An Introduction to
Geophysical Exploration

GEOSCIENCE TEXTS

SERIES EDITOR

A. HALLAM
Lapworth Professor of Geology
University of Birmingham

GEOSCIENCE TEXTS

An Introduction to Geophysical Exploration

PHILIP KEAREY
Department of Geology
University of Bristol

MICHAEL BROOKS
Department of Geology
University of Wales
College of Cardiff

SECOND EDITION

OXFORD

BLACKWELL SCIENTIFIC PUBLICATIONS

LONDON EDINBURGH BOSTON

MELBOURNE PARIS BERLIN VIENNA

First published 1984
Reprinted 1987, 1989
Second edition 1991
Reprinted 1992, 1993

Set by Setrite Typesetters, Hong Kong
Printed and bound in Great Britain at
the University Press, Cambridge

DISTRIBUTORS

Marston Book Services Ltd
PO Box 87
Oxford OX2 0DT
(*Orders*: Tel: 0865-791155
          Fax: 0865-791927
          Telex: 837515)

USA
Blackwell Scientific Publications, Inc.
238 Main Street
Cambridge, MA 02142
(*Orders*: Tel: 800 759-6102
          617 876-7000)

Canada
Oxford University Press
70 Wynford Drive
Don Mills
Ontario M3C 1J9
(*Orders*: Tel: (416) 441-2941)

Australia
Blackwell Scientific Publications Pty Ltd
54 University Street
Carlton, Victoria 3053
(*Orders*: Tel: (03) 347-5552)

British Library
Cataloguing in Publication Data

Kearey, P.
  An introduction to geophysical exploration – 2nd ed.
  1. Mineral deposits. Prospecting. Applications of
  geophysics
  I. Title II. Brooks, M. (Michael) *1936*– III. Series
  622.15
  ISBN 0-632-02921-8
  ISBN 0-632-02923-4 pbk

Library of Congress
Cataloging in Publication Data

Kearey, P.
  An introduction to geophysical exploration/Philip
  Kearey, Michael Brooks. – 2nd ed.
  p. cm. – (Geoscience texts)
  Includes bibliographical references and index.
  ISBN 0-632-02921-1 (hardback).
  ISBN 0-632-02923-4 (limp)
  1. Prospecting – Geophysical methods.
  I. Brooks, M. II. Title. III. Series.
  TN269.K37 1991
  622′.15 – dc20

Contents

Plates 1, 2 and 4 appear between pages 70 and 71

Plates 3 and 5 appear between pages 86 and 87

Preface

This book provides a general introduction to the most important methods of geophysical exploration. These methods represent a primary tool for investigation of the subsurface and are applicable to a very wide range of problems. Although their main application is in prospecting for natural resources the methods are also used, for example, as an aid to geological surveying, as a means of deriving information on the Earth's internal physical properties, and in engineering or archaeological site investigations. Consequently, geophysical exploration is of importance not only to geophysicists but also to geologists, physicists, engineers and archaeologists. The book covers the physical principles, methodology, interpretational procedures and fields of application of the various survey methods. The main emphasis has been placed on seismic methods because these represent the most extensively used techniques, being routinely and widely employed by the oil industry in prospecting for hydrocarbons. Since this is an introductory text we have not attempted to be completely comprehensive in our coverage of the subject. Readers seeking further information on any of the survey methods described should refer to the more advanced texts listed at the end of each chapter.

We hope that the book will serve as an introductory course text for students in the above-mentioned disciplines and also as a useful guide for specialists who wish to be aware of the value of geophysical surveying to their own disciplines. In preparing a book for such a wide possible readership it is inevitable that problems arise concerning the level of mathematical treatment to be adopted. Geophysics is a highly mathematical subject and although we have attempted to show that no great mathematical expertise is necessary for a broad understanding of geophysical surveying, a full appreciation of the more advanced data processing and interpretational techniques does require a reasonable mathematical ability. Our approach to this problem has been to keep the mathematics as simple as possible and to restrict full mathematical analysis to relatively simple cases. We consider it important, however, that any user of geophysical surveying should be aware of the more advanced techniques of analysing and interpreting geophysical data since these can greatly increase the amount of useful information obtained from the data. In discussing such techniques we have adopted a semi-quantitative or qualitative approach which allows the reader to assess their scope and importance, without going into the details of their implementation.

In the second edition of this book we have attempted to incorporate suggestions from readers of the first edition, and recent developments in exploration techniques. There are new chapters on Radiometric Surveying and Geophysical Borehole Logging, and new sections on vertical seismic profiling, marine gravimeters, time−domain electromagnetic methods, non-contacting resistivity measurements and ground-penetrating radar. The treatment of three-dimensional seismic surveying and seismic stratigraphy has been expanded. We have incorporated the application of the surveying methods into the relevant chapters on techniques rather than dealing with applications in a separate chapter. We have also provided a set of problems at the end of each main chapter. Since, however, much geophysical interpretation involves an integration of data, several of the problems make reference to more than one exploration method.

ACKNOWLEDGEMENTS

We thank our friends and colleagues Dr P.F. Ellis, Dr J. Shaw and Dr R.G. Pearce for their helpful comments on early versions of the manuscript. The text figures were drafted by Mrs J. Bees, Mrs A. Gregory, Ms T. Lanigan and Mrs M. Millen. Mr D.G. Hilton is thanked for extensive photographic assistance.

1 / The principles and limitations of geophysical exploration methods

1.1 INTRODUCTION

This chapter is provided for readers with no prior knowledge of geophysical exploration methods and is pitched at an elementary level. It may be passed over by readers already familiar with the basic principles and limitations of geophysical surveying.

The science of geophysics applies the principles of physics to the study of the Earth. Geophysical investigations of the interior of the Earth involve taking measurements at or near the Earth's surface that are influenced by the internal distribution of physical properties. Analysis of these measurements can reveal how the physical properties of the Earth's interior vary vertically and laterally.

By working at different scales, geophysical methods may be applied to a wide range of investigations from studies of the entire Earth (global geophysics) to exploration of a localized region of the upper crust. In the geophysical exploration methods (also referred to as geophysical surveying) discussed in this book, measurements within geographically restricted areas are used to determine the distributions of physical properties at depths that reflect the local subsurface geology.

An alternative method of investigating subsurface geology is, of course, by drilling boreholes, but these are expensive and provide information only at discrete locations. Geophysical surveying, although sometimes prone to major ambiguities or uncertainties of interpretation, provides a relatively rapid and cost-effective means of deriving areally distributed information on subsurface geology. In the exploration for subsurface resources the methods are capable of detecting and delineating local features of potential interest that could not be discovered by any realistic drilling programme. Geophysical surveying does not dispense with the need for drilling but, properly applied, it can optimize exploration programmes by maximizing the rate of ground coverage and minimizing the drilling requirement. The importance of geophysical exploration as a means of deriving subsurface geological information is so great that the basic principles and scope of the methods and their main fields of application should be ap-

preciated by any practising earth scientist. This book provides a general introduction to the main geophysical methods in widespread use.

1.2 THE SURVEY METHODS

There is a broad division of geophysical surveying methods into those that make use of natural fields of the Earth and those that require the input into the ground of artificially generated energy. The natural field methods utilize the gravitational, magnetic, electrical and electromagnetic fields of the Earth, searching for local perturbations in these naturally occurring fields that may be caused by concealed geological features of economic or other interest. Artificial source methods involve the generation of local electrical or electromagnetic fields that may be used analogously to natural fields, or, in the most important single group of geophysical surveying methods, the generation of seismic waves whose propagation velocities and transmission paths through the subsurface are mapped to provide information on the distribution of geological boundaries at depth. Generally, natural field methods can provide information on Earth properties to significantly greater depths and are logistically more simple to carry out than artificial source methods. The latter, however, are capable of producing a more detailed and better resolved picture of the subsurface geology.

Several geophysical surveying methods can be used at sea or in the air. The higher capital and operating costs associated with marine or airborne work and the problems of accurate position fixing are offset by the increased speed of operation and the benefit of being able to survey areas where ground access is difficult or impossible.

A wide range of geophysical surveying methods exists, for each of which there is an 'operative' physical property to which the method is sensitive. The methods are listed in Table 1.1.

The type of physical property to which a method responds clearly determines its range of applications. Thus, for example, the magnetic method is very suitable for locating buried magnetite ore bodies

Table 1.1. Geophysical surveying methods.

Method	Measured parameter	'Operative' physical property
Seismic	Travel times of reflected/refracted seismic waves	Density and elastic moduli, which determine the propagation velocity of seismic waves
Gravity	Spatial variations in the strength of the gravitational field of the Earth	Density
Magnetic	Spatial variations in the strength of the geomagnetic field	Magnetic susceptibility and remanence
Electrical		
Resistivity	Earth resistance	Electrical conductivity
Induced polarization	Polarization voltages or frequency-dependent ground resistance	Electrical capacitance
Self potential	Electrical potentials	Electrical conductivity
Electromagnetic	Response to electromagnetic radiation	Electrical conductivity and inductance
Radar	Travel times of reflected radar pulses	Dielectric constant

because of their high magnetic susceptibility. Similarly, seismic or electrical methods are suitable for the location of a buried water table because saturated rock may be distinguished from dry rock by its higher seismic velocity and higher electrical conductivity.

Other considerations also determine the type of methods employed in a geophysical exploration programme. For example, reconnaissance surveys are often carried out from the air because of the high speed of operation. In such cases the electrical or seismic methods are not applicable, since these require physical contact with the ground for the direct input of energy.

Geophysical methods are often used in combination. Thus, the initial search for metalliferous mineral deposits often utilizes airborne magnetic and electromagnetic surveying. Similarly, routine reconnaissance of continental shelf areas often includes simultaneous gravity, magnetic and seismic surveying. At the interpretation stage, ambiguity arising from the results of one survey method may often be removed by consideration of results from a second survey method.

Geophysical exploration commonly takes place in a number of stages. For example, in the offshore search for oil and gas, an initial gravity reconnaissance survey may reveal the presence of a large sedimentary basin that is subsequently explored using seismic methods. A first round of seismic exploration may highlight areas of particular interest where further detailed seismic work needs to be carried out.

The main fields of application of geophysical surveying, together with an indication of the most appropriate surveying methods for each application, are listed in Table 1.2.

Exploration for hydrocarbons and for metalliferous minerals represents the main application of geophysical surveying. In terms of the amount of money expended annually, seismic methods are the most important technique because of their routine and widespread use in the exploration for hydrocarbons. Seismic methods are particularly well suited to the investigation of the layered sequences in sedimentary basins that are the primary targets for oil or gas. On the other hand, seismic methods are quite unsuited to the exploration of igneous and metamorphic terrains for the near-surface, irregular ore bodies that represent the main source of metalliferous minerals. Exploration for ore bodies is mainly carried out using electromagnetic and magnetic surveying methods.

In several geophysical survey methods it is the local variation in a measured parameter, relative to some normal background value, that is of primary interest. Such variation is attributable to a localized subsurface zone of distinctive physical property and possible geological importance. A local variation of this type is known as a *geophysical anomaly*. For example, the Earth's gravitational field, after the application of certain corrections, would everywhere

Table 1.2. Geophysical surveying applications.

Application	Appropriate survey methods*
Exploration for fossil fuels (oil, gas, coal)	S, G, M, (EM)
Exploration for metalliferous mineral deposits	M, EM, E, SP, IP, R
Exploration for bulk mineral deposits (sand & gravel)	S, (E), (G)
Exploration for underground water supplies	E, S, (G), (Rd)
Engineering/construction site investigation	E, S, Rd, (G), (M)

* G = gravity; M = magnetic; S = seismic; E = electrical resistivity; SP = self potential; IP = induced polarization; EM = electromagnetic; R = radiometric; Rd = ground-penetrating radar. Subsidiary methods in brackets.

be constant if the subsurface were of uniform density. Any lateral density variation associated with a change of subsurface geology results in a local deviation in the gravitational field. This local deviation from the otherwise constant gravitational field is referred to as a gravity anomaly.

Although many of the geophysical methods require complex methodology and relatively advanced mathematical treatment in interpretation, much information may be derived from a simple assessment of the survey data. This is illustrated in the following section where a number of geophysical surveying methods are applied to the problem of detecting and delineating a specific geological feature, namely a salt dome. No terms or units are defined here, but the examples serve to illustrate the way in which geophysical surveys can be applied to the solution of a particular geological problem.

Salt domes are emplaced when a buried salt layer, because of its buoyancy, rises through overlying denser strata in a series of approximately cylindrical bodies. The rising columns of salt pierce the overlying strata or arch them into a domed form. A salt dome has physical properties that are different from the surrounding sediments and which enable its detection by geophysical methods. These properties are (**1**) a relatively low density, (**2**) a negative magnetic susceptibility, (**3**) a relatively high propagation velocity for seismic waves, and (**4**) a high electrical resistivity (specific resistance).

1 The relatively low density of salt with respect to its surroundings renders the salt dome a zone of anomalously low mass. The Earth's gravitational field is perturbed by subsurface mass distributions and the salt dome therefore gives rise to a gravity

anomaly that is negative with respect to surrounding areas. Fig. 1.1 presents a contour map of gravity anomalies measured over the Grand Saline Salt Dome in east Texas, USA. The gravitational readings have been corrected for effects which result from the Earth's rotation, irregular surface relief and regional geology so that the contours reflect only variations in the shallow density structure of the area resulting from the local geology. The location of the salt dome is known from both drilling and mining operations and its subcrop is indicated. It is readily apparent that there is a well-defined negative gravity anomaly centred over the salt dome and the circular gravity contours reflect the circular outline of the dome. Clearly, gravity surveys provide a powerful method for the location of features of this type.

2 A less familiar characteristic of salt is its negative magnetic susceptibility, full details of which must be deferred to Chapter 7. This property of salt causes a local decrease in the strength of the Earth's magnetic field in the vicinity of a salt dome. Fig. 1.2 presents a contour map of the strength of the magnetic field over the Grand Saline Salt Dome covering the same area as Fig. 1.1. Readings have been corrected for the large-scale variations of the magnetic field with latitude, longitude and time so that, again, the contours reflect only those variations resulting from variations in the magnetic properties of the subsurface. As expected, the salt dome is associated with a negative magnetic anomaly although the magnetic low is displaced slightly from the centre of the dome. This example illustrates that salt domes may be located by magnetic surveying but the technique is not widely used as the associated anomalies are

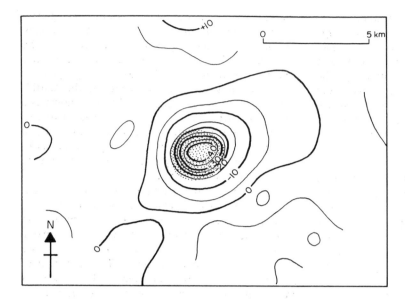

Fig. 1.1 The gravity anomaly over the Grand Saline Salt Dome, Texas, USA (contours in gravity units — see Chapter 6). The stippled area represents the subcrop of the dome. (Redrawn from Peters & Dugan 1945.)

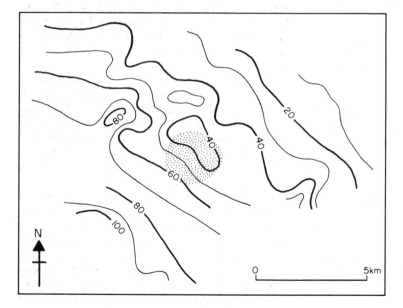

Fig. 1.2 Magnetic anomalies over the Grand Saline Salt Dome, Texas, USA (contours in nT — see Chapter 7). The stippled area represents the subcrop of the dome. (Redrawn from Peters & Dugan 1945.)

usually very small and therefore difficult to detect.
3 Seismic rays normally propagate through salt at a higher velocity than through the surrounding sediments. A consequence of this velocity difference is that any seismic energy incident on the boundary of a salt body is partitioned into a refracted phase that is transmitted through the salt and a reflected phase that travels back through the surrounding sediments (Chapter 3). These two seismic phases provide alternative means of locating a concealed salt body.

For a series of seismic rays travelling from a single shot point into a fan of seismic detectors (see Fig. 5.21), rays transmitted through any intervening salt dome will travel at a higher average velocity than in the surrounding medium and, hence, will arrive relatively early at the recording site. By means of this 'fan-shooting' it is possible to delineate sections of ground which are associated with anomalously short travel times and which may therefore be underlain by a salt body.

An alternative, and more effective, approach to the seismic location of salt domes utilizes energy reflected off the salt, as shown schematically in Fig. 1.3. A survey configuration of closely-spaced shots and detectors is moved systematically along a profile line and the travel times of rays reflected back from any subsurface geological interfaces are measured. If a salt dome is encountered, rays reflected off its top surface will delineate the shape of the concealed body.

4 Earth materials with anomalous electrical resistivity may be located using either electrical or electromagnetic geophysical techniques. Shallow features are normally investigated using artificial field methods in which an electrical current is introduced into the ground and potential differences between points on the surface are measured to reveal anomalous material in the subsurface (Chapter 8). However, this method is restricted in its depth of penetration by the limited power that can be introduced into the ground. Much greater penetration can be achieved by making use of the natural Earth currents (telluric currents) generated by the motions of charged particles in the ionosphere. These currents extend to great depths within the Earth and, in the absence of any electrically anomalous material, flow parallel to the surface. A salt dome, however, possesses an anomalously high electrical resistivity and electric currents preferentially flow around and over the top of such a structure rather than through it. This pattern of flow causes distortion of the constant potential gradient at the surface that would be associated with a homogeneous subsurface and indicates the presence of the high resistivity salt. Fig. 1.4 presents the results of a telluric current survey of the Haynesville Salt Dome, Texas, USA. The contour values represent quantities describing the extent to which the telluric currents are distorted by subsurface phenomena and their configuration reflects the shape of the subsurface salt dome with some accuracy.

1.3 THE PROBLEM OF AMBIGUITY IN GEOPHYSICAL INTERPRETATION

If the internal structure and physical properties of the Earth were precisely known, the magnitude of any particular geophysical measurement taken at the Earth's surface could be predicted uniquely. Thus, for example, it would be possible to predict the travel time of a seismic wave reflected off any

buried layer or to determine the value of the gravity or magnetic field at any surface location. In geophysical surveying the problem is the converse of the above, namely, to deduce some aspect of the Earth's internal structure on the basis of geophysical measurements taken at (or near to) the Earth's surface. The former type of problem is known as a *direct* problem, the latter as an *inverse* problem. Whereas direct problems are theoretically capable of unambiguous solution, inverse problems suffer from an inherent ambiguity, or non-uniqueness, in the conclusions that can be drawn.

To exemplify this point a simple analogy to geophysical surveying may be considered. In *echo-sounding*, high frequency acoustic pulses are transmitted by a transducer mounted on the hull of a ship and echoes returned from the sea bed are detected by the same transducer. The travel time of the echo is measured and converted into a water depth, multiplying the travel time by the velocity with which sound waves travel through water, i.e. $1500\,\mathrm{m\,s^{-1}}$. Thus an echo time of $0.10\,\mathrm{s}$ indicates a path length of $0.10 \times 1500 = 150\,\mathrm{m}$, or a water depth of $150/2 = 75\,\mathrm{m}$, since the pulse travels down to the sea bed and back up to the ship.

Using the same principle, a simple seismic survey may be used to determine the depth of a buried geological interface (e.g. the top of a limestone layer). This would involve generating a seismic pulse at the Earth's surface and measuring the travel time of a pulse reflected back to the surface from the top of the limestone. However, the conversion of this travel time into a depth requires knowledge of the velocity with which the pulse travelled along the reflection path and, unlike the velocity of sound in water, this information is generally not known. If a velocity is assumed, a depth estimate can be derived but it represents only one of many possible solutions. And since rocks differ significantly in the velocity with which they propagate seismic waves, it is by no means a straightforward matter to translate the travel time of a seismic pulse into an accurate depth to the geological interface from which it was reflected.

The solution to this particular problem, as discussed in Chapter 4, is to measure the travel times of reflected pulses at several offset distances from a seismic source because the variation of travel time as a function of range provides information on the velocity distribution with depth. However, although the degree of uncertainty in geophysical interpretation can often be reduced to an acceptable level by the general expedient of taking additional (and in

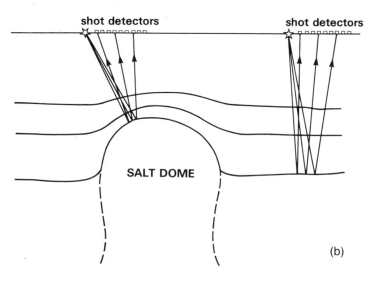

(a)

(b)

Fig. 1.3 (a) Seismic reflection section across a buried salt dome (courtesy Prakla-Seismos GMBH). (b) Simple structural interpretation of the seismic section, illustrating some possible ray paths for reflected rays.

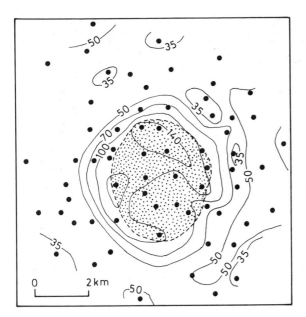

Fig. 1.4 Perturbation of telluric currents over the Haynesville Salt Dome, Texas, USA (for explanation of units see Chapter 9). The stippled area represents the subcrop of the dome. (Redrawn from Boissonas & Leonardon 1948).

some cases different kinds of) field measurements, the problem of inherent ambiguity cannot be circumvented.

The general problem is that significant differences from an actual subsurface geological situation may give rise to insignificant, or immeasurably small, differences in the quantities actually measured during a geophysical survey. Thus ambiguity arises because many different geological configurations could reproduce the observed measurements. This basic limitation results from the unavoidable fact that geophysical surveying attempts to solve a difficult inverse problem. It should also be noted that experimentally derived quantities are never exactly determined and experimental error adds a further degree of indeterminacy to that caused by the incompleteness of the field data and the ambiguity associated with the inverse problem. Since a unique solution cannot, in general, be recovered from a set of field measurements, geophysical interpretation is concerned either to determine properties of the subsurface that all possible solutions share, or to introduce assumptions to restrict the number of admissable solutions (Parker 1977). In spite of these inherent problems, however, geophysical surveying is an invaluable tool for the investigation of subsurface geology and occupies a key role in exploration programmes for geological resources.

1.4 THE STRUCTURE OF THE BOOK

The above introductory sections illustrate in a simple way the very wide range of approaches to the geophysical investigation of the subsurface and warn of inherent limitations in geophysical interpretations.

Chapter 2 provides a short account of the more important data processing techniques of general applicability to geophysics. In Chapters 3 to 10 the individual survey methods are treated systematically in terms of their basic principles, survey procedures, interpretation techniques and major applications. Chapter 11 describes the application of these methods to specialized surveys undertaken in boreholes. All these chapters contain suggestions for further reading which provide a more extensive treatment of the material covered in this book. A set of problems is given for all the major geophysical methods.

2 / Geophysical data processing

2.1 INTRODUCTION

Much of geophysical surveying is concerned with the measurement and analysis of waveforms that express the variation of some measurable quantity as a function of distance or time. The quantity may, for example, be the strength of the Earth's gravitational or magnetic field along a profile line across a geological structure; or it may be the displacement of the ground surface as a function of time associated with the passage of seismic waves from a nearby explosion. The analysis of waveforms such as these represents an essential aspect of geophysical data processing and interpretation. The fundamental principles on which the various methods of data analysis are based are brought together in this chapter, along with a discussion of the techniques of digital data processing by computer that are routinely used by geophysicists.

Throughout this chapter waveforms are referred to as functions of time, but all the principles discussed, relating to spectral analysis and digital filtering, are equally applicable to functions of distance. In the latter case, frequency (number of waveform cycles per unit time) is replaced by spatial frequency or *wavenumber* (number of waveform cycles per unit distance).

2.2 DIGITIZATION OF GEOPHYSICAL DATA

Waveforms of geophysical interest generally represent continuous (analogue) functions of time or distance. The quantity of information and, in some cases, the complexity of data processing to which these waveforms are subjected are such that the processing can only be accomplished effectively and economically by digital computers. Consequently, the data often need to be expressed in digital form for input to a computer, whatever the form in which they were originally recorded.

A continuous, smooth function of time or distance can be expressed digitally by sampling the function at a fixed interval and recording the instantaneous value of the function at each sampling point. Thus the analogue function of time $f(t)$ shown in Fig. 2.1(a) can be represented as the digital function $g(t)$ shown in Fig. 2.1(b) where the continuous function has been replaced by a series of discrete values at fixed intervals of time τ.

The two basic parameters of a digitizing system are the sampling precision (dynamic range) and the sampling frequency.

Dynamic range is an expression of the ratio of the largest measurable amplitude A_{max} to the smallest measurable amplitude A_{min} in a sampled function. The higher the dynamic range, the more faithfully will amplitude variations in the analogue waveform be represented in the digitized version of the waveform. Dynamic range is normally expressed in the *decibel* (dB) scale used to define electrical power ratios: the ratio of two power values P_1 and P_2 is given by $10 \log_{10} (P_1/P_2)$ dB. Since electrical *power* is proportional to the square of *signal amplitude* A, $10 \log_{10} (P_1/P_2) = 10 \log_{10} (A_1/A_2)^2 = 20 \log_{10} A_1/A_2$). Thus, if a digital sampling scheme measures amplitudes over the range from 1 to 1024 units of amplitude, the dynamic range is given by $20 \log_{10} (A_{max}/A_{min}) = 20 \log_{10} 1024 \approx 60$ dB.

For convenience of handling in digital computers, digital samples are expressed in binary form (i.e. they are composed of a sequence of digits that have the value of either 0 or 1). Each binary digit is known as a *bit* and the sequence of bits representing the sample value is known as a *word*. The dynamic range of a digitized waveform is determined by the number of bits in each word. For example, a dynamic range of 60 dB requires 11-bit words since the appropriate amplitude ratio of 1024 ($=2^{10}$) is rendered as 10 000 000 000 in binary form. A dynamic range of 84dB represents an amplitude ratio of 2^{14} and, hence, requires sampling with 15-bit words. Thus, increasing the number of bits in each word in digital sampling increases the dynamic range of the digital function.

Intuitively, it may appear that the digital sampling of a continuous function inevitably leads to a loss of fidelity in the resultant digital function, since the latter is only specified by discrete values at a series of spaced points. In fact, as discussed below, there is

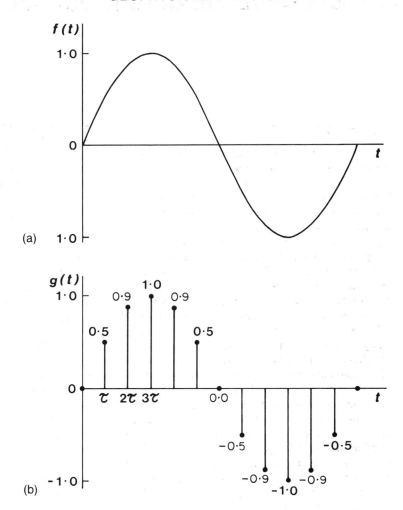

Fig. 2.1 (a) Analogue representation of a sinusoidal function. (b) Digital representation of the same function.

no significant loss of information content as long as the frequency of sampling is at least twice as high as the highest frequency component in the sampled function.

Sampling frequency is the number of sampling points in unit time or unit distance. Thus if a waveform is sampled every two milliseconds (sampling interval), the sampling frequency is 500 samples per second (or 500 Hz). Sampling at this rate will preserve all frequencies up to 250 Hz in the sampled function. This frequency of half the sampling frequency is known as the *Nyquist frequency* (f_N) and the *Nyquist interval* is the frequency range from zero up to f_N.

$$f_N = 1/2 \, \Delta t \qquad (2.1)$$

where Δt = sampling interval.

If frequencies above the Nyquist frequency are present in the sampled function, a serious form of distortion results known as *aliasing*, in which the higher frequency components are 'folded back' into the Nyquist interval. Consider the example illustrated in Fig. 2.2 in which a sine wave is sampled at different sampling frequencies. At the higher sampling rate (Fig. 2.2(a)) the waveform is accurately reproduced but at the lower rate (Fig. 2.2(b)) it is rendered as a fictitious frequency within the Nyquist interval. The relationship between input and output frequencies in the case of a sampling frequency of 500 Hz is shown in Fig. 2.2(c). It is apparent that an input frequency of 125 Hz, for example, is retained in the output but that an input frequency of 625 Hz is folded back to be output at 125 Hz also.

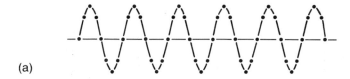

(a)

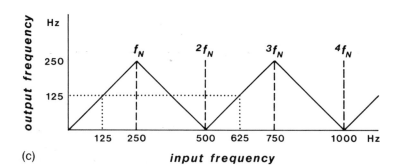

(b)

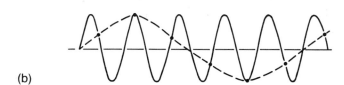

(c)

Fig. 2.2 (a) Sine wave frequency less than Nyquist frequency. (b) Sine wave frequency greater than Nyquist frequency showing the fictitious frequency that is generated by aliasing. (c) Relationship between input and output frequencies for a sampling frequency of 500 Hz (Nyquist frequency $f_N = 250$ Hz).

To overcome the problem of aliasing, either the sampling frequency must be at least twice as high as the highest frequency component present in the sampled function, or the function must be passed through an *antialias filter* prior to digitization. The antialias filter is a low pass frequency filter with a sharp cut-off that removes frequency components above the Nyquist frequency, or attenuates them to an insignificant amplitude level.

2.3 SPECTRAL ANALYSIS

A distinction may be made between *periodic waveforms* (Fig. 2.3(a)), that repeat themselves at a fixed time period T, and *transient waveforms* (Fig. 2.3(b)), that are non-repetitive.

By means of *Fourier analysis* any periodic waveform, however complex, may be decomposed into a series of sine (or cosine) waves whose frequencies are integer multiples of the basic repetition frequency $1/T$, known as the fundamental frequency. The higher frequency components, at frequencies of n/T ($n = 1, 2, 3 \ldots$), are known as harmonics. Thus the complex waveform of Fig. 2.4(a) is built up from the addition of the two individual sine wave components shown. To express any waveform in terms of its constituent sine wave components, it is necessary to define not only the frequency of each component but also its amplitude and phase. If in the above example the relative amplitude and phase relations of the individual sine waves are altered, summation can produce the quite different waveform illustrated in Fig. 2.4(b).

From the above it follows that a periodic waveform can be expressed in two different ways: in the *time domain*, expressing wave amplitude as a function of time, and in the *frequency domain*, expressing the amplitude and phase of its constituent sine waves as a function of frequency. The waveforms shown in Fig. 2.4(a) and (b) are represented in Fig. 2.5(a) and (b) in terms of their *amplitude* and *phase spectra*. These spectra, being composed of a series of discrete amplitude and phase components, are known as *line spectra*.

Transient waveforms do not repeat themselves, that is, they have an infinitely long period. They may

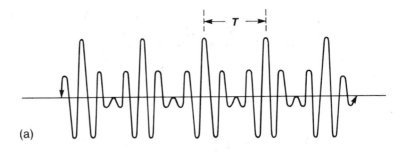

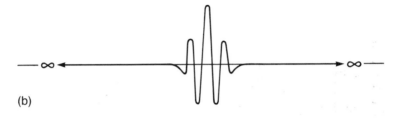

Fig. 2.3 (a) Periodic and (b) transient waveforms.

(a)

(b)

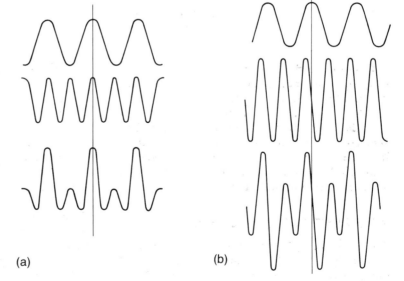

Fig. 2.4 Complex waveforms resulting from the summation of two sine wave components of frequency f and $2f$. (a) The two sine wave components are of equal amplitude and in phase. (b) The higher frequency component has twice the amplitude of the lower frequency component and is $\pi/2$ out of phase. (After Anstey 1965.)

(a)

(b)

thus be regarded, by analogy with a periodic waveform, as having an infinitesimally small fundamental frequency $(1/T \rightarrow 0)$ and, consequently, harmonics that occur at infinitesimally small frequency spacings to give continuous amplitude and phase spectra rather than the line spectra of periodic waveforms. Digitization provides a means of dealing with the continuous spectra of transient waveforms. Clearly it is impossible to cope analytically with a spectrum containing an infinite number of sine wave components and the continuous amplitude and phase spectra are therefore subdivided into a number of thin frequency slices, giving each slice a frequency equal to the mean frequency of the slice and an

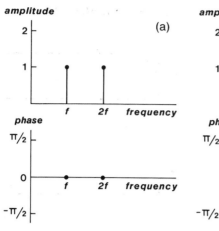

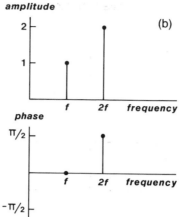

Fig. 2.5 Representation in the frequency domain of the waveforms illustrated in Fig. 2.4, showing their amplitude and phase spectra.

amplitude and phase proportional to the area of the slice of the appropriate spectrum (Fig. 2.6). This digital expression of a continuous spectrum in terms of a finite number of discrete frequency components provides an approximate representation in the frequency domain of a transient waveform in the time domain. Increasing the number of frequency slices improves the accuracy of the approximation.

Fourier transformation may be used to convert a time function $g(t)$ into its equivalent amplitude and phase spectra $A(f)$ and $\phi(f)$, or into a complex function of frequency $G(f)$ known as the *frequency spectrum*, where

$$G(f) = A(f)e^{i\phi(f)} \tag{2.2}$$

The time and frequency domain representations of a waveform, $g(t)$ and $G(f)$, are known as a *Fourier pair*, represented by the notation

$$g(t) \leftrightarrow G(f) \tag{2.3}$$

Components of a Fourier pair are interchangeable, such that, if $G(f)$ is the Fourier transform of $g(t)$, then $g(t)$ is the Fourier transform of $G(f)$.

Fig. 2.7 illustrates Fourier pairs for various waveforms of geophysical significance. All the examples illustrated have *zero phase spectra*, that is, the individual sine wave components of the waveforms are in phase at zero time. In this case $\phi(f) = 0$ for all values of f.

Fig. 2.7(a) shows a spike function (also known as a Dirac function), which is the shortest possible transient waveform. Fourier transformation shows that the spike function has a continuous frequency spectrum of constant amplitude from zero to infinity;

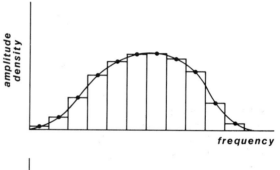

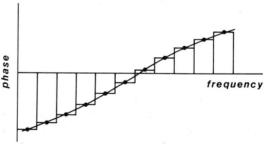

Fig. 2.6 Digital representation of the continuous amplitude and phase spectra associated with a transient waveform.

thus, a spike function contains all frequencies from zero to infinity at equal amplitude. The 'DC bias' waveform of Fig. 2.7(b) has, as would be expected, a line spectrum comprising a single component at zero frequency. Note that Fig. 2.7(a) and (b) demonstrate the principle of interchangeability of Fourier pairs stated above (equation (2.3)).

Fig. 2.7(c) and (d) illustrate transient waveforms

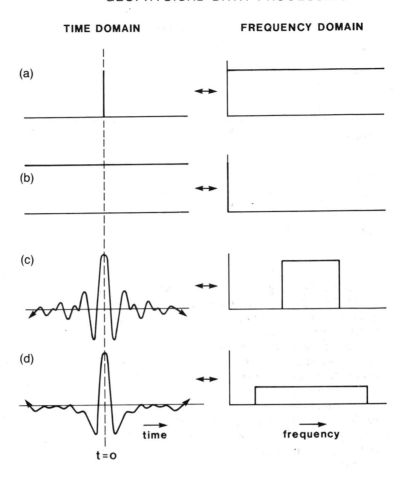

Fig. 2.7 Fourier transform pairs for various waveforms. (a) A spike function. (b) A 'DC bias'. (c) and (d) Transient waveforms approximating seismic pulses.

approximating the shape of seismic pulses, together with their amplitude spectra. Both have a band-limited amplitude spectrum, the spectrum of narrower bandwidth being associated with the longer transient waveform. In general, the shorter a time pulse the wider is its frequency bandwidth and in the limiting case a spike pulse has an infinite bandwidth.

Waveforms with zero phase spectra such as those illustrated in Fig. 2.7 are symmetrical about the time axis and, for any given amplitude spectrum, produce the maximum peak amplitude in the resultant waveform. If phase varies linearly with frequency, the waveform remains unchanged in shape but is displaced in time; if the phase variation with frequency is non-linear the shape of the waveform is altered. A particularly important case in seismic data processing is the phase spectrum associated with *minimum delay* in which there is a maximum concentration of energy at the front end of the waveform. Analysis of seismic

pulses sometimes assumes that they exhibit minimum delay (see Chapter 4).

Fourier transformation of digitized waveforms is readily enacted by computers, using a 'fast Fourier transform' (FFT) algorithm as in the Cooley-Tukey method (Brigham 1974). FFT subroutines can thus be routinely built into data processing programmes in order to carry out spectral analysis of geophysical waveforms.

Fourier transformation can be extended into two dimensions (Rayner 1971), and can thus be applied to areal distributions of data such as gravity and magnetic contour maps. In this case the time variable is replaced by horizontal distance and the frequency variable by wavenumber (number of waveform cycles per unit distance). The application of two-dimensional Fourier techniques to the interpretation of potential field data is discussed in Chapters 6 and 7.

2.4 **WAVEFORM PROCESSING**

The principles of convolution, deconvolution and correlation form the common basis for many methods of geophysical data processing, especially in the field of seismic reflection surveying. They are introduced here in general terms and are referred to extensively in later chapters.

2.4.1 **Convolution**

Convolution (Kanasewich 1981) is a mathematical operation defining the change of shape of a waveform resulting from its passage through a filter. Filtering modifies a waveform by discriminating against its constituent sine wave components to alter their relative amplitudes or phase relations or both. Filtering is an inherent characteristic of any transmission system. Thus, for example, a seismic pulse generated by an explosion is altered in shape by filtering effects, both in the ground and in the recording system, so that the recorded pulse (the filtered output) differs significantly from the initial pulse (the input).

As a simple example of filtering, consider a weight suspended from the end of a vertical spring. If the top of the spring is perturbed by a sharp up-and-down movement (the input), the response of the weight (the filtered output) is a series of damped oscillations out of phase with the initial perturbation (Fig. 2.8).

The effect of a filter may be categorized by its *impulse response* which is defined as the output of the filter when the input is a spike function (Fig. 2.9). The Fourier transform of the impulse response is known as the *transfer function* and this specifies the amplitude and phase response of the filter, thus defining its operation completely.

The effect of a filter is described mathematically by a convolution operation such that, if the input signal $g(t)$ to the filter is *convolved with* the impulse response $f(t)$ of the filter, known as the *convolution operator*, the filtered output $y(t)$ is obtained.

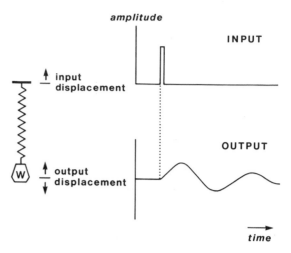

Fig. 2.8 The principle of filtering illustrated by the perturbation of a suspended weight system.

$$y(t) = g(t) * f(t) \tag{2.4}$$

where the asterisk denotes the convolution operation.

Fig. 2.10(a) shows a spike function input to a filter whose impulse response is given in Fig. 2.10(b). Clearly the latter is also the filtered output since, by definition, the impulse response represents the output for a spike input. Fig. 2.10(c) shows an input comprising two separate spike functions and the filtered output (Fig. 2.10(d)) is now the superimposition of the two impulse response functions offset in time by the separation of the spikes and scaled according to the individual spike amplitudes. Since any transient wave can be represented as a series of spike functions (Fig. 2.10(e)), the general form of a filtered output (Fig. 2.10(f)) can be regarded as the summation of a set of impulse responses related to a succession of spikes simulating the overall shape of the input wave.

In Fig. 2.11 the individual steps in the convolution process are shown for two digital functions, a double spike function given by $g_i = g_1, g_2, g_3 = 2, 0, 1$ and an

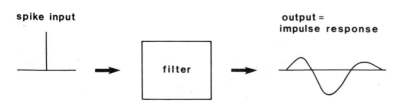

Fig. 2.9 The impulse response of a filter.

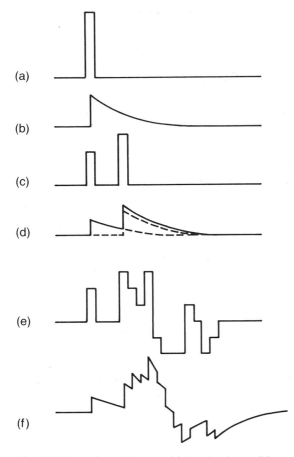

Fig. 2.10 Examples of filtering. (a) A spike input. (b) Filtered output equivalent to impulse response of filter. (c) An input comprising two spikes. (d) Filtered output given by summation of two impulse response functions offset in time. (e) A complex input represented by a series of contiguous spike functions. (f) Filtered output given by the summation of a set of impulse responses. (After Anstey 1965.)

impulse response function given by $f_i = f_1, f_2, f_3, f_4 =$ 4, 3, 2, 1, where the numbers refer to discrete amplitude values at the sampling points of the two functions.

From Fig. 2.11 it can be seen that the convolved output $y_i = y_1, y_2, y_3, y_4, y_5, y_6 = 8, 6, 8, 5, 2, 1$. Note that the convolved output is longer than the input waveforms: if the functions to be convolved have lengths of m and n, the convolved output has a length of $(m + n - 1)$.

Convolution involves time inversion (or folding) of one of the functions and its progressive sliding past the other function, the individual terms in the convolved output being derived by summation of the cross-multiplication products over the overlapping parts of the two functions.

In general, if $g_i(i = 1, 2, \ldots, m)$ is an input function and $f_j(j = 1, 2, \ldots, n)$ is a convolution operator, then the convolution output function y_k is given by

$$y_k = \sum_{i=1}^{m} g_i f_{k-i} \ (k = 1, 2, \ldots, m + n - 1) \qquad (2.5)$$

It can be shown that the convolution of two functions in the time domain is mathematically equivalent to multiplication of their amplitude spectra and addition of their phase spectra in the frequency domain. The operation of convolution can thus be performed by transforming the time functions into the frequency domain, multiplying their amplitude spectra, summing their phase spectra and taking the inverse transform of the resultant frequency spectrum. Thus, digital filtering can be enacted in either the time domain or the frequency domain. With large data sets, filtering by computer is more efficiently carried out in the frequency domain as less mathematical operations are involved.

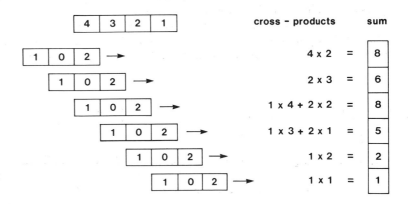

Fig. 2.11 Convolution of two digital functions.

Convolution, or its equivalent in the frequency domain, finds very wide application in geophysical data processing, notably in the digital filtering of seismic and potential field data and the construction of synthetic seismograms for comparison with field seismograms (see Chapters 4 and 6).

2.4.2 Deconvolution

Deconvolution, or *inverse filtering* (Kanasewich 1981) is a process that counteracts a previous convolution (or filtering) action. Consider the convolution operation given in equation (2.4):

$$y(t) = g(t) * f(t)$$

$y(t)$ is the filtered output derived by passing the input waveform $g(t)$ through a filter of impulse response $f(t)$. Knowing $y(t)$ and $f(t)$, the recovery of $g(t)$ represents a deconvolution operation. Suppose that $f'(t)$ is the function that must be convolved with $y(t)$ to recover $g(t)$

$$g(t) = y(t) * f'(t) \tag{2.6}$$

Substituting for $y(t)$ as given by equation (2.4)

$$g(t) = g(t) * f(t) * f'(t) \tag{2.7}$$

Now

$$g(t) = g(t) * \delta(t) \tag{2.8}$$

where $\delta(t)$ is a spike function (a unit amplitude spike at zero time); that is, a time function $g(t)$ convolved with a spike function produces an unchanged convolution output function $g(t)$. From equations (2.7) and (2.8) it follows that

$$f(t) * f'(t) = \delta(t)$$

Thus, $f'(t)$ can be derived for application in equation (2.6) to recover the input signal $g(t)$. The function $f'(t)$ represents the *deconvolution operator*.

Deconvolution is an essential aspect of seismic data processing, being used to improve seismic records by removing the adverse filtering effects encountered by seismic waves during their passage through the ground. In the seismic case, referring to equation (2.4), $y(t)$ is the seismic record resulting from the passage of a seismic wave $g(t)$ through a portion of the Earth, which acts as a filter with an impulse response $f(t)$. The particular problem with deconvolving a seismic record is that the input waveform $g(t)$ and the impulse response $f(t)$ of the Earth filter are in general unknown. Thus the 'deterministic' approach to deconvolution outlined above

cannot be employed and the deconvolution operator has to be designed using statistical methods. This special approach to the deconvolution of seismic records, known as predictive deconvolution, is discussed further in Chapter 4.

2.4.3 Correlation

Cross-correlation of two digital waveforms involves cross-multiplication of the individual waveform elements and summation of the cross-multiplication products over the common time interval of the waveforms. The cross-correlation function involves progressively sliding one waveform past the other and, for each time shift, or lag, summing the cross-multiplication products to derive the cross-correlation as a function of lag value. The cross-correlation operation is similar to convolution but does not involve folding of one of the waveforms.

Given two digital waveforms of finite length, x_i and y_i ($i = 1, 2, \ldots, n$), the cross-correlation function is given by

$$\phi_{xy}(\tau) = \sum_{i=1}^{n-\tau} x_{i+\tau} y_i \quad (-m < \tau < +m)$$

where τ is the lag and m is known as the maximum lag value of the function.

It can be shown that cross-correlation in the time domain is mathematically equivalent to multiplication of amplitude spectra and subtraction of phase spectra in frequency domain.

Clearly if two identical non-periodic waveforms are cross-correlated (Fig. 2.12) all the cross-multiplication products will sum at zero lag to give a maximum positive value. When the waveforms are displaced in time, however, the cross-multiplication products will tend to cancel out to give small values. The cross-correlation function therefore peaks at zero lag and reduces to small values at large time shifts. Two closely similar waveforms will likewise produce a cross-correlation function that is strongly peaked at zero lag. On the other hand, if two dissimilar waveforms are cross-correlated the sum of cross-multiplication products will always be near to zero due to the tendency for positive and negative products to cancel out at all values of lag. In fact, for two waveforms containing only random noise the cross-correlation function $\phi_{xy}(\tau)$ is zero for all values of τ. Thus, the cross-correlation function measures the degree of similarity of waveforms.

An important application of cross-correlation is in

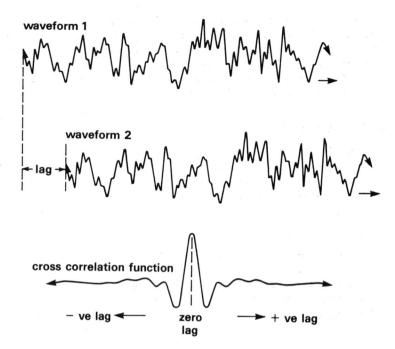

waveform 1

waveform 2

← lag →

cross correlation function

− ve lag ←   zero lag   → + ve lag

Fig. 2.12 Cross-correlation of two identical waveforms.

the detection of weak signals embedded in noise. If a waveform contains a known signal concealed in noise at unknown time, cross-correlation of the waveform with the signal function will produce a cross-correlation function centred on the time value at which the signal function and its concealed equivalent in the waveform are in phase (Fig. 2.13).

A special case of correlation is that in which a waveform is cross-correlated with itself, to give the *autocorrelation function* $\phi_{xx}(\tau)$. This function is symmetrical about a zero lag position, so that

$$\phi_{xx}(\tau) = \phi_{xx}(-\tau)$$

The autocorrelation function of a periodic waveform is also periodic, with a frequency equal to the repetition frequency of the waveform. Thus, for example, the autocorrelation function of a cosine wave is also a cosine wave. For a transient waveform, the autocorrelation function decays to small values at large values of lag. These differing properties of the autocorrelation function of periodic and transient waveforms determine one of its main uses in geophysical data processing, namely, the detection of hidden periodicities in any given waveform. Side lobes in the autocorrelation function (Fig. 2.14) are an indication of the existence of periodicities in the original waveform, and the spacing of the side lobes

defines the repetition period. This property is particularly useful in the detection and suppression of multiple reflections in seismic records (see Chapter 4).

The autocorrelation function contains all the frequency information of the original waveform but none of the phase information, the original phase relationships being replaced by a zero phase spectrum. In fact, the autocorrelation function and the square of the amplitude spectrum $A(f)$ can be shown to form a Fourier pair

$$\phi_{xx}(\tau) \leftrightarrow A(f)^2$$

Since the square of the amplitude represents the power term (energy contained in the frequency component) the autocorrelation function can be used to compute the *power spectrum* of a waveform.

2.5 DIGITAL FILTERING

In waveforms of geophysical interest, the signal is almost invariably superimposed on unwanted noise. In favourable circumstances the signal/noise ratio (SNR) is high, so that the signal is readily identified and extracted for subsequent analysis. Often the SNR is low and special processing is necessary to enhance the information content of the waveforms.

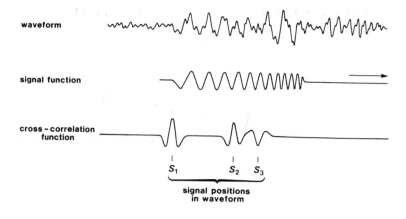

waveform

signal function

cross-correlation function

S_1 S_2 S_3

signal positions in waveform

Fig. 2.13 Cross-correlation to detect occurrences of a known signal concealed in noise. (After Sheriff 1973.)

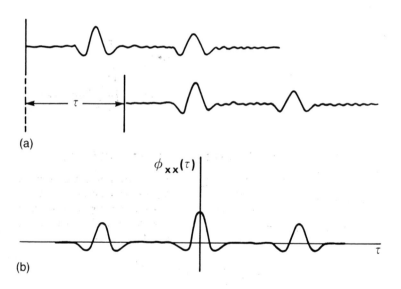

(a)

$\phi_{xx}(\tau)$

τ

(b)

Fig. 2.14 Autocorrelation of the waveform exhibiting periodicity shown in (a) produces the autocorrelation function with side lobes shown in (b). The spacing of the side lobes defines the repetition period of the original waveform.

Digital filtering is widely employed in geophysical data processing to improve SNR or otherwise improve the signal characteristics.

A very wide range of digital filters is in routine use in geophysical, and especially seismic, data processing. The two main types of digital filter are frequency filters and inverse (deconvolution) filters.

2.5.1 Frequency filters

Frequency filters discriminate against selected frequency components of an input waveform and may be low-pass (LP), high-pass (HP), band-pass (BP) or band-reject (BR) in terms of their frequency response. Frequency filters are employed when the signal and noise components of a waveform have different frequency characteristics and can therefore be separated on this basis.

Analogue frequency filtering is still in widespread use and analogue antialias (LP) filters are an essential component of analogue-to-digital conversion systems (see Section 2.2). Nevertheless, digital frequency filtering by computer offers much greater flexibility of filter design and facilitates filtering of much higher performance than can be obtained with analogue filters.

To illustrate the design of a digital frequency filter, consider the case of a LP filter whose cut-off frequency is f_c. The desired output characteristics of the ideal LP filter are represented by the amplitude spectrum shown in Fig. 2.15(a). The spectrum has a constant unit amplitude between 0 and f_c and zero

FREQUENCY DOMAIN   **TIME DOMAIN**

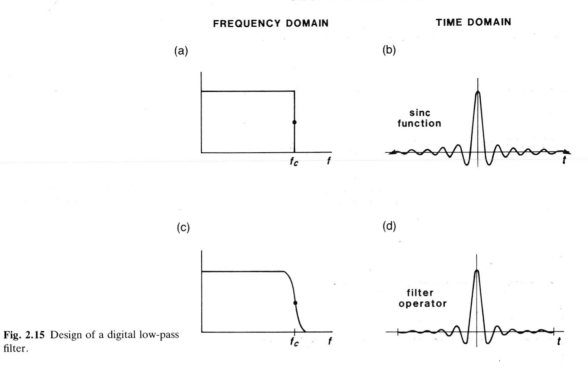

Fig. 2.15 Design of a digital low-pass filter.

amplitude outside this range: the filter would therefore pass all frequencies between 0 and f_c without attenuation and would totally suppress frequencies above f_c. This amplitude spectrum represents the transfer function of the ideal LP filter.

Inverse Fourier transformation of the transfer function into the time domain yields the impulse response of the ideal LP filter (see Fig. 2.15(b)). However, this impulse response (a sinc function) is infinitely long and must therefore be truncated for practical use as a convolution operator in a digital filter. Fig. 2.15(c) represents the frequency response of a practically realizable LP filter operator of finite length (Fig. 2.15(d)). Convolution of the input waveform with the latter will result in LP filtering with a ramped cut-off (Fig. 2.15(c)) rather than the instantaneous cut-off of the ideal LP filter.

HP, BP and BR time-domain filters can be designed in a similar way by specifying a particular transfer function in the frequency domain and using this to design a finite-length impulse response function in the time domain. As with analogue filtering, digital frequency filtering generally alters the phase spectrum of the waveform and this effect may be undesirable. However, *zero-phase filters* can be designed that facilitate digital filtering without altering the phase spectrum of the filtered signal.

2.5.2 **Inverse (deconvolution) filters**

The main applications of inverse filtering to remove the adverse effects of a previous filtering operation lie in the field of seismic data processing. A discussion of inverse filtering in the context of deconvolving seismic records is given in Chapter 4.

2.6 **PROBLEMS**

1 Over the distance between two recording sites at different ranges from a seismic source, seismic waves are found to have attenuated by 5 dB. What is the ratio of the wave amplitudes observed at the two sites?

2 In a geophysical survey, time-series data are sampled at 4 ms intervals for digital recording. (a) What is the Nyquist frequency? (b) In the absence of antialias filtering, at what frequency would noise at 200 Hz be aliased back into the Nyquist interval?

3 If a digital recording of a geophysical time series is required to have a dynamic range of 120 dB, what number of bits is required in each binary word?

4 If the digital signal $(-1, 3, -2, -1)$ is convolved with the filter operator $(2, 3, 1)$, what is the convolved output?

5 Cross-correlate the signal function $(-1, 3, -1)$ with the waveform $(-2, -4, -4, -3, 3, 1, 2, 2)$ containing signal and noise, and indicate the likely position of the signal in the waveform on the basis of the cross-correlation function.

6 A waveform is composed of two in-phase components of equal amplitude at frequencies f and $3f$. Represent the waveform in the time domain and the frequency domain.

FURTHER READING

Brigham, E.O. (1974) *The Fast Fourier Transform*. Prentice-Hall, New Jersey.

Camina, A.R. & Janacek, G.J. (1984) *Mathematics for Seismic Data Processing and Intepretation*. Graham & Trotman, London.

Claerbout, J.F. (1985) *Fundamentals of Geophysical Data Processing*. McGraw-Hill, New York.

Dobrin, M.B. & Savit, C.H. (1988) *Introduction to Geo-physical Prospecting* (4th edn). McGraw-Hill, New York.

Kanasewich, E.R. (1981) *Time Sequence Analysis in Geophysics* (3rd edn). University of Alberta Press.

Kulhánek, O. (1976) *Introduction to Digital Filtering in Geophysics*. Elsevier, Amsterdam.

Menke, W. (1989) *Geophysical Data Analysis: Discrete Inverse Theory*. Academic Press, London.

Rayner, J.N. (1971) *An Introduction to Spectral Analysis*. Pion, England.

Robinson, E.A. & Trietel, S. (1980) *Geophysical Signal Analysis*. Prentice-Hall, New Jersey.

Sheriff, R.E. & Geldart, L.P. (1983) *Exploration Seismology Vol 2: Data-processing and Interpretation*. Cambridge University Press, Cambridge.

3 / Elements of seismic surveying

3.1 INTRODUCTION

In seismic surveying, seismic waves are propagated through the Earth's interior and the travel times are measured of waves that return to the surface after refraction or reflection at geological boundaries within the ground. These travel times may be converted into depth values and, hence, the distribution of subsurface interfaces of geological interest may be systematically mapped.

Seismic surveying was first carried out in the early 1920s. It represented a natural development of the already long-established methods of earthquake seismology in which the travel times of earthquake waves recorded at seismological observatories are used to derive information on the internal structure of the Earth. Earthquake seismology provides information on the gross internal layering of the Earth, and measurement of the velocity of earthquake waves through the various Earth layers provides major clues as to their composition and constitution. In the same way, but on a smaller scale, seismic surveying provides a clear and, indeed, uniquely detailed picture of subsurface geology. It undoubtedly represents the single most important geophysical surveying method, in terms of the amount of survey activity and the very wide range of its applications.

Many of the principles of earthquake seismology are applicable to seismic surveying. However, the latter is concerned solely with the structure of the ground down to several kilometres at most and utilizes artificial seismic sources such as explosions, whose location, timing and source characteristics are, unlike earthquakes, under the direct control of the geophysicist. Seismic surveying also utilizes specialized recording systems and associated data processing and interpretation techniques.

Seismic methods are widely applied to exploration problems involving the detection and mapping of subsurface boundaries of, normally, simple geometry. The methods are particularly well suited to the mapping of layered sedimentary sequences and are therefore widely used in the search for oil and gas. The methods are also well suited, on a smaller scale, to the mapping of near-surface sediment layers, the location of the water table and, in an engineering context, site investigation of foundation conditions including the determination of depth to bedrock. Seismic surveying can be carried out on land or at sea, and it is used extensively in offshore geological surveys and the exploration for offshore resources.

In this chapter the physical principles on which seismic methods are based are reviewed at an elementary level, starting with a discussion of the nature of seismic waves and going on to consider their mode of propagation through the ground, with particular reference to reflection and refraction at subsurface interfaces between different rock types. To understand the different types of seismic wave that propagate through the ground away from a seismic source, some elementary concepts of stress and strain need to be considered.

3.2 STRESS AND STRAIN

When external forces are applied to a body, balanced internal forces are set up within it. *Stress* is a measure of the intensity of these balanced internal forces. The stress acting on an area of any surface within the body may be resolved into a component of normal stress perpendicular to the surface and a component of shearing stress in the plane of the surface.

At any point in a stressed body three orthogonal planes can be defined on which the components of stress are wholly normal stresses, that is, no shearing stresses act along them. These planes define three orthogonal axes known as the principal axes of stress, and the normal stresses acting in these directions are known as the *principal stresses*. Each principal stress represents a balance of equal-magnitude but oppositely-directed force components. The stress is said to be compressive if the forces are directed towards each other and tensile if they are directed away from each other.

If the principal stresses are all of equal magnitude within a body the condition of stress is said to be *hydrostatic*, since this is the state of stress throughout a fluid body at rest. No shearing stresses exist in a hydrostatic stress field since these cannot be sustained by a fluid body. If the principal stresses are unequal, shearing stresses exist along all surfaces within the stressed body except for the three orthogonal planes intersecting in the principal axes.

A body subjected to stress undergoes a change of shape and/or size known as *strain*. Up to a certain limiting value of stress, known as the yield strength of a material, the strain is linearly related to the applied stress (Hooke's Law). This elastic strain is reversible so that removal of stress leads to a removal of strain. If the yield strength is exceeded the strain becomes non-linear and partly irreversible (i.e. permanent strain results), and is known as plastic or ductile strain. If the stress is increased still further the body fails by fracture. A typical stress–strain curve is illustrated in Fig. 3.1.

The linear relationship between stress and strain in the elastic field is specified for any material by its various *elastic moduli*, each of which expresses the ratio of a particular type of stress to the resultant strain. Consider a rod of original length l and cross-sectional area A which is extended by an increment Δl through the application of a stretching force F to its end faces (Fig. 3.2(a)). The relevant elastic modulus is Young's modulus E, defined by

$$\text{Young's modulus } E = \frac{\text{longitudinal stress } F/A}{\text{longitudinal strain } \Delta l/l}$$

Note that extension of such a rod will be accompanied by a reduction in its diameter, i.e. the rod will suffer lateral as well as longitudinal strain.

The ratio of the lateral to the longitudinal strain is known as *Poisson's ratio* (σ).

The *bulk modulus K* expresses the stress–strain ratio in the case of a simple hydrostatic pressure P applied to a cubic element (Fig. 3.2(b)), the resultant volume strain being the change of volume Δv divided by the original volume v

$$\text{Bulk modulus } K = \frac{\text{volume stress } P}{\text{volume strain } \Delta v/v}$$

In a similar manner the *shear modulus* (μ) is defined as the ratio of shearing stress (τ) to the resultant shear strain $\tan \theta$ (Fig. 3.2(c))

$$\text{Shear modulus } \mu = \frac{\text{shearing stress } \tau}{\text{shear strain } \tan \theta}$$

Finally, the *axial modulus ψ* defines the ratio of longitudinal stress to longitudinal strain in the case when there is no lateral strain, i.e. when the material is constrained to deform uniaxially (Fig. 3.2(d))

$$\text{Axial modulus } \psi = \frac{\text{longitudinal stress } F/A}{\substack{\text{longitudinal strain} \\ \text{(uniaxial) } \Delta l/l}}$$

3.3 SEISMIC WAVES

Seismic waves are parcels of elastic strain energy that propagate outwards from a seismic source such as an earthquake or an explosion. Sources suitable for seismic surveying generate shortlived wave trains, known as pulses, that typically contain a wide range

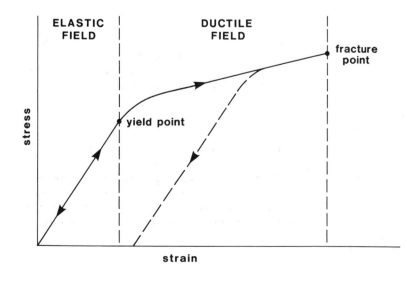

Fig. 3.1 A typical stress–strain curve for a solid body.

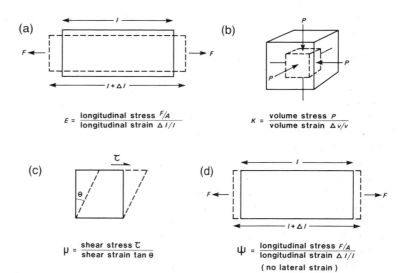

Fig. 3.2 The elastic moduli. (a) Young's modulus E. (b) Bulk modulus K. (c) Shear modulus μ. (d) Axial modulus ψ.

of frequencies. Except in the immediate vicinity of the source, the strains associated with the passage of a seismic pulse are minute and may be assumed to be elastic. On this assumption the propagation velocities of seismic pulses are determined by the elastic moduli and densities of the materials through which they pass. There are two groups of seismic waves, *body waves* and *surface waves*.

3.3.1 Body waves

Body waves of two types can propagate through the body of an elastic solid. *Compressional waves* (the longitudinal, primary or *P*-waves of earthquake seismology) propagate by compressional and dilatational uniaxial strains in the direction of wave travel. Particle motion associated with the passage of a compressional wave involves oscillation, about a fixed point, in the direction of wave propagation (Fig. 3.3(a)). *Shear waves* (the transverse, secondary or *S*-waves of earthquake seismology) propagate by a pure shear strain in a direction perpendicular to the direction of wave travel. Individual particle motions involve oscillation, about a fixed point, in a plane at right angles to the direction of wave propagation (Fig. 3.3(b)). If all the particle oscillations are confined to a plane, the shear wave is said to be plane-polarized.

The velocity of propagation of a body wave in any material is given by

$$v = \left[\frac{\text{appropriate elastic modulus of material}}{\text{density } \rho \text{ of material}}\right]^{1/2}$$

Hence the velocity v_p of a compressional body wave, which involves a uniaxial compressional strain, is given by

$$v_p = \left[\frac{\psi}{\rho}\right]^{1/2}$$

or, since $\psi = K + 4/3\mu$, by

$$v_p = \left[\frac{K + 4/3\mu}{\rho}\right]^{1/2}$$

and the velocity v_s of a shear body wave, which involves a pure shear strain, is given by

$$v_s = \left[\frac{\mu}{\rho}\right]^{1/2}$$

It will be seen from these equations that compressional waves always travel faster than shear waves in the same medium. The ratio v_p/v_s in any material is determined solely by the value of Poisson's ratio (σ) for that material

$$v_p/v_s = \left[\frac{2(1-\sigma)}{(1-2\sigma)}\right]^{1/2}$$

and since Poisson's ratio for consolidated rocks is typically about 0.25, $v_p \approx 1.7v_s$.

Body waves are non-dispersive, i.e. all frequency components in a wave train or pulse travel through any material at the same velocity, determined only by the elastic moduli and density of the material.

(a)   P - wave

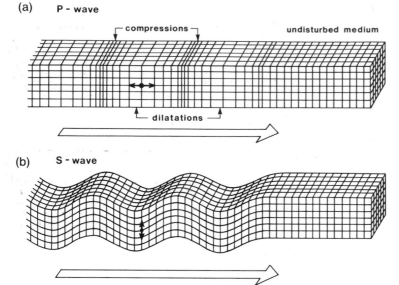

Fig. 3.3 Elastic deformations and ground particle motions associated with the passage of body waves. (a) A *P*-wave. (b) An *S*-wave. (From Bolt 1982.)

3.3.2 Surface waves

In a bounded elastic solid, seismic waves known as surface waves can propagate along the boundary of the solid.

Rayleigh waves propagate along a free surface, or along the boundary between two dissimilar solid media, the associated particle motions being elliptical in a plane perpendicular to the surface and containing the direction of propagation (Fig. 3.4(a)). The orbital particle motion is in the opposite sense to the circular particle motion associated with an oscillatory water wave, and is therefore sometimes described as retrograde. A further major difference between Rayleigh waves and oscillatory water waves is that the former involve a shear strain and are thus restricted to solid media. The amplitude of Rayleigh waves decreases exponentially with distance below the surface. They have a propagation velocity lower than that of shear body waves and in a homogeneous half space they would be non-dispersive. In practice, Rayleigh waves travelling round the surface of the Earth are observed to be dispersive, their waveform undergoing progressive change during propagation as a result of the different frequency components travelling at different velocities. This dispersion is directly attributable to velocity variation with depth in the Earth's interior and, indeed, analysis of the observed pattern of dispersion is a powerful method of studying the velocity structure of the lithosphere and asthenosphere (Knopoff 1983).

In a layered solid a second set of surface waves, known as *Love waves*, appears in the surface layer if its shear body wave velocity v_s is lower than that of the underlying layer. Love waves are polarized shear waves with an associated oscillatory particle motion parallel to the free surface and perpendicular to the direction of wave motion (Fig. 3.4(b)). The velocity of Love waves is intermediate between the shear wave velocity of the surface layer and that of deeper layers, and Love waves are inherently dispersive. The observed pattern of Love wave dispersion can be used in a similar way to Rayleigh wave dispersion to study the structure of the lithosphere and asthenosphere (Knopoff 1983).

Although recent experimental surveys of shallow structure have been carried out using shear waves and surface waves (for example, local surface wave dispersion patterns can be used to study the thickness and structure of sedimentary basins), the vast bulk of seismic surveying utilizes only compressional waves and in the following account attention will be concentrated on these waves.

3.3.3 Waves and rays

A seismic pulse propagates outwards from a seismic source at a velocity determined by the physical

(a)

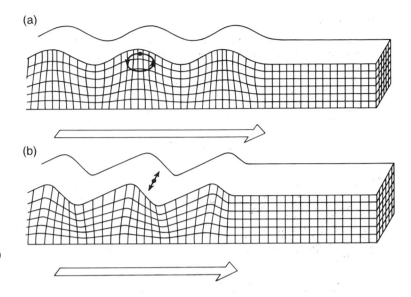

(b)

Fig. 3.4 Elastic deformations and
ground particle motions associated
with the passage of surface waves. (a)
A Rayleigh wave. (b) A Love wave.
(From Bolt 1982.)

properties of the surrounding rocks. If the pulse
travels through a homogeneous rock it will travel at
the same velocity in all directions away from the
source so that at any subsequent time the wavefront,
defined as the locus of all points which the pulse has
reached, will be a sphere. *Seismic rays* are defined as
thin pencils of seismic energy travelling along ray
paths that, in isotropic media, are everywhere perpendicular to wavefronts (Fig. 3.5). Rays have no
physical significance but represent a useful concept
in discussing travel paths of seismic energy through
the ground.

It should be noted that the propagation velocity of
a seismic wave is the velocity with which the seismic
energy travels through a medium. This is *not* the
same as the velocity of a particle of the medium
perturbed by the passage of the wave. In the case of
compressional body waves, for example, their propagation velocity through rocks is typically a few
thousand metres per second. The associated oscillatory ground motions involve *particle velocities* that
depend on the amplitude of the wave. For the weak
seismic events routinely recorded in seismic surveys
particle velocities may be as small as $10^{-8}\,\mathrm{m\,s^{-1}}$ and
involve ground displacements of only about $10^{-10}\,\mathrm{m}$.
The detection of seismic waves involves measuring
these very small particle velocities.

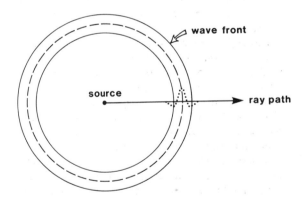

Fig. 3.5 The relationship of a ray path to the associated
wavefront.

3.4 COMPRESSIONAL WAVE VELOCITIES OF ROCKS

By virtue of their various compositions, textures
(e.g. grain shape and degree of sorting), porosities
and contained pore fluids, rocks differ in their elastic
moduli and densities and, hence, in their seismic
velocities. Information on the compressional wave
velocity v_p of rock layers encountered by seismic

surveys is important for two main reasons: firstly it is necessary for the conversion of seismic wave travel times into depths; secondly, it provides an indication of the lithology of a rock or, in some cases, the nature of the pore fluids contained within it.

Rock velocities may be measured *in situ* by field measurements, or in the laboratory using suitably prepared rock samples. In the field, seismic surveys yield estimates of velocity for rock layers delineated by reflecting or refracting interfaces, as discussed in detail in Chapters 4 and 5. If boreholes exist in the vicinity of a seismic survey, it may be possible to correlate velocity values so derived with individual rock units encountered within borehole sequences. As discussed in Chapter 11, velocity may also be measured directly in boreholes using a sonic probe, which emits high frequency pulses and measures the travel time of the pulses through a small vertical interval of wall rock. Drawing the probe up through the borehole yields a *sonic log*, or continuous velocity log (CVL), which is a record of velocity variation through the borehole section (Section 11.8, Fig. 11.14).

In the laboratory, velocities are determined by measuring the travel time of high frequency (about 1 MHz) acoustic pulses transmitted through cylindrical rock specimens. By this means, the effect on velocity of varying temperature, confining pressure, pore fluid pressure or composition may be quantitatively assessed. It is important to note that laboratory measurements at low confining pressures are of doubtful validity. The intrinsic velocity of a rock is not normally attained in the laboratory below a confining pressure of about 100 MPa (megapascals), or 1 kbar, at which pressure the original solid contact between grains characteristic of the pristine rock is re-established.

The following general findings of velocity studies are noteworthy:
1 Compressional wave velocity increases with confining pressure (very rapidly over the first 100 MPa).
2 Sandstone and shale velocities show a systematic increase with depth of burial and with age, due to the combined effects of progressive compaction and cementation.
3 For a wide range of sedimentary rocks the compressional wave velocity is related to density, and well-established velocity–density curves have been published (Sheriff & Geldart 1983; see Section 6.9, Fig. 6.16). Hence, the densities of inaccessible subsurface layers may be predicted if their velocity is known from seismic surveys.

4 The presence of gas in sedimentary rocks reduces the elastic moduli, Poisson's ratio and the v_p/v_s ratio. v_p/v_s ratios greater than 2.0 are characteristic of unconsolidated sand, whilst values less than 2.0 may indicate either a consolidated sandstone or a gas-filled unconsolidated sand. The potential value of v_s in detecting gas-filled sediments accounts for the current interest in shear wave seismic surveying.

Typical compressional wave velocity values and ranges for a wide variety of Earth materials are given in Table 3.1.

3.5 ATTENUATION OF SEISMIC ENERGY ALONG RAY PATHS

As a seismic pulse propagates, the original energy E transmitted outwards from the source becomes distributed over a spherical shell of expanding radius. If the radius of the shell is r, the amount of energy contained within a unit area of the shell is $E/4\pi r^2$. Along a ray path, therefore, the energy contained in the ray falls off as r^{-2} due to the effect of the *geometrical spreading* of the energy. Wave amplitude which, within a homogeneous material, is proportional to the square root of the wave energy, therefore falls off as r^{-1}.

A further cause of energy loss along a ray path arises because the ground is imperfectly elastic in its response to the passage of seismic waves. Elastic energy is gradually absorbed into the medium by internal frictional losses, leading eventually to the total disappearance of the seismic disturbance. *The absorption coefficient* (α) expresses the proportion of energy lost during transmission through a distance equivalent to a complete wavelength λ. Values of α for common Earth materials range from 0.25 to 0.75 dB λ^{-1}.

Over the range of frequencies utilized in seismic surveying the absorption coefficient is normally assumed to be independent of frequency. If the amount of absorption per wavelength is constant, it follows that higher frequency waves attenuate more rapidly than lower frequency waves as a function of time or distance. To illustrate this point, consider two waves with frequencies of 10 Hz and 100 Hz to propagate through a rock in which $v_p = 2.0$ km s^{-1} and $\alpha = 0.5$ dB λ^{-1}. The 100 Hz wave ($\lambda = 20$ m) will be attenuated due to absorption by 5 dB over a distance of 200 m, whereas the 10 Hz wave ($\lambda = 200$ m) will be attenuated by only 0.5 dB over the same distance. The shape of a seismic pulse with a broad frequency content therefore changes con-

Table 3.1. Compressional wave velocities in Earth materials.

	v_p (km s^{-1})
Unconsolidated materials	
Sand (dry)	0.2−1.0
Sand (water saturated)	1.5−2.0
Clay	1.0−2.5
Glacial till (water saturated)	1.5−2.5
Permafrost	3.5−4.0
Sedimentary rocks	
Sandstones	2.0−6.0
Tertiary sandstone	2.0−2.5
Pennant sandstone (Carboniferous)	4.0−4.5
Cambrian quartzite	5.5−6.0
Limestones	2.0−6.0
Cretaceous chalk	2.0−2.5
Jurassic oolites and bioclastic limestones	3.0−4.0
Carboniferous limestone	5.0−5.5
Dolomites	2.5−6.5
Salt	4.5−5.0
Anhydrite	4.5−6.5
Gypsum	2.0−3.5
Igneous/Metamorphic rocks	
Granite	5.5−6.0
Gabbro	6.5−7.0
Ultramafic rocks	7.5−8.5
Serpentinite	5.5−6.5
Pore fluids	
Air	0.3
Water	1.4−1.5
Ice	3.4
Petroleum	1.3−1.4
Other materials	
Steel	6.1
Iron	5.8
Aluminium	6.6
Concrete	3.6

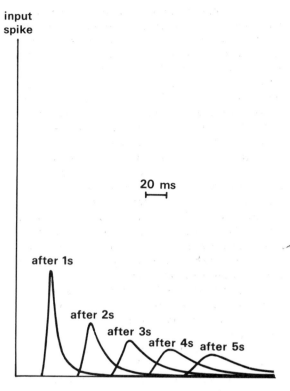

Fig. 3.6 The progressive change of shape of an original spike pulse during its propagation through the ground due to the effects of absorption. (After Anstey 1977.)

tinuously during propagation due to the progressive loss of the higher frequencies. In general, the effect of absorption is to produce a progressive lengthening of the seismic pulse (Fig. 3.6).

3.6 RAY PATHS IN LAYERED MEDIA

At an interface between two rock layers there is generally a change of propagation velocity resulting from the difference in physical properties of the two layers. At such an interface the energy within an incident seismic pulse is partitioned into transmitted and reflected pulses. The relative amplitudes of the transmitted and reflected pulses, in terms of the velocities and densities of the two layers, are given by Zoeppritz's equations (Telford *et al.* 1976).

3.6.1 Reflection and transmission of normally incident seismic rays

Consider a compressional ray of amplitude A_0 normally incident on an interface between two media of differing velocity and density (Fig. 3.7). A transmitted ray of amplitude A_2 travels on through the interface in the same direction as the incident ray and a reflected ray of amplitude A_1 returns back along the path of the incident ray.

The total energy of the transmitted and reflected rays must, of course, equal the energy of the incident ray. The relative proportions of energy transmitted

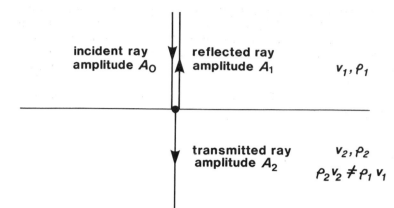

Fig. 3.7 Reflected and transmitted rays associated with a ray normally incident on an interface of acoustic impedance contrast.

and reflected are determined by the contrast in *acoustic impedance Z* across the interface. The acoustic impedance of a rock is the product of its density and its compressional wave velocity, i.e. $Z = \rho v$. It is difficult to relate acoustic impedance to a tangible rock property but, in general, the harder a rock the higher is its acoustic impedance.

Acoustic impedance is closely analogous to electrical impedance and, just as the maximum transmission of electrical energy requires a matching of electrical impedances, so the maximum transmission of seismic energy requires a matching of acoustic impedances. Hence, the smaller the contrast in acoustic impedance across a rock interface the greater is the proportion of energy transmitted through the interface.

The *reflection coefficient R* is the ratio of the amplitude A_1 of the reflected ray to the amplitude A_0 of the incident ray

$$R = A_1/A_0$$

For a normally incident ray this is given, from solution of Zoeppritz's equations, by

$$R = \frac{\rho_2 v_2 - \rho_1 v_1}{\rho_2 v_2 + \rho_1 v_1} = \frac{Z_2 - Z_1}{Z_2 + Z_1}$$

where ρ_1, v_1, Z_1 and ρ_2, v_2, Z_2 are the density, P-wave velocity and acoustic impedance values in the first and second layers, respectively. From this equation it follows that $-1 \le R \le +1$. A negative R-value signifies a phase change of π (180°) in the reflected ray.

The *transmission coefficient T* is the ratio of the amplitude A_2 of the transmitted ray to the amplitude A_0 of the incident ray

$$T = A_2/A_0$$

For a normally incident ray this is given, from solution of Zoeppritz's equations, by

$$T = \frac{2Z_1}{Z_2 + Z_1}$$

Reflection and transmission coefficients are sometimes expressed in terms of energy rather than wave amplitude. If energy intensity I is defined as the amount of energy flowing through a unit area normal to the direction of wave propagation in unit time, so that I_0, I_1 and I_2 are the intensities of the incident, reflected and transmitted rays respectively, then

$$R' = I_1/I_0 = \left[\frac{Z_2 - Z_1}{Z_2 + Z_1}\right]^2$$

and

$$T' = I_2/I_0 = \frac{4Z_1Z_2}{(Z_2 + Z_1)^2}$$

where R' and T' are the reflection and transmission coefficients expressed in terms of energy.

If R or $R' = 0$, all the incident energy is transmitted. This is the case when there is no contrast of acoustic impedance across an interface (i.e. $Z_1 = Z_2$), even if the density and velocity values are different in the two layers. If R or $R' = +1$ or -1, all the incident energy is reflected. A good approximation to this situation occurs at the free surface of a water layer: rays travelling upwards from an explosion in a water layer are almost totally reflected back from the water surface with a phase change of π ($R = -0.9995$).

Values of reflection coefficient R for interfaces between different rock types rarely exceed ± 0.5 and are typically less than ± 0.2. Thus, normally, the bulk of seismic energy incident on a rock interface is transmitted and only a small proportion is reflected.

By use of an empirical relationship between velocity and density (see also Section 6.9), it is possible to calculate the reflection coefficient from velocity information alone (Gardner *et al.* 1974, Meckel & Nath 1977)

$$R = 0.625 \ln (v_1/v_2)$$

3.6.2 Reflection and refraction of obliquely incident rays

When a P-ray is obliquely incident on an interface of acoustic impedance contrast, reflected and transmitted P-rays are generated as in the case of normal incidence. Additionally, some of the incident compressional energy is converted into reflected and transmitted S-rays (Fig. 3.8) that are polarized in a vertical plane. Zoeppritz's equations show that the amplitudes of the four phases are a function of the angle of incidence θ. The converted rays may attain a significant magnitude at large angles of incidence; they are, however, of only minor interest in seismic surveying and are not considered further here.

In the case of oblique incidence, the transmitted P-ray travels through the lower layer with a changed direction of propagation (Fig. 3.9) and is referred to as a *refracted ray*. The situation is directly analogous to the behaviour of a light ray obliquely incident on the boundary between, say, air and water and Snell's

Law of Optics applies equally to the seismic case. The generalized form of Snell's Law states that for any ray the quantity $\sin i/v$ remains a constant, known as the *ray parameter p*, where i is the angle of inclination of the ray in a layer in which it is travelling with a velocity v.

For the refracted P-ray shown in Fig. 3.9, therefore

$$\frac{\sin \theta_1}{v_1} = \frac{\sin \theta_2}{v_2}$$

or

$$\frac{\sin \theta_1}{\sin \theta_2} = \frac{v_1}{v_2}$$

Note that if $v_2 > v_1$ the ray is refracted away from the normal to the interface, i.e. $\theta_2 > \theta_1$.

Snell's Law applies to the reflected ray also, from which it follows that the angle of reflection equals the angle of incidence (Fig. 3.9).

3.6.3 Critical refraction

When the velocity is higher in the underlying layer there is a particular angle of incidence, known as the *critical angle θ_c*, for which the angle of refraction is 90°. This gives rise to a critically refracted ray that travels along the interface at the higher velocity v_2. At any greater angle of incidence there is total internal reflection of the incident energy (apart from converted S-rays over a further range of angles). The critical angle is given by

$$\frac{\sin \theta_c}{v_1} = \frac{\sin 90°}{v_2} = \frac{1}{v_2}$$

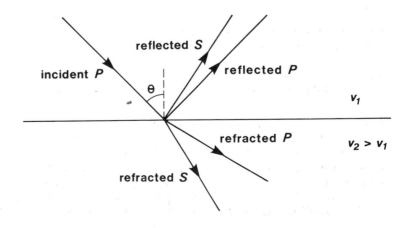

Fig. 3.8 Reflected and refracted P-
and S-rays generated by a P-ray
obliquely incident on an interface of
acoustic impedance contrast.

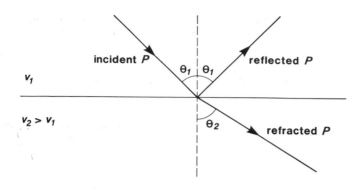

Fig. 3.9 Reflected and refracted P-rays associated with a P-ray obliquely incident on an interface of acoustic impedance contrast.

so that

$$\theta_c = \sin^{-1}(v_1/v_2)$$

The passage of the critically refracted ray along the top of the lower layer causes a perturbation in the upper layer that travels forward at the velocity v_2, which is greater than the seismic velocity v_1 of the layer. The situation is analogous to that of a projectile travelling through air at a velocity greater than the velocity of sound in air and the result is the same, namely, the generation of a shock wave. This wave is known as a *head wave* in the seismic case, and it passes up obliquely through the upper layer towards the surface (see Fig. 3.10). Any ray associated with the head wave is inclined at the critical angle θ_c. By means of the head wave, seismic energy is returned to the surface after critical refraction in an underlying layer of higher velocity.

3.6.4 Diffraction

In the above discussion of the reflection and transmission of seismic energy at interfaces of acoustic impedance contrast it was implicitly assumed that the interfaces were continuous and of low curvature. At abrupt discontinuities in interfaces, or structures whose radius of curvature is shorter than the wavelength of incident waves, the laws of reflection and refraction no longer apply. Such phenomena give rise to a radial scattering of incident seismic energy known as *diffraction*. Common sources of diffraction in the ground include the edges of faulted layers (Fig. 3.11) and small isolated objects, such as boulders, in an otherwise homogeneous layer.

Diffracted phases are commonly observed in seismic recordings and are sometimes difficult to discriminate from reflected and refracted phases, as discussed in Chapter 4.

3.7 REFLECTION AND REFRACTION SURVEYING

Consider the simple geological section shown in Fig. 3.12 involving two homogeneous layers of seismic velocities v_1 and v_2 separated by a horizontal interface at a depth z, the compressional wave velocity being higher in the underlying layer (i.e. $v_2 > v_1$).

From a near-surface seismic source S there are three types of ray path by which energy reaches the surface at a distance from the source, where it may be recorded by a suitable detector as at D, a horizontal distance x from S. The *direct ray* travels along a straight line through the top layer from source to detector at velocity v_1. The *reflected ray* is obliquely incident on the interface and is reflected back through the top layer to the detector, travelling along its entire path at the top layer velocity v_1. The *refracted ray* travels obliquely down to the interface at velocity v_1, along a segment of the interface at the higher velocity v_2, and back up through the upper layer at v_1.

The travel time of a direct ray is given simply by

$$t_{DIR} = x/v_1$$

which defines a straight line of slope $1/v_1$ passing through the time–distance origin.

The travel time of a reflected ray is given by

$$t_{RFL} = (x^2 + 4z^2)^{1/2}/v_1$$

which, as discussed in Chapter 4, is the equation of an hyperbola.

The travel time of a refracted ray (for derivation see Chapter 5) is given by

$$t_{RFR} = x/v_2 + \frac{2z(v_2^2 - v_1^2)^{1/2}}{v_1 v_2}$$

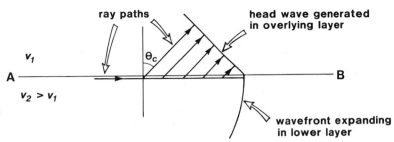

Fig. 3.10 Generation of a head wave in the upper layer by a wave propagating through the lower layer.

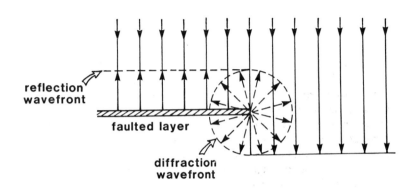

Fig. 3.11 Diffraction caused by the truncated end of a faulted layer.

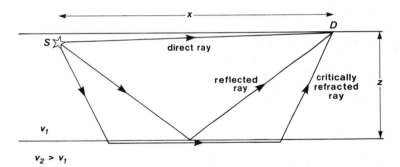

Fig. 3.12 Direct, reflected and refracted ray paths from a near surface source to a surface detector in the case of a simple two-layer model.

which is the equation of a straight line having a slope of $1/v_2$ and an intercept on the time axis of $2z(v_2^2 - v_1^2)^{1/2}/v_1 v_2$.

Travel-time curves, or time–distance curves, for direct, refracted and reflected rays are illustrated in Fig. 3.13. By suitable analysis of the travel-time curve for reflected *or* refracted rays it is possible to compute the depth to the underlying layer. This provides two independent seismic surveying methods for locating and mapping subsurface interfaces, namely, *reflection surveying* and *refraction surveying*. These have their own distinctive methodologies and fields of application and they are discussed separately in detail in Chapters 4 and 5. However, some general

remarks about the two methods may be made here with reference to the travel-time curves of Fig. 3.13. The curves are more complicated in the case of a multilayered model, but the following remarks still apply.

The first arrival of seismic energy at a surface detector offset from a surface source is always a direct ray or a refracted ray. The direct ray is overtaken by a refracted ray at the *crossover distance* x_{cros}. Beyond this offset distance the first arrival is always a refracted ray. Since critically refracted rays travel down to the interface at the critical angle there is a certain distance, known as the *critical distance* x_{crit}, within which refracted energy will not

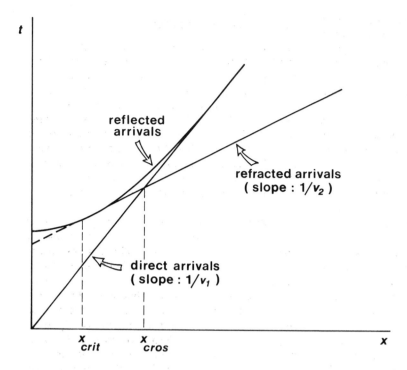

Fig. 3.13 Travel-time curves for direct, reflected and refracted rays in the case of a simple two-layer model.

be returned to surface. At the critical distance, the travel times of reflected rays and refracted rays coincide because they follow effectively the same path. Reflected rays are never first arrivals: they are always preceded by direct rays and, beyond the critical distance, by refracted rays also.

The above characteristics of the travel-time curves determine the methodology of refraction and reflection surveying. In refraction surveying, recording ranges are chosen to be sufficiently large to ensure that the crossover distance is well exceeded in order that refracted rays may be detected as first arrivals of seismic energy. Indeed, some types of refraction survey consider only these first arrivals, which can be detected with unsophisticated field recording systems. In general this approach means that the deeper a refractor, the greater is the range over which recordings of refracted arrivals need to be taken.

In reflection surveying, by contrast, reflected phases are sought that are never first arrivals and are normally of very low amplitude because geological reflectors tend to have small reflection coefficients. Consequently, reflections are normally concealed in seismic records by higher amplitude events such as direct or refracted body waves and surface waves.

Reflection surveying methods therefore have to be capable of discriminating between reflected energy and many types of synchronous noise. Recordings are normally restricted to small offset distances, well within the critical distance for the reflecting interfaces of main interest. However, in multichannel reflection surveying recordings are conventionally taken over a significant range of offset distances, for reasons that are discussed fully in Chapter 4.

3.8 SEISMIC SOURCES AND THE SEISMIC/ACOUSTIC SPECTRUM

A seismic source is a localized region within which the sudden release of energy leads to a rapid stressing of the surrounding medium. Most seismic sources preferentially generate the compressional wave energy that is mainly utilized in seismic surveying.

There is a very wide variety of seismic sources, characterized by differing energy levels and frequency characteristics. In general a seismic source contains a wide range of frequency components within the range from 1 Hz to a few hundred hertz, though the energy is often concentrated in a narrower frequency band. In addition to the seismic sources there are also several acoustic sources that generate acoustic waves (i.e. sound waves in water or air) which are useful in marine seismic surveying. The complete seismic/acoustic spectrum is shown in Fig. 3.14.

Many considerations govern the selection of a suitable seismic source for a particular survey application. The general problem in seismic surveying is to recognize a seismic signal that has been markedly attenuated by propagation through the ground and which is embedded in the general background level of seismic noise that characterizes any recording site. There are many sources of noise in the seismic spectrum including microseisms (caused by weak natural sources such as wind or water waves), industrial activity and traffic vibration. There is, therefore, an inherent problem of signal:noise ratio (SNR) in seismic surveying. This problem becomes extreme when the SNR reduces below unity.

Source characteristics can be modified by the use of several sources in an array designed, for example, to improve the frequency spectrum of the transmitted pulse. This matter is taken up in Chapter 4 when discussing the design parameters of seismic reflection surveys. In this section the various types of seismic/acoustic source in common use are introduced.

3.8.1 Explosive sources

On land, explosives are normally detonated in shallow shot holes to improve the coupling of the energy source with the ground and to minimize surface damage. An inherent problem of explosions at sea is the generation of *bubble pulses* caused by oscillation of the high-pressure gas bubble resulting from the initial explosion. Bubble pulses have the effect of unduly lengthening the seismic pulse (Fig. 3.15). Steps can, however, be taken to suppress the effect of the bubble pulse by detonating near to the water surface so that the gas bubble escapes into the air.

Explosives offer a reasonably cheap and highly efficient seismic source with a wide frequency spectrum, but their use normally requires special permission and presents logistical difficulties of storage and transportation. They are slow to use on land because of the need to drill shot holes. Their main shortcoming, however, is that they do not provide the type of precisely repeatable source signature required by some modern processing techniques, nor can the detonation of explosives be repeated at fixed and precise time intervals as required for efficient reflection profiling at sea carried out by survey vessels underway.

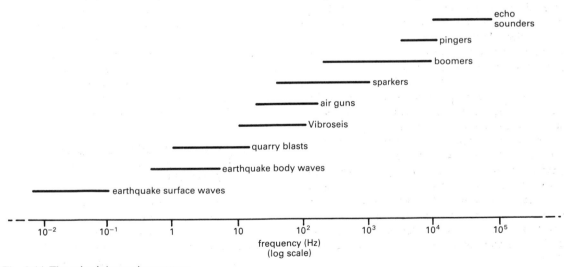

Fig. 3.14 The seismic/acoustic spectrum.

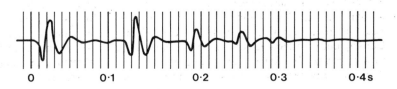

Fig. 3.15 The train of bubble pulses associated with the detonation of an explosive charge in water. (After Sheriff 1973.)

3.8.2 **Non-explosive sources**

LAND SOURCES

The most common method of reflection surveying on land utilizing a non-explosive source is the Vibroseis® method. This uses truck-mounted vibrators to pass into the ground an extended vibration of low amplitude and continuously varying frequency, known as a *sweep signal*. A typical sweep signal lasts from several seconds up to a few tens of seconds and varies progressively in frequency between limits of about 10 and 80 Hz. The field recordings consist of overlapping reflected wave trains of very low amplitude concealed in the ambient seismic noise. In order both to increase the signal-to-noise ratio and to shorten the pulse length, each recorded seismogram is cross-correlated (see Section 2.4.3) with the known sweep signal to produce a correlated seismogram or *correlogram*. The correlogram has a similar appearance to the type of seismogram that would be obtained with a high-energy impulsive source such as an explosion, but the seismic arrivals appear as symmetrical (zero-phase) wavelets known as *Klauder wavelets* (Fig. 3.16).

The Vibroseis® source is quick and convenient to use and produces a precisely known and repeatable signal. The vibrator unit needs a firm base on which to operate, such as a tarmac road, and it will not work well on soft ground. The peak force of a vibrator is only about 10^5 N and to increase the transmitted energy for deep penetration surveys, vibrators are typically employed in groups with a phase-locked response. Multiple sweeps are commonly employed, the recordings from individual sweeps being added together (stacked) to increase the signal:noise ratio. A particular advantage of vibrators is that they can be used in towns since they cause no damage or significant disturbance to the environment. The cross-correlation method of extracting the signal is also capable of coping with the inherently high noise levels of urban areas.

The principle of using a precisely known source signature of long duration is extended with the *Mini-Sosie* source. A pneumatic hammer delivers a random sequence of impacts to a base plate, thus transmitting a pulse-encoded signal of low amplitude into the ground. The source signal is recorded by a detector on the base plate and used to cross-correlate with the field recordings of reflected arrivals of the pulse-encoded signal from buried interfaces. Peaks in the cross-correlation function reveal the positions of reflected signals in the recordings.

The horizontal impact of a weight on to one side

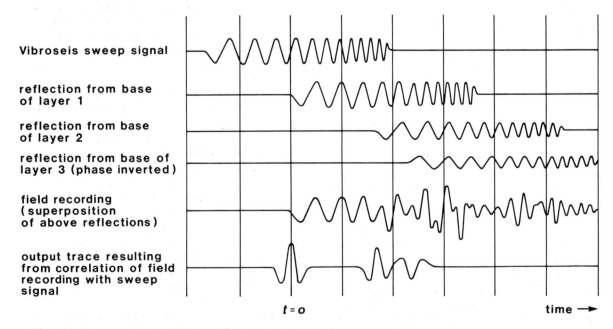

Vibroseis sweep signal

reflection from base of layer 1

reflection from base of layer 2

reflection from base of layer 3 (phase inverted)

field recording (superposition of above reflections)

output trace resulting from correlation of field recording with sweep signal

t = o   time →

Fig. 3.16 Cross-correlation of a Vibroseis® seismogram with the input sweep signal to locate the positions of occurrence of reflected arrivals.

of a vertical plate partially embedded in the ground can be used as a source for shear wave seismology, in which special shear wave detectors are used. One application of shear wave seismology is in engineering site investigation where the separate measurement of v_p and v_s for near-surface layers allows direct calculation of Poisson's ratio and estimation of the elastic moduli, which provide valuable information on the *in situ* geotechnical properties of the ground.

MARINE SOURCES

Several sources, having different energy levels and frequency characteristics, are available for marine reflection surveying. Before describing these sources it should be noted that seismic reflection surveys are normally required to achieve a specific depth of penetration and a specific resolution (i.e. ability to resolve individual, closely-spaced reflectors) and that a source must be chosen appropriate to the specified task.

The resolution is basically determined by the pulse length: for a pulse of any particular length there is a minimum separation of reflectors below which the reflected pulses will overlap in time in the seismic recording. Although the pulse length can be shortened at the processing stage by deconvolution (see Section 4.7.2) many seismic sources are used in conjunction with simple seismic profiling systems in which the analogue signal from the receiver is amplified, band-pass filtered and fed directly to a chart recorder (see Chapter 4). In such systems the resolution of the resulting seismic record is inherently limited by the recorded pulse length. Since the higher energy sources necessary for deeper penetration are characterized by lower dominant frequencies and longer pulse lengths there is generally a trade-off between penetration and resolution: the deeper the penetration the lower will be the resolving power. The common types of marine sources are described briefly below.

Air guns (Fig. 3.17(a)) are pneumatic sources in which a chamber is charged with compressed air fed through a hose from a shipboard compressor and the air is released, by electrical triggering, through side vents into the water in the form of a high-pressure bubble. The operating pressure is typically 10–15 MPa. A wide range of chamber sizes is available, leading to different energy outputs and frequency characteristics. For deep penetration surveys the total energy transmitted may be increased by the use of arrays of air guns mounted on a frame that is towed behind the survey vessel. The primary pulse generated by an air gun is followed by a train of bubble pulses that increase the overall length of the pulse. With some loss of peak energy output, the growth of bubble pulses can be effectively suppressed by reducing the rate at which the compressed air is released into the water layer. Arrays of guns of differing dimensions and, therefore, different bubble pulse periods can be used to produce a high-energy source in which primary pulses interfere constructively whilst bubble pulses interfere destructively (Fig. 3.18).

Whilst vibrator and air gun sources were developed for land and marine surveys respectively, it is of interest to note that both have subsequently been modified for operation in another environment to that for which they were originally designed. Thus experiments have been carried out using marine vibrator units, with special baseplates, deployed in fixtures attached to a survey vessel (Baeten *et al* 1988). In a similar way, air guns enclosed in large water-filled bags that can be lowered on to the ground surface have been installed in truck-mounted systems for use in land surveys. Such applications have not become widespread.

Water guns (Fig. 3.17(b)) are pneumatic sources in which the compressed air, rather than being released into the water layer, is used to drive a piston that ejects a water jet into the surrounding water. A vacuum cavity is created behind the advancing water jet and this implodes under the influence of the ambient hydrostatic pressure generating a strong acoustic pulse free of bubble oscillations. Since the implosion represents collapse into a vacuum, no gaseous material is compressed to 'bounce back' as a bubble pulse. The resulting short pulse length offers a potentially higher resolution than is achieved with air guns.

Several marine sources utilize explosive mixtures of gases. In *sleeve exploders*, propane and oxygen are piped into a submerged flexible rubber sleeve where the gaseous mixture is fired by means of a spark plug. The products of the resultant explosion cause the sleeve to expand rapidly generating a shock wave in the surrounding water. The exhaust gases are vented to surface through a valve that opens after the explosion, thus attenuating the growth of bubble pulses.

Sparkers, *boomers* and *pingers* are devices for converting electrical energy into acoustic energy. The sparker pulse is generated by the discharge of a

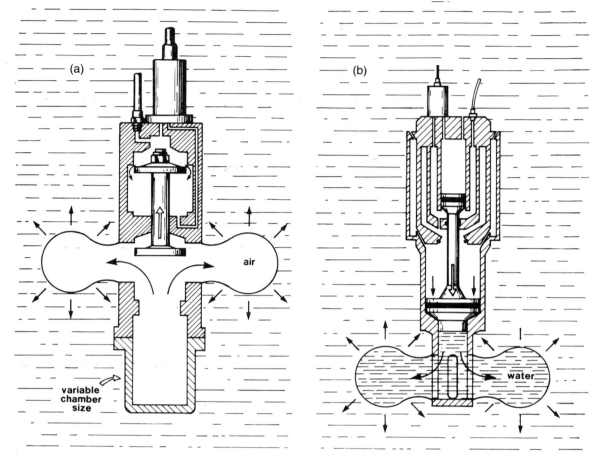

Fig. 3.17 Schematic cross-sections through (a) a Bolt air gun and (b) a Sodera water gun to illustrate the principles of operation. (Redrawn with permission of Bolt Associates and Sodera Ltd.)

large capacitor bank directly into the sea water through an array of electrodes towed in a frame behind the survey vessel. Operating voltages are typically 3.5 to 4.0 kV and peak currents may exceed 200 A. This electrical discharge leads to the formation and rapid growth of a plasma bubble and the consequent generation of an acoustic pulse. The boomer source comprises a rigid aluminium plate attached by a spring-loaded mounting to a resin block in which is embedded a heavy-duty spiral coil. A capacitor bank is discharged through the coil and the electromagnetic field thus generated sets up eddy currents in the aluminium plate. These currents generate a secondary field that opposes the primary field and the plate is rapidly repulsed, setting up a compressional wave in the water. The device is

typically towed behind the survey vessel in a catamaran mounting. Sparkers and boomers generate broad band acoustic pulses and can be operated over a wide range of energy levels so that the source characteristics can to some extent be tailored to the needs of a particular survey. In general, boomers offer better resolution (down to 0.5 m) but more restricted depth penetration (a few hundred metres maximum).

Pingers consist of small ceramic piezoelectric transducers, mounted in a towing fish, which when activated by an electrical impulse emit a very short, high-frequency acoustic pulse of low energy. They offer a very high resolving power (down to 0.1 m) but limited penetration (a few tens of metres in mud, much less in sand or rock). They are useful in

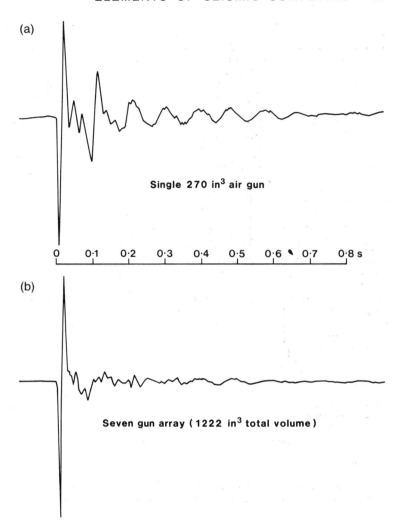

Fig. 3.18 Comparison of the source signatures of (a) a single air gun (peak pressure: 4.6 bar metres) and (b) a seven-gun array (peak pressure: 19.9 bar metres). Note the effective suppression of bubble pulses in the latter case. (Redrawn with permission of Bolt Associates.)

offshore engineering applications such as surveys of proposed routes for submarine pipelines.

Further discussion of the use of air guns, sparkers, boomers and pingers in single-channel seismic reflection profiling systems is given in Section 4.10.

3.9 SEISMIC DATA ACQUISITION SYSTEMS

The basic field activity in seismic surveying is the collection of *seismograms* which may be defined as analogue or digital time series that register the amplitude of ground motions as a function of time during the passage of a seismic wave train. The acquisition of seismograms involves conversion of the seismic ground motions into electrical signals, amplification and filtering of the signals and their registration on a chart recorder and/or tape recorder. The conventional seismic survey procedure is to monitor ground motions at a large number of surface locations; thus multichannel recording systems are usually employed with, exceptionally, up to several hundred separate recording channels. Except in the simplest recording systems the data are tape recorded to facilitate subsequent processing. Modern recording systems utilize digital tape recording so that the data are available in a suitable form for input to computers. A block diagram of a seismic recording system is shown in Fig. 3.19.

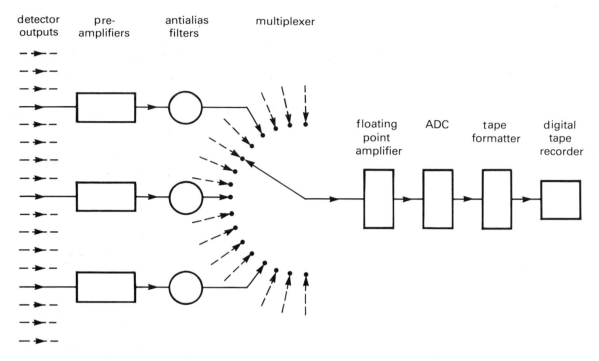

detector outputs   pre-amplifiers   antialias filters   multiplexer

floating point amplifier   ADC   tape formatter   digital tape recorder

Fig. 3.19 Schematic block diagram of a multichannel seismic recording system.

3.9.1 **Seismic detectors**

The detectors used in seismic surveying are electro-mechanical transducers that convert a mechanical input (the seismic pulse) into an electrical output. Devices used on land to detect seismic ground motions are known as *seismometers* or *geophones*. In water, the passage of a compressional seismic wave is marked by transient pressure changes and these are detected by *hydrophones* towed or suspended in the water column or, in very shallow water, laid on the sea bed. Hydrophones may also be used in the water-saturated ground conditions encountered in swamps or marshland. Detectors may comprise individual geophones or hydrophones, or arrays of these devices connected together in series/parallel to provide a summed output.

Although there are several types of geophone, the most common is the *moving-coil* geophone (Fig. 3.20). In this instrument a cylindrical coil is suspended from a spring support in the field of a permanent magnet which is attached to the instrument casing. The magnet has a cylindrical pole piece inside the coil and an annular pole piece surrounding the coil. The suspended coil represents an oscillatory

system with a resonant frequency determined by the mass of the coil and the stiffness of its spring suspension.

The geophone is fixed by a spike base into soft ground or mounted firmly on hard ground. It moves in sympathy with the ground surface during the passage of a seismic wave, causing relative motion between the suspended coil and the fixed magnet. Movement of the coil in the magnetic field generates a voltage across the terminals of the coil. The oscillatory motion of the coil is inherently damped because the current flowing in the coil induces a magnetic field that interacts with the field of the magnet to oppose the motion of the coil. The amount of this damping can be altered by connecting a shunt resistance across the coil terminals to control the amount of current flowing in the coil. Additional damping is arranged by winding the coil on a metal former. The magnetic field induced by eddy currents flowing in the metal former also opposes the coil motion.

Ideally, the output waveform of a geophone closely mirrors the ground motion and this is arranged by careful selection of the amount of damping. Too little damping results in an oscillatory

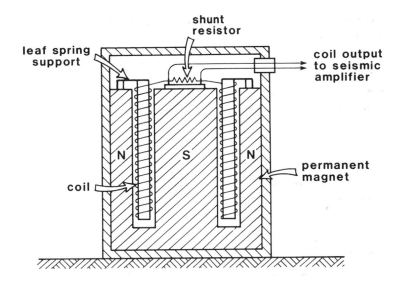

Fig. 3.20 Schematic cross-section through a moving-coil geophone.

output at the resonant frequency, whilst over-damping leads to a reduction of sensitivity. Damping is typically arranged to be about 0.7 of the critical value at which oscillation would just fail to occur for an impulsive mechanical input such as a sharp tap. With this amount of damping the frequency response of the geophone is effectively flat above the resonant frequency. The effect of differing amounts of damping on the frequency and phase response of a geophone is shown in Fig. 3.21.

To preserve the shape of the seismic waveform, geophones should have a flat frequency response and minimal phase distortion within the frequency range of interest. Consequently, geophones should be arranged to have a resonant frequency well below the main frequency band of the seismic signal to be recorded. Most commercial seismic reflection surveys employ geophones with a resonant frequency between 4 and 15 Hz.

Above the resonant frequency, the output of a moving-coil geophone is proportional to the velocity of the coil. Note that the coil velocity is related to the very low particle velocity associated with a seismic ground motion and not to the much higher propagation velocity of the seismic energy (see p. 25). The sensitivity of a geophone, measured in output volts per unit of velocity, is determined by the number of windings in the coil and the strength of the magnetic field, hence, instruments of larger and heavier construction are required for higher sensitivity. The miniature geophones used in

commercial reflection surveying typically have a sensitivity of about $10\,\text{V}$ per $\text{m}\,\text{s}^{-1}$.

Moving-coil geophones are sensitive only to the component of ground motion along the axis of the coil. Vertically travelling compressional waves from subsurface reflectors cause vertical ground motions and are therefore best detected by geophones with an upright coil as illustrated in Fig. 3.20. The optimal recording of seismic phases that involve mainly horizontal ground motions, such as horizontally-polarized shear waves, requires geophones in which the coil is mounted and constrained to move horizontally. As discussed in Chapter 4, geophones are typically deployed in linear or areal arrays containing several geophones whose individual outputs are summed. Such arrays provide detectors with a directional response that facilitates the enhancement of signal and the suppression of certain types of noise (see p. 53).

In seismic surveying the outputs of several detectors are fed to a multichannel recording system mounted in a recording vehicle. The individual detector outputs may be fed along a multicore cable or multiplexed at the detector location and transmitted along a lighter cable containing far fewer conductors. Some modern systems utilize lightweight fibre-optic cables or telemetry links to transmit the detector outputs to the recording vehicle.

Hydrophones are composed of ceramic piezo-electric elements which produce an output voltage proportional to the pressure variations associated

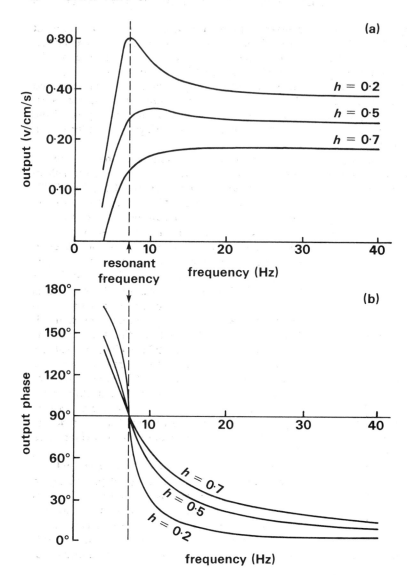

Fig. 3.21 Amplitude and phase responses of a geophone with a resonant frequency of 7 Hz, for different damping factors h. Output phase is expressed relative to input phase. (After Telford *et al.* 1976.)

with the passage of a compressional seismic wave through water. The sensitivity is typically 0.1 mV Pa^{-1}. For multichannel seismic surveying at sea, large numbers of individual hydrophones are made up into hydrophone *streamers* by distributing them along an oil-filled plastic tube. The tube is arranged to have neutral buoyancy and is manufactured from materials with an acoustic impedance close to that of water to ensure good transmission of seismic energy to the hydrophone elements. Since piezoelectric elements are sensitive to accelerations, hydrophones are often composed of two elements mounted back to back and connected in series so that the effects of

accelerations of the streamer as it is towed through the water are cancelled out in the hydrophone outputs. As with geophone deployment in land seismic surveying, groups of hydrophones may be connected together into linear arrays to produce detectors with a directional response.

3.9.2 Seismic amplifiers and tape recorders

Seismic amplifiers are required to amplify signals in the frequency range from a few hertz to a few hundred hertz (or, in some marine systems, up to a few thousand hertz), and have to cope with a very wide range of signal amplitudes. The amplitude of ground motions near to a seismic source may reduce by a factor of a million or more between the early arrivals of strong direct waves and surface waves and the later arrivals of very weak waves reflected back to the surface from deep interfaces. An amplitude ratio of one million is equivalent to a dynamic range of 120 dB. A maximum dynamic range for geophones of about 140 dB and an inherent minimum noise level in seismic amplifiers of about 1 microvolt effectively limits the maximum dynamic range of a seismic recording to 120 dB.

Most seismic amplifier systems contain frequency filters for high-pass, low-pass, band-pass or band-reject filtering. Filtering is commonly employed to produce a suitable visual record in the field for monitoring purposes, either at the time of the original recording or subsequently, by playback of a tape recording. The tape is normally recorded broad band (except for antialias filtering in the case of digital recording; see Chapter 2) in order to retain the maximum amount of information in the seismic recording. Optimal frequency filtering can then be carried out digitally as an aspect of the subsequent computer processing of the data.

The approach to seismic amplifier design depends upon whether analogue or digital tape recording is to be employed. The maximum dynamic range of analogue tape recording is about 50 dB so that in analogue recording systems the dynamic range of the seismic signal needs to be reduced prior to the recording stage. This can be accomplished by various means. *Automatic gain control* (AGC) alters the amplification factor of the amplifier in accordance with the amplitude of the input signal (Fig. 3.22). Up to a certain input level the gain is approximately

constant but it reduces progressively for higher input levels. Thus the stronger signals are relatively attenuated and the overall dynamic range is markedly reduced. Time variable gain can be used to suppress the gain when strong signals are being received (known as *initial suppression* or *presuppression*), and to increase the gain in the later part of the recording, when the seismic signal has reduced to a very low level.

In digital recording systems, the analogue output of the seismic amplifiers has to be passed through an *analogue-to-digital converter* (ADC) (Fig. 3.19). Conventionally, a single ADC is used to digitize all the seismic channels by means of multiplexing. This involves electronic switching of the ADC sequentially through all the channels and, for each channel, sampling the instantaneous output value and registering it in digital form as a binary word (see Chapter 2). Thus one full scan of all the channels produces a sequence of binary words each representing a sample value for an individual channel. The required switching rate of the ADC is determined by the required digital sampling interval and the number of channels to be multiplexed. For example, if each channel of a 50-channel amplifier system is to be sampled every 2.5 ms, the ADC must scan all 50 channels in less than 2.5 ms, which requires a switching rate of faster than 0.05 ms. In fact, ADCs operate at much faster switching rates than this.

Tape recording of the seismic data in a multiplexed form means that the initial stage of processing on playback is demultiplexing to recover the form of the outputs of the individual seismic amplifier channels. Demultiplexing by computer is a simple matter of reordering the sequence of binary words recorded on tape into separate one-dimensional

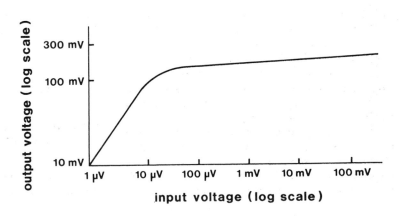

Fig. 3.22 The non-linear amplification factor of an automatic gain control (AGC) system.

arrays representing the outputs of each recording channel.

The multiplexed digital data are recorded in a standard tape format. A common, internationally accepted format is 9-track recording on half-inch tape with a data packing density of 1600 bits per inch, but a higher packing density of 6250 bits per inch is becoming increasingly common.

Since the dynamic range of a digitized waveform is determined solely by the length (i.e. number of bits) of each binary word (see Chapter 2), the only limitation on the dynamic range of a digital recording is the number of bits recorded. However, there are practical limits to the recorded word length because the greater the number of bits the faster the required tape speed and the greater the data storage problem. In addition to the digital word defining the amplitude of the ground motion it is necessary to record an extra bit, known as the sign-bit, to register the associated direction of ground motion (up or down). In conventional seismic recording, 16 to 20 bits are acquired per sample point.

For any given number of bits, the effective range of a digital recording can be increased by the use of *floating point* amplifiers. These measure the magnitude of any channel output sampled by the multiplexer and represent the digital output by two numbers, one giving the value of the output to the required number of significant places, the other giving the power of two to which this number has to be raised.

Floating point amplifiers have largely replaced *binary gain* amplifiers which automatically adjust the gain level of the recording in binary steps (6 dB) through the recording period on the basis of the amplitude of the signal output from the multiplexer. Tape recording the binary gain level as a function of time enables true signal amplitudes to be recovered during subsequent computer processing.

For the visual display of seismic records for monitoring purposes, a multichannel oscillographic recorder is used either to display the filtered output of the seismic amplifiers or for the field playback of tape recorded data. In some seismic recording systems without tape recording facilities the chart recording produced by the oscillographic recorder represents the only permanent record of the seismic data.

Some oscillographic recorders contain a facility to store records digitally in an internal memory and to sum the results obtained from successive shots prior to display of the seismograms. Summing of the results from a number of shots results in an improved SNR in the resultant seismograms. In such recorders, the content of the memory can typically be inspected on a display screen before being played out as a chart recording to produce a permanent record.

3.10 PROBLEMS

1 How does the progressive loss of higher frequencies in a propagating seismic pulse lead to an increase in pulse length?

2 A 10 Hz seismic wave travelling at 5 km s^{-1} propagates for 1000 m through a medium with an absorption coefficient of 0.2 dBλ^{-1}. What is the wave attenuation in decibels due solely to absorption?

3 A wave component with a wavelength of 100 m propagates through an homogeneous medium from a seismic source at the bottom of a borehole. Between two detectors, located in boreholes at radial distances of 1 km and 2 km from the source, the wave amplitude is found to be attenuated by 10 dB. Calculate the contribution of geometrical spreading to this value of attenuation and, thus, determine the absorption coefficient of the medium.

4 What is the crossover distance for direct and critically refracted rays in the case of a horizontal interface at a depth of 200 m separating a top layer of velocity 3.0 km s^{-1} from a lower layer of velocity 5.0 km s^{-1}?

5 A seismic pulse generated by a surface source is returned to surface after reflection at the tenth of a series of horizontal interfaces, each of which has a reflection coefficient R of 0.1. What is the attenuation in amplitude of the pulse caused by energy partitioning at all interfaces encountered along its path?

6 At what frequency would a 150 Hz signal be recorded by a digital recording system with a sampling rate of 100 Hz?

FURTHER READING

Al-Sadi, H.N. (1980) *Seismic Exploration.* Birkhäuser Verlag, Basel.

Anstey, N.A. (1977) *Seismic Interpretation: The Physical Aspects.* IHRDC, Boston.

Anstey, N.A. (1981) *Seismic Prospecting Instruments. Vol 1: Signal Characteristics and Instrument Specifications.* Gebrüder Borntraeger, Berlin.

Dobrin, M.B. & Savit, C.H. (1988) *Introduction to Geophysical Prospecting* (4th edn). McGraw-Hill, New York.

Gregory, A.R. (1977) Aspects of rock physics from laboratory and log data that are important to seismic interpretation. *In*: Payton, C.E. (ed.), *Seismic Stratigraphy— Applications to Hydrocarbon Exploration.* Memoir 26, American Association of Petroleum Geologists, Tulsa.

Lavergne, M. (1989) *Seismic Methods*. Editions Technip, Paris.

Sheriff, R.E. & Geldart, L.P. (1982) *Exploration Seismology Vol 1: History, Theory and Data Acquisition*. Cambridge University Press, Cambridge.

Sheriff, R.E. & Geldart, L.P. (1983) *Exploration Seismology Vol 2: Data-processing and Interpretation*. Cambridge University Press, Cambridge.

Waters, K.H. (1978) *Reflection Seismology—A Tool For Energy Resource Exploration*. Wiley, New York.

4 / Seismic reflection surveying

4.1 INTRODUCTION

In seismic reflection surveys the travel times are measured of arrivals reflected from subsurface interfaces between media of different acoustic impedance. Reflection surveys are most commonly carried out in areas of shallowly dipping sedimentary sequences. In such situations, velocity varies much more as a function of depth, due to the differing physical properties of the individual layers, than horizontally, due to lateral facies changes within the individual layers. For the purposes of initial consideration, the horizontal variations of velocity may be ignored.

Fig. 4.1 shows a horizontally-layered ground with vertical-reflected ray paths from the various layer boundaries. This model assumes the subsurface to be composed of a series of depth intervals each characterized by an *interval velocity* v_i, which may be the uniform velocity within a homogeneous geological unit or the average velocity over a depth interval containing more than one unit. If z_i is the thickness of such an interval and τ_i is the one-way travel time of a ray through it, the interval velocity is given by

$$v_i = \frac{z_i}{\tau_i}$$

The interval velocity may be averaged over several depth intervals to yield a *time-average velocity* or, simply, *average velocity* $\overline{V}$. Thus the average velocity of the top n layers in Fig. 4.1 is given by

$$\overline{V} = \frac{\sum\limits_{i=1}^{n} z_i}{\sum\limits_{i=1}^{n} \tau_i} = \frac{\sum\limits_{i=1}^{n} v_i \tau_i}{\sum\limits_{i=1}^{n} \tau_i}$$

or, if Z_n is the total thickness of the top n layers and T_n is the total one-way travel time through the n layers

$$\overline{V} = \frac{Z_n}{T_n}$$

4.2 GEOMETRY OF REFLECTED RAY PATHS

4.2.1 Single horizontal reflector

The basic geometry of the reflected ray path is shown in Fig. 4.2(a) for the simple case of a single horizontal reflector lying at a depth z beneath an homogeneous top layer of velocity V. The equation for the travel time t of the reflected ray from a shot point to a detector at a horizontal offset, or shot–detector separation, x is given by the ratio of the travel path length to the velocity

$$t = (x^2 + 4z^2)^{1/2}/V \qquad (4.1)$$

In a reflection survey, reflection times t are measured at offset distances x and it is required to determine z and V. If reflection times are measured

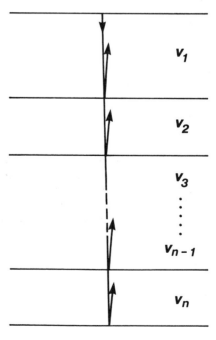

Fig. 4.1 Vertical reflected ray paths in a horizontally-layered ground.

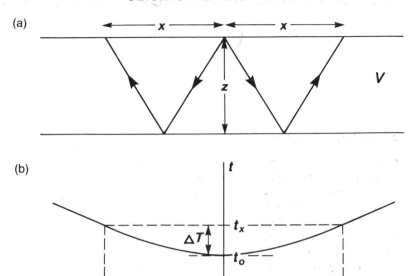

Fig. 4.2 (a) Geometry of reflected ray paths and (b) time–distance curve for reflected rays from a horizontal reflector. ΔT=normal moveout.

at different offsets x, equation (4.1) can be solved for these unknowns.

Equation (4.1) can be arranged into the normal hyperbolic form to give

$$\frac{V^2 t^2}{4z^2} - \frac{x^2}{4z^2} = 1 \qquad (4.2)$$

Thus the graph of travel time of reflected rays plotted against offset distance (the *time–distance curve*) is an hyperbola whose axis of symmetry is the time axis (Fig. 4.2(b)). Substituting $x = 0$ in equation (4.1), the travel time t_0 of a vertically reflected ray is obtained. $t_0 = 2z/V$ represents the intercept on the time axis of the time–distance curve (see Fig. 4.2(b)). Equation (4.1) can be written

$$t^2 = \frac{4z^2}{V^2} + \frac{x^2}{V^2} \qquad (4.3)$$

Thus

$$t^2 = t_0^2 + \frac{x^2}{V^2} \qquad (4.4)$$

From equation (4.3)

$$t = \frac{2z}{V}\left\{1 + \left(\frac{x}{2z}\right)^2\right\}^{1/2}$$

$$= t_0\left\{1 + \left(\frac{x}{Vt_0}\right)^2\right\}^{1/2} \qquad (4.5)$$

Binomial expansion of equation (4.5) gives

$$t = t_0\left\{1 + \frac{1}{2}\left(\frac{x}{Vt_0}\right)^2 - \frac{1}{8}\left(\frac{x}{Vt_0}\right)^4 + \dots\right\}$$

For small offset–depth ratios (i.e. $x/Vt_0 \ll 1$), which is the normal case in reflection surveying, this equation may be truncated after the first term to obtain

$$t \approx t_0\left\{1 + \frac{1}{2}\left(\frac{x}{Vt_0}\right)^2\right\} \qquad (4.6)$$

This is the most convenient form of the time–distance equation for reflected rays and it is used in various ways in the processing and interpretation of reflection data.

Moveout is defined as the difference between the travel times t_1 and t_2 of reflected ray arrivals recorded at two offset distances x_1 and x_2. From equation (4.6)

$$t_2 - t_1 \approx \frac{x_2^2 - x_1^2}{2V^2 t_0}$$

Normal moveout (NMO) at an offset distance x is the difference in travel time ΔT between reflected arrivals at x and at zero offset (see Fig. 4.2)

$$\Delta T \approx t_x - t_0 \approx \frac{x^2}{2V^2 t_0} \tag{4.7}$$

Note that NMO is a function of offset, velocity and reflector depth z (since $z = Vt_0/2$). The concept of moveout is fundamental to the recognition, correlation and enhancement of reflection events, and to the calculation of velocities using reflection data. It is implicitly or explicitly used at several stages in the processing and interpretation of reflection data.

To exemplify its use, consider the $T-\Delta T$ *method* of velocity analysis. Rearranging the terms of equation (4.7) yields

$$V \approx \frac{x}{(2t_0 \Delta T)^{1/2}} \tag{4.8}$$

Using this relationship, the velocity V above the reflector can be computed from knowledge of the zero-offset reflection time t_0 $(= T)$ and the NMO ΔT. In practice, such velocity values are obtained by computer analysis which produces a statistical estimate based upon many such calculations using large numbers of reflected ray paths (see Section 4.6). Once the velocity has been derived, it can be used in conjunction with t_0 to compute the depth z to the reflector using $z = Vt_0/2$.

4.2.2 Sequence of horizontal reflectors

In a multilayered ground, inclined rays reflected from the nth interface undergo refraction at all higher interfaces to produce a complex travel path (Fig. 4.3(a)). At offset distances that are small compared to reflector depths, the travel-time curve is still essentially hyperbolic but the homogeneous top layer velocity V in equations (4.1) and (4.7) is replaced by the *average velocity* $\overline{V}$ or, to a closer approximation (Dix 1955), the *root-mean-square velocity* V_{rms} of the layers overlying the reflector. As the offset increases, the departure of the actual

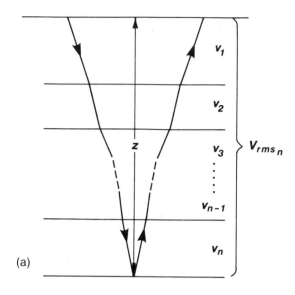

(a)

(b)

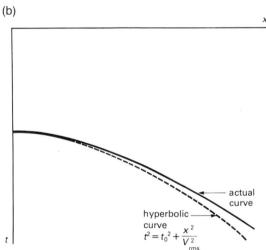

Fig. 4.3 (a) The complex travel path of a reflected ray through a multilayered ground. (b) The time−distance curve for reflected rays following the above type of path. Note that the divergence from the hyperbolic travel-time curve for a homogeneous overburden of velocity V_{rms} increases with offset.

travel-time curve from a hyperbola becomes more marked (Fig. 4.3(b)).

The root-mean-square velocity of the section of ground down to the nth interface is given by

$$V_{rms_n} = \left[\sum_{i=1}^{n} v_i^2 \, \tau_i \Big/ \sum_{i=1}^{n} \tau_i \right]^{1/2}$$

where v_i is the interval velocity of the ith layer and τ_i is the one-way travel time of the reflected ray through the ith layer.

Thus at small offsets x $(x \ll z)$, the total travel time t_n of the ray reflected from the nth interface at depth z is given to a close approximation by

$$t_n \approx (x^2 + 4z^2)^{1/2}/V_{rms_n} \qquad \text{(cf. equation (4.1))}$$

and the NMO for the nth reflector is given by

$$\Delta T_n \approx \frac{x^2}{2V_{rms_n}^2 t_0} \qquad \text{(cf. equation (4.7))}.$$

The individual NMO value associated with each reflection event may therefore be used to derive a root-mean-square velocity value for the layers above the reflector. Values of V_{rms} down to different reflectors can then be used to compute interval velocities using the *Dix formula*. To compute the interval velocity v_n for the nth interval

$$v_n = \left[\frac{V_{rms_n}^2 t_n - V_{rms_{n-1}}^2 t_{n-1}}{t_n - t_{n-1}} \right]^{1/2}$$

where $V_{rms_{n-1}}$, t_{n-1} and V_{rms_n}, t_n are, respectively, the root-mean-square velocity and reflected ray travel times to the $n-1$th and nth reflectors (Dix 1955).

4.2.3 Dipping reflector

In the case of a dipping reflector (Fig. 4.4(a)) the value of dip θ enters the time–distance equation as an additional unknown

$$t = (x^2 + 4z^2 + 4xz \sin \theta)^{1/2}/V$$
(cf. equation (4.1))

or, in the hyperbolic form

$$\frac{V^2 t^2}{4z^2 \cos^2 \theta} - \frac{(x + 2z \sin \theta)^2}{4z^2 \cos^2 \theta} = 1$$
(cf equation (4.2)).

The axis of symmetry of the hyperbola is now no longer the time axis (Fig. 4.4(b)).

Proceeding as in the case of a horizontal reflector the following truncated binomial expansion is obtained

$$t \approx t_0 \left\{ 1 + \frac{(x^2 + 4xz \sin \theta)}{2V^2 t_0^2} \right\} \qquad (4.9)$$

Consider two receivers at equal offsets x updip and downdip from a central shot point (Fig. 4.4).

Because of the dip of the reflector, the reflected ray paths are of different length and the two rays will therefore have different travel times. *Dip moveout* ΔT_d is defined as the difference in travel times t_x and t_{-x} of rays reflected from the dipping interface to receivers at equal and opposite offsets x and $-x$

$$\Delta T_d = t_x - t_{-x}$$

Using the individual travel times defined by equation (4.9)

$$\Delta T_d = \frac{2x \sin \theta}{V}$$

Rearranging terms, and for small angles of dip (when $\sin \theta \approx \theta$)

$$\theta \approx V \Delta T_d / 2x$$

Hence the *dip moveout* ΔT_d may be used to compute the reflector dip θ if V is known. V can be derived via equation (4.8) using the NMO ΔT which, for small dips, may be obtained with sufficient accuracy by averaging the updip and downdip moveouts

$$\Delta T \approx (t_x + t_{-x} - 2t_0)/2$$

4.2.4 Ray paths of multiple reflections

In addition to rays that return to the surface after reflection at a single interface, known as *primary reflections*, there are many paths in a layered subsurface by which rays may return to the surface after reflection at more than one interface. Such rays are called *reverberations*, *multiple reflections* or simply *multiples*. A variety of possible ray paths involving multiple reflection is shown in Fig. 4.5(a).

Multiple reflections tend to have lower amplitudes than primary reflections because of the loss of energy at each reflection. However, there are two types of multiple that are reflected at interfaces of high reflection coefficient and therefore tend to have amplitudes comparable with primary reflections: (1) *ghost reflections*, where rays from a buried explosion on land are reflected back from the surface or the base of the weathered layer (see p. 59) to produce a reflection event, known as a ghost reflection, that arrives a short time after the primary; and (2) *water layer reverberations*, where rays from a marine source are repeatedly reflected at the sea bed and sea surface.

Multiple reflections that involve only a short additional path length arrive so soon after the

(a)

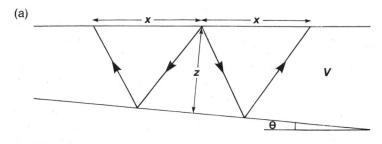

(b)

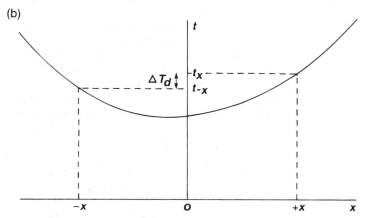

Fig. 4.4 (a) Geometry of reflected ray paths and (b) time−distance curve for reflected rays from a dipping reflector. ΔT_d = dip moveout.

primary event that they merely extend the overall length of the recorded pulse. Such multiples are known as *short-path multiples* (or short-period reverberations) and these may be contrasted with *long-path multiples* (long-period reverberations) whose additional path length is sufficiently long that the multiple reflection is a distinct and separate event in the seismic record (Fig. 4.5(b)).

The correct recognition of multiples is essential. Misidentification of a long-path multiple as a primary event, for example, would lead to serious interpretation error. The arrival times of multiple reflections are predictable, however, from primary reflection times and multiples can therefore be suppressed by suitable data processing techniques to be described later (Section 4.7).

4.3 MULTICHANNEL REFLECTION SURVEYING

The basic requirement of a multichannel reflection survey is to obtain recordings of reflected pulses at several offset distances from a shot point. As discussed in Chapter 3 this requirement is complicated in practice by the fact that the reflected pulses are never the first arrivals of seismic energy, and

they are generally of very low amplitude. Moreover, reflected pulses are typically concealed in noise which includes other, unwanted, seismic phases such as direct and refracted body waves and surface waves. Special procedures to enhance reflected arrivals have to be adopted during field recording and subsequent data reduction and processing.

The requirement to record reflected arrivals at more than one offset distance is met by multichannel recording of seismic arrivals at a spread of detectors located in the vicinity of a shot point. In *two-dimensional surveys* (*reflection profiling*), data are collected along survey lines that nominally contain all shot points and receivers. For the purpose of data processing, reflected ray paths are assumed to lie in the vertical plane containing the survey line. Thus, in the presence of cross-dip the resultant seismic sections do not provide a true representation of the subsurface structure, since actual reflection points then lie outside the vertical plane. Two-dimensional survey methods are adequate for the mapping of structures (such as cylindrical folds, or faults) which maintain uniform geometry along strike. They may also be used to investigate three-dimensional structures by mapping lateral changes across a series of closely-spaced survey lines or around a grid of lines.

(a)

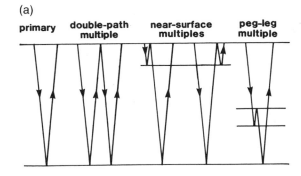

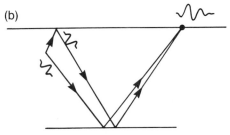

short-path multiples extend pulse length

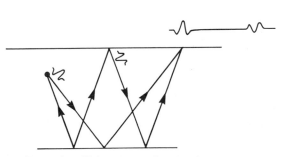

long-path multiples generate discrete pulse

Fig. 4.5 (a) Various types of multiple reflection in a layered ground. (b) The difference between short-path and long-path multiples.

However, as discussed below, *three-dimensional surveys* provide a much better means of mapping three-dimensional structures and, in areas of structural complexity, they may provide the only means of obtaining reliable structural interpretations.

Reflection profiling is normally carried out along profile lines with the shot point and its associated spread of detectors being moved progressively along the line to build up lateral coverage of the underlying geological section. This progression is carried out in a stepwise fashion on land but continuously, by a ship underway, at sea. The gathering of reflection survey data along profile lines permits an imaging, in reflection times, of the geological structure underlying the survey line. The third dimension of geological structure may be studied by implementing an intersecting network of reflection lines.

The two most common shot-detector configurations in multichannel reflection profiling surveys are the *split spread* (or *straddle spread*) and the *single-ended spread* (Fig. 4.6), with the number of detectors in a spread being normally 24 or more. In split spreads, the detectors are distributed on either side of a central shot point; in single-ended spreads, the shot point is located at one end of the detector spread. Surveys on land are commonly carried out with a split spread geometry, but in marine reflection surveys single-ended spreads are the normal configuration, with a marine source towed ahead of a hydrophone streamer.

The general aim of three-dimensional surveys is to achieve a higher degree of resolution of the subsurface geology than is achievable by two-dimensional surveys. Three-dimensional survey methods involve collecting field data in such a way that recorded arrivals are not restricted to rays that have travelled in a single vertical plane. In a three-dimensional survey, the disposition of shots and receivers is such that groups of recorded arrivals can be assembled that represent rays reflected from an *area* of each reflecting interface. Three-dimensional surveying therefore samples a volume of the subsurface rather than an area contained in a vertical plane, as in two-dimensional surveying.

On land, three-dimensional data are normally collected using the *crossed array method* in which shots and detectors are distributed along orthogonal sets of lines (in-lines and cross-lines) to establish a grid of recording points. For a single pair of lines, the areal coverage of a subsurface reflector is illustrated in Fig. 4.7.

At sea, three-dimensional data may be collected along closely-spaced parallel tracks with the hydrophone streamer feathered to tow obliquely to the ship's track such that it sweeps across a swathe of the sea floor as the vessel proceeds along its track. By ensuring that the swathes associated with adjacent tracks overlap, data may be assembled to provide areal coverage of subsurface reflectors. In the alternative *dual source array method*, sources are deployed on side gantries to port and starboard of the hydrophone streamer and fired alternately

(a)

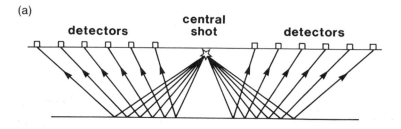

(b)

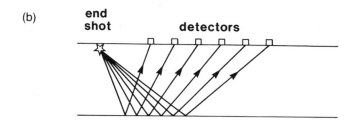

Fig. 4.6 Shot-detector configurations used in multichannel seismic reflection profiling. (a) Split spread, or straddle spread. (b) Single-ended spread.

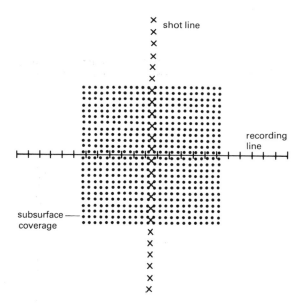

Fig. 4.7 The areal coverage derived from a single pair of crossing lines in a three-dimensional survey. Each dot represents the mid-point between a shot and a detector.

(Fig. 4.8). Dual streamers may similarly be deployed to obtain three-dimensional data.

High quality position fixing is a prerequisite of three-dimensional marine surveys in order that the locations of all shot-detector midpoints are accurately determined. Position fixing is normally achieved in nearshore areas using radio navigation systems, in which a location is determined by calculation of range from onshore radio transmitters. Beyond the range of such systems satellite navigation is used, with Doppler sonar being employed to determine the velocity of the vessel along the survey track for interpolation of position during the time interval between individual satellite fixes (Lavergne 1989).

4.4 THE REFLECTION SEISMOGRAM (SEISMIC TRACE)

The oscillographic recording of the amplified output of each detector in a reflection spread is a visual representation of the local pattern of vertical ground motion (on land) or pressure variation (at sea) over a short interval of time following the triggering of a nearby seismic source. This *reflection seismogram* or *seismic trace* represents the combined response of the layered ground and the recording system to a seismic pulse.

At each layer boundary a proportion of the incident energy in the pulse is reflected back towards the detector. The detector therefore receives a series of reflected pulses, scaled in amplitude according to the distance travelled and the reflection coefficients of the various layer boundaries. The pulses arrive at times determined by the depths to the boundaries and the velocities of propagation between them.

Assuming that the pulse shape remains unchanged

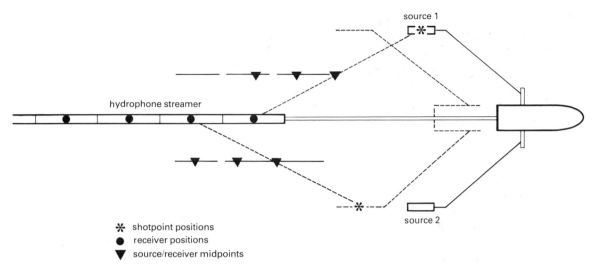

Fig. 4.8 The dual source array method of collecting three-dimensional seismic data at sea. Alternate firing of sources 1 and 2 into the hydrophone streamer produces two parallel sets of source-detector midpoints.

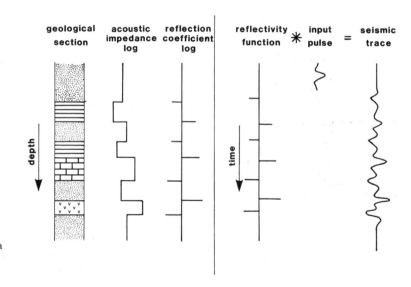

Fig. 4.9 The reflection seismogram viewed as the convolved output of a reflectivity function with an input pulse.

as it propagates through such a layered ground, the resultant seismic trace may be regarded as the convolution of the input pulse with a time series known as a *reflectivity function* composed of a series of spikes. Each spike has an amplitude related to the reflection coefficient of a boundary and a travel time equivalent to the two-way reflection time for that boundary. This time series represents the *impulse response* of the layered ground (i.e. the output for a spike input). The convolution model is illustrated

schematically in Fig. 4.9. Since the pulse has a finite length, individual reflections from closely-spaced boundaries are seen to overlap in time on the resultant seismogram.

In practice, as the pulse propagates it lengthens due to the progressive loss of its higher frequency components by absorption. The basic reflection seismogram may then be regarded as the convolution of the reflectivity function with a *time-varying* seismic pulse. The seismogram will be further complicated by the superimposition of various types of noise such as multiple reflections, direct and refracted body waves, surface waves (ground roll), air waves and coherent and incoherent noise unconnected with the seismic source. In consequence of these several effects, reflection seismograms generally have a complex appearance and reflection events are often not recognizable without the application of suitable processing techniques.

The initial display of seismic profile data is normally in groups of seismic traces recorded from a common shot, known as *common shot point gathers* or, simply, *shot gathers*. The playout of shot gathers at the time of field recording provides a means of checking that a satisfactory recording has been achieved from any particular shot. In shot gathers, the seismic traces are plotted side by side in their correct relative positions and the records are commonly displayed with their time axes arranged vertically in a draped fashion. In these seismic records, recognition of reflection events and their correlation from trace to trace is much assisted if one half of the normal 'wiggly-trace' waveform is blocked out. Fig. 4.10 shows a draped section with this mode of display, derived from a split spread multichannel survey. A short time after the shot instant the first arrival of seismic energy reaches the innermost phones (the central traces) and this energy passes out symmetrically through the two arms of the split spread. The first arrivals are followed by a series of reflection events revealed by their hyperbolic moveout.

4.5 MULTICHANNEL REFLECTION SURVEY DESIGN

4.5.1 Vertical and horizontal resolution

Reflection surveys are normally designed to provide a specified depth of penetration and a particular degree of resolution of the subsurface geology in both the vertical and horizontal dimensions. The vertical resolution is a measure of the ability to recognize individual, closely-spaced reflectors and is determined by the pulse length on the recorded seismic section. For a reflected pulse represented by a simple wavelet, the maximum resolution possible is between one quarter and one eighth of the dominant wavelength of the pulse (Sheriff 1985). Thus, for a reflection survey involving a signal with a dominant frequency of 50 Hz propagating in sedimentary strata with a velocity of $2.0 \, \mathrm{km \, s^{-1}}$, the

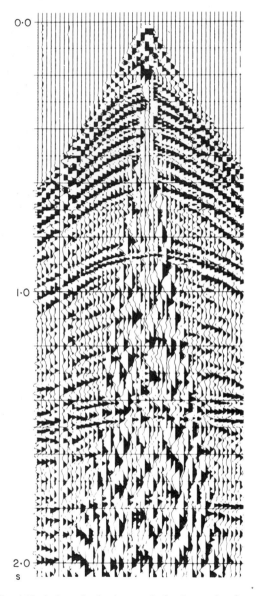

Fig. 4.10 A draped seismic record of a shot gather from a split spread (courtesy Prakla-Seismos GMBH). Sets of reflected arrivals from individual interfaces are recognizable by their characteristic hyperbolic form. The late-arriving, high-amplitude, low-frequency events, defining a triangular-shaped central zone within which reflected arrivals are masked, represent surface waves (ground roll).

dominant wavelength would be 40 m and the vertical resolution may therefore be no better than about 10 m. Since deeper travelling seismic waves tend to have a lower dominant frequency due to the progressive loss of higher frequencies by absorption (Section 3.5), vertical resolution decreases as a function of depth. It should be noted that the vertical resolution of a seismic survey may be improved at the data processing stage by a shortening of the recorded pulse length using inverse filtering (deconvolution) (Section 4.7).

There are two main controls on the horizontal resolution of a reflection survey, one being intrinsic to the physical process of reflection and the other being determined by the detector spacing. To deal with the latter point first, the horizontal resolution is clearly determined by the spacing of the individual depth estimates from which the reflector geometry is reconstructed. From Fig. 4.11 it can be seen that, for a flat-lying reflector, the horizontal resolution is equal to half the detector spacing. Note, also, that the length of reflector sampled by any detector spread is half the spread length. The spacing of detectors must be kept small to ensure that reflections from the same interface can be correlated reliably from trace to trace in areas of complex geology.

Notwithstanding the above, there is an absolute limit to the achievable horizontal resolution in consequence of the actual process of reflection. The path by which energy from a source is reflected back to a detector may be expressed geometrically by a simple ray path. However, such a ray path has only geometrical significance and the actual reflection process is best described by considering any reflecting interface to be composed of an infinite number of point scatterers, each of which contributes energy to the reflected signal (Fig. 4.12). The actual reflected pulse then results from interference of an infinite number of backscattered rays.

Energy that is returned to a detector within half a wavelength of the initial reflected arrival interferes constructively to build up the reflected signal, and the part of the interface from which this energy is returned is known as the first *Fresnel zone* (Fig. 4.12) or, simply, the Fresnel zone. Around the first Fresnel zone are a series of annular zones from which the reflected energy tends, overall, to interfere destructively and cancel out. The width of the Fresnel zone represents an absolute limit on the horizontal resolution of a reflection survey since reflectors separated by a distance smaller than this cannot be individually distinguished. The width w of the Fresnel zone is related to the dominant wavelength λ of the source and the reflector depth z by

$$w \approx (2z\,\lambda)^{1/2} \qquad (\text{for } z \gg \lambda)$$

Since, as noted above, deeper travelling reflected energy tends to have a lower dominant frequency due to the effects of absorption, the size of the first Fresnel zone increases as a function of reflector depth. Hence horizontal resolution, like vertical resolution, reduces with increasing reflector depth.

4.5.2 Design of detector arrays

Each detector in a conventional reflection spread consists of an *array* (or *group*) of several geophones or hydrophones arranged in a specific pattern and connected together in series/parallel to produce a single channel of output. The effective offset of an array is taken to be the distance from the shot to the centre of the array. Arrays of phones provide a directional response and are used to enhance the near-vertically travelling reflected pulses and to suppress several types of horizontally travelling *coherent* noise, i.e. noise that can be correlated from trace to trace as opposed to random noise. To exemplify

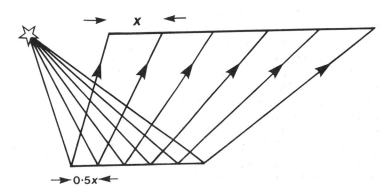

Fig. 4.11 The horizontal resolution of a seismic reflection survey is half the detector spacing.

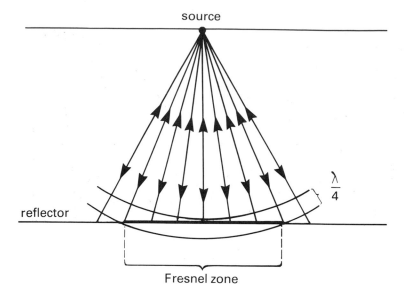

Fig. 4.12 Energy is returned to source from all points of a reflector. The part of the reflector from which energy is returned within half a wavelength of the initial reflected arrival is known as the Fresnel zone.

this, consider a Rayleigh surface wave (a vertically polarized wave travelling along the surface) and a vertically travelling compressional wave reflected from a deep interface to pass simultaneously through two geophones connected in series/parallel and spaced at half the wavelength of the Rayleigh wave. At any given instant, ground motions associated with the Rayleigh wave will be in opposite directions at the two phones and the individual outputs of the phones will therefore be out of phase and cancelled by summing. Ground motions associated with the reflected compressional wave will, however, be in phase at the two phones and the summed outputs of the phones will therefore be twice their individual outputs.

The directional response of any linear array is governed by the relationship between the apparent wavelength λ_a of a wave in the direction of the array, the number of elements n in the array and their spacing Δx. The response is given by a response function R

$$R = \frac{\sin n\beta}{\sin \beta}$$

where

$$\beta = \pi \Delta x / \lambda_a$$

R is a periodic function that is fully defined in the interval $0 \le \Delta x / \lambda_a \le 1$ and is symmetrical about $\Delta x / \lambda_a = 0.5$. Typical array response curves are shown in Fig. 4.13.

Arrays comprising areal rather than linear patterns of phones may be used to suppress horizontal noise travelling along different azimuths.

The initial stage of a reflection survey involves field trials in the survey area to determine the most suitable combination of source, offset recording range, array geometry and detector spacing (the horizontal distance between the centres of adjacent phone arrays, often referred to as the *group interval*) to produce good seismic data in the prevailing conditions.

Source trials involve tests of the effect of varying, for example, the shot depth and charge size of an explosive source or the number, chamber sizes and trigger delay times of individual guns in an air gun array. The detector array geometry needs to be designed to suppress the prevalent coherent noise events (mostly source-generated). On land, the local noise is investigated by means of a *noise test* in which shots are fired into a spread of closely-spaced detectors (*noise spread*) consisting of individual phones, or arrays of phones clustered together to eliminate their directional response. A series of shots is fired with the noise spread being moved progressively out to large offset distances. The purpose of the noise test is to determine the characteristics of the coherent noise, in particular, the velocity across the spread and dominant frequency of the air waves, surface waves (ground roll), direct and shallow refracted arrivals, that together tend to conceal the low-amplitude reflections. A typical

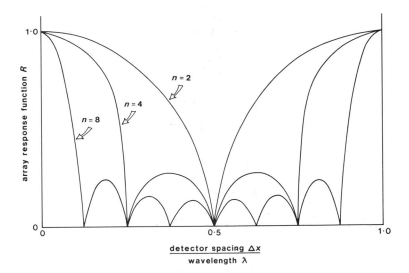

Fig. 4.13 Response functions for different detector arrays. (After Al-Sadi 1980.)

noise section derived from such a test is shown in Fig. 4.14(a) and clearly reveals a number of coherent noise events that need to be suppressed to enhance the SNR of reflected arrivals. Such noise sections provide the necessary information for the optimal design of detector phone arrays. Fig. 4.14(b) shows a time section obtained with a suitable array geometry designed to suppress the local noise events and reveals the presence of reflection events that were totally concealed in the noise section.

It is apparent from the above account that the use of suitably designed arrays can markedly improve the SNR of reflection events on field seismic recordings. Further improvements in SNR and survey resolution are achievable by various types of data processing discussed later in the chapter.

4.5.3 Common depth point (CDP) surveying

If the shot-detector spread in a multichannel reflection survey is moved forward in such a way that no two reflected ray paths sample the same point on a subsurface reflector, the survey coverage is said to be *single-fold*. Each seismic trace then represents a unique sampling of some point on the reflector. In common depth point profiling, which has become the standard method of two-dimensional multichannel seismic surveying, it is arranged that a set of traces recorded at different offsets contains reflections from a common depth point (CDP) on the reflector (Fig. 4.15(a)). The shot points and detector

locations for such a set of traces, known as a *CDP gather*, have a *common midpoint* (CMP) below which the common depth point is assumed to lie (Fig. 4.15(a)).

In three-dimensional surveying the common depth point principle applies similarly, but each CDP gather involves an areal rather than a linear distribution of shot points and detector locations (Fig. 4.16). Thus, for example, a 20-fold coverage is obtained in a crossed-array three-dimensional survey if reflected ray paths from five shots along different shot lines to four detectors along different recording lines all have a common reflection point.

In two-dimensional CDP surveying, known as *CDP profiling*, the common depth points are all assumed to lie within the vertical section containing the survey line; in three-dimensional surveying, the common depth points are distributed across an area of any subsurface reflector.

The advantages of CDP surveying are that (1) the CDP gather represents the best possible data set for computing velocity from the normal moveout (NMO) effect; and (2) with accurate velocity information the moveout can be removed from each trace of a CDP gather to produce a set of traces that may be summed algebraically (i.e. *stacked*) to produce a *CDP stack* in which reflected arrivals are enhanced relative to the seismic noise.

Strictly, the common depth point principle breaks down in the presence of dip because the common depth point then no longer directly underlies the shot-detector midpoint and the reflection point

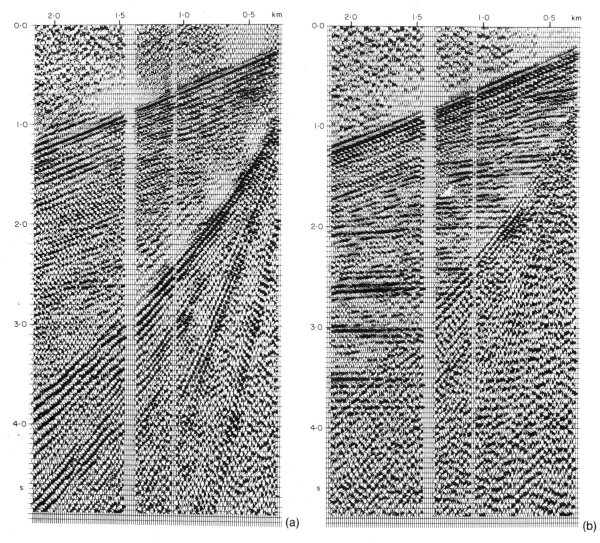

Fig. 4.14 Noise test to determine the appropriate detector array for a seismic reflection survey. (a) Draped seismic record obtained with a noise spread composed of clustered (or 'bunched') geophones. (b) Seismic record obtained over the same ground with a spread composed of 140-m-long geophone arrays. (From Waters 1978.)

differs for rays travelling to different offsets (see Fig. 4.15(b)). Nevertheless, the method is sufficiently robust that marked improvements in SNR almost invariably result from CDP stacks as compared with single traces.

The *fold* of the stacking refers to the number of traces in the CDP gather and may conventionally be 6, 12, 24, 48 or, exceptionally, over 1000. The fold is alternatively expressed as a percentage: single-fold = 100% coverage, six-fold = 600% coverage and so on. The theoretical improvement in SNR brought about by stacking n traces containing a mixture of coherent in-phase signals and random (incoherent) noise is $\sqrt{n}$. Stacking attenuates or even totally suppresses the long-path multiples that have a significantly different moveout from the primary reflections: thus when the latter are stacked in phase the former are not in phase and do not sum.

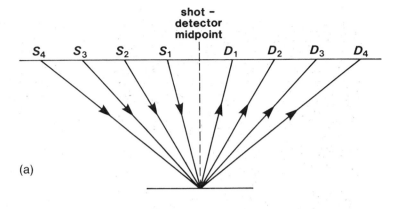

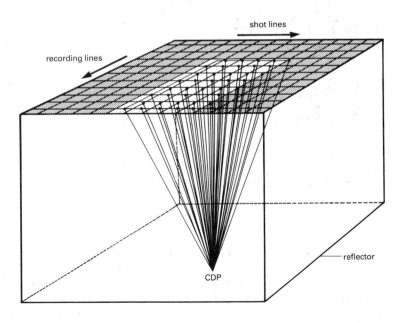

Fig. 4.15 Common depth point (CDP) reflection profiling. (a) A set of rays from different shots to detectors all reflected off a common point on a horizontal reflector. (b) The common depth point is not achieved in the case of a dipping reflector.

Fig. 4.16 Reflected ray paths defining a common depth point from an areal distribution of shot points and detector locations in a three-dimensionl survey.

The fold of a CDP profile is determined by the quantity $N/2n$, where N is the number of phone arrays along a spread and n is the number of phone array spacings by which the spread is moved forward between shots (the *move-up rate*). Thus with a 24-channel spread ($N = 24$) and a move-up rate of two array spacings per shot interval $(n = 2)$, the coverage would be 24/4 = 6-fold. A field procedure for the routine collection of 6-fold CDP coverage using a single-ended 12-channel spread configuration progressively moved forward along a profile line is shown in Fig. 4.17

4.5.4 Display of seismic reflection data

CDP profiling data from two-dimensional surveys are conventionally displayed as *seismic sections* in which the individual stacked seismograms are plotted side by side, in close proximity, with their time axes arranged vertically. Reflection events may then be traced across the section by correlating pulses from seismogram to seismogram and in this way the distribution of subsurface reflectors beneath the survey line may be mapped. However, whilst it is tempting to envisage seismic sections as straightforward images of geological cross-sections it must not be forgotten that the vertical dimension of the sections is time, not depth.

The product of three-dimensional seismic surveying is a volume of data (Fig. 4.18, Plate 1) representing reflection coverage from an area of each subsurface reflector. From this reflection data volume, conventional two-dimensional seismic sections may be constructed not only along the actual shot lines and recording lines employed but also along any other vertical slice through the data volume. Hence, seismic sections may be simulated for any azimuthal direction across the survey area by taking a vertical slice through the data volume, and this enables optimal two-dimensional representation of any recorded structural features.

More importantly, horizontal slices may be taken through the data volume to display the pattern of reflections intersected by any time plane. Such a representation of the three-dimensional data is known as a *time slice* or *seiscrop*, and analysis of reflection patterns displayed in time slices provides a powerful means of mapping three-dimensional structures (see Plates 1 & 2). In particular, structures may be traced laterally through the data volume, rather than having to be interpolated between adjacent lines as is the case in two-dimensional surveys. The manipulation of data volumes obtained from three-dimensional surveys is carried out at computer work stations using software routines that enable seismic sections and time slices to be displayed as required. Automatic event picking and contouring are also facilitated (Brown 1986).

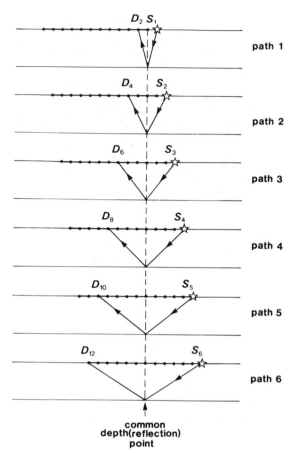

Fig. 4.17 A field procedure for obtaining 6-fold CDP coverage with a single-ended 12-channel detector spread moved progressively along the survey line.

4.6 TIME CORRECTIONS APPLIED TO SEISMIC TRACES

Two main types of correction need to be applied to reflection times on individual seismic traces in order that the resultant seismic sections give a true representation of geological structure. These are the *static* and *dynamic* corrections, so-called because the former is a fixed time correction applied to an entire trace whereas the latter varies as a function of reflection time.

4.6.1 Static correction

Reflection times on seismic traces recorded on land have to be corrected for time differences introduced by near-surface irregularities. These irregularities have the effect of shifting reflection

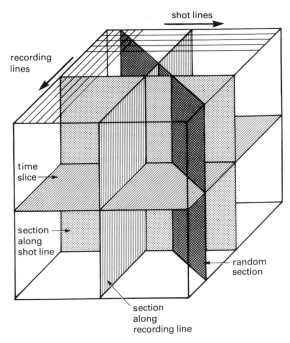

shot lines

recording lines

time slice

section along shot line

random section

section along recording line

Fig. 4.18 The reflection data volume obtained from a three-dimensional seismic survey. By taking vertical slices through this data volume, it is possible to generate seismic sections in any azimuthal direction; by taking horizontal slices (time slices), the areal distribution of reflection events can be studied at any two-way reflection time.

events on adjacent traces out of their true time relationships. The two major sources of irregularity (Fig. 4.19) are: (1) elevation differences between individual shots and detectors; and (2) the presence of a *weathered layer*, which is a heterogeneous surface layer, a few metres to several tens of metres thick, of abnormally low seismic velocity. The weathered layer is mainly caused by the presence within the surface zone of open joints and microfractures and by the unsaturated state of the zone. Although it may be only a few metres thick, its abnormally low velocity causes large time delays to rays passing through it. Thus variations of thickness in the weathered layer may, if not corrected for, lead to false indications of significant structural relief features on underlying reflectors being portrayed on seismic sections. In marine surveys there is no elevation difference between individual shots and detectors but the water layer represents a surface layer of anomalously low velocity in some ways analogous to the weathered layer on land.

The *static correction* is a combined weathering and

elevation correction that removes the effects of the low velocity surface layer and reduces all reflection times to a common height datum (Fig. 4.19). The correction is calculated on the assumption that the reflected ray path is effectively vertical immediately beneath any shot or detector. The travel time of the ray is then corrected for the time taken to travel the vertical distance between the shot or detector elevation and the survey datum, there being a component of correction for each end of the ray path. Survey datum may lie above the local base of the weathered layer, or even above the local land surface. In adjusting travel times to datum in such cases the height interval between the base of the weathered layer and datum is effectively replaced by material with the velocity of the main top layer.

Calculation of the static correction requires knowledge of the velocity and thickness of the weathered layer and the velocity of the underlying layer under all shot and detector locations. The first arrivals of energy at detectors in a reflection spread are normally rays that have been refracted along the top of the layer underlying the weathered layer and these arrivals can be used in a refraction interpretation of the top layer geometry using methods discussed in Chapter 5. If the normal reflection spread does not contain recordings at sufficiently small offsets to detect these shallow refracted rays and the direct rays defining the weathered layer velocity v_w, special short refraction spreads may be established for this purpose.

Direct measurements of the weathered layer velocity may also be obtained by *uphole surveys* in which small shots are fired at various depths down boreholes penetrating through the weathered layer and the velocities are measured of rays travelling from the shots to a surface detector. Conversely, a surface shot may be recorded by down-hole detectors. In reflection surveys utilizing buried shots, a surface detector is routinely located at the surface close to the shot hole to measure the *vertical time* (*VT*) or *uphole time*, from which the velocity of the surface layer above the shot may be calculated.

A purely empirical but often very effective approach to the computation of the static correction is to assume that the weathered layer and surface relief are the only cause of irregularities in the travel times of rays reflected from a shallow interface and to apply appropriate time adjustments to the individual traces to produce a smooth reflection profile in the time section.

Due to the fact that the velocity and thickness of

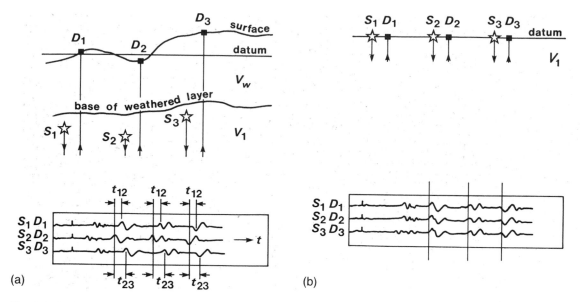

Fig. 4.19 Static corrections. (a) Seismograms showing time differences between reflection events on adjacent seismograms due to the different elevations of shots and detectors and the presence of a weathered layer. (b) The same seismograms after the application of elevation and weathering corrections, showing good alignment of the reflection events. (After O'Brien 1974.)

the weathered layer can never be precisely defined, the static correction always contains errors, or residuals, which have the effect of diminishing the SNR of CDP stacks and reducing the coherence of reflection events on time sections. These residuals can be calculated by computer in a *residual static analysis* by searching for systematic residual effects associated with individual shot and detector locations and applying these as corrections to the time sections. Fig. 4.20 shows the marked improvement in SNR and reflection coherence achievable by the application of these automatically computed residual static corrections.

In marine reflection surveys the static correction is commonly restricted to a conversion of travel times to mean sea level datum, without removing the overall effect of the water layer. Travel times are increased by $(d_s + d_h)v_w$, where d_s and d_h are the depths below mean sea level of the source and hydrophone array and v_w is the velocity of sea water.

4.6.2 Dynamic correction and velocity analysis

The *dynamic correction* is applied to reflection times to remove the effect of normal moveout. The correction is therefore numerically equal to the NMO and, as such, is a function of offset, velocity and reflector depth. Consequently, the correction has to be calculated separately for each time increment of a seismic trace.

Adequate correction for normal moveout is dependent on the use of accurate velocities. In common depth point surveys the appropriate velocity is derived by computer analysis of moveout in the groups of traces (*CDP gathers*) that contain reflections from a common depth point. Prior to this *velocity analysis*, static corrections must be applied to the individual traces to remove the effect of the low velocity surface layer and to reduce travel times to a common height datum. The method is exemplified with reference to Fig. 4.21 which illustrates a set of statically corrected traces containing a reflection event with a zero offset travel time of t_o. Dynamic corrections are calculated for a range of velocity values and the dynamically corrected traces are stacked. The *stacking velocity* V_{st} is defined as that velocity value which produces the maximum amplitude of the reflection event in the stack of traces. This clearly represents the condition of successful removal of NMO. Since the stacking velocity is that which removes NMO, it is given by the equation

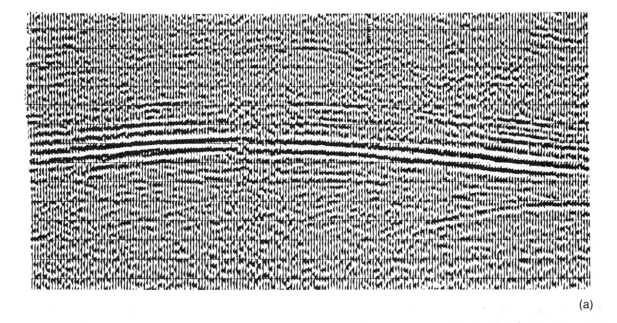

(a)

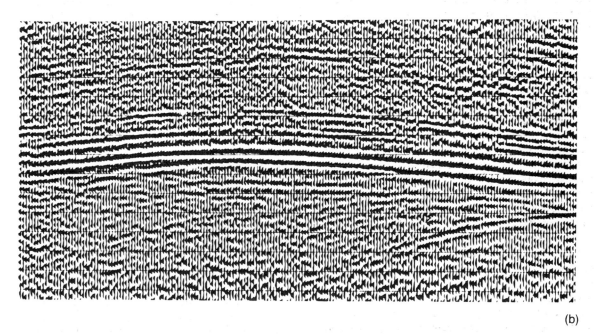

(b)

Fig. 4.20 Major improvement to a seismic section resulting from residual static analysis. (a) Manual statics only. (b) After residual static correction. (Courtesy Prakla Seismos GMBH.)

$$t^2 = t_o^2 + \frac{x^2}{V_{st}^2}$$

(cf equation (4.4)).

As previously noted, the travel–time curve for re-

flected rays in a multilayered ground is not an hyperbola (see Fig. 4.3(b)). However, if the maximum offset value x is small compared with reflector depth, the stacking velocity closely approximates

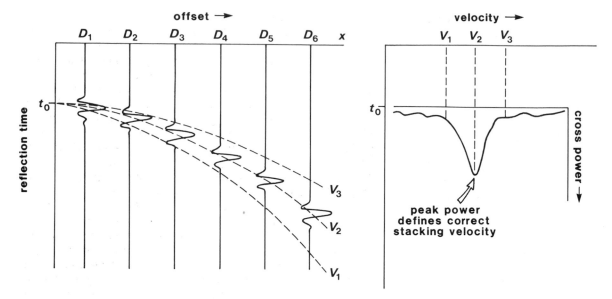

Fig. 4.21 A set of reflection events in a CDP gather is corrected for NMO using a range of velocity values. The stacking velocity is that which produces peak cross-power from the stacked events, i.e. the velocity that most successfully removes the NMO. In the case illustrated, V_2 represents the stacking velocity. (After Taner & Koehler 1969.)

the root-mean-square velocity V_{rms}, though it is obviously also affected by any reflector dip. Values of V_{st} for different reflectors can therefore be used in a similar way to derive interval velocities using the Dix formula (see p. 47). In practice, stacking velocities are computed for narrow time intervals along the entire trace to produce a *velocity spectrum* (Fig. 4.22). Velocity spectra are produced for several locations along a CDP profile to provide stacking velocity values for use in the dynamic correction of local traces.

In three-dimensional surveying, stacking velocities and velocity spectra may be derived in a similar manner from the CDP gathers involving areally distributed shotpoints and detectors.

4.7 REFLECTION DATA PROCESSING: FILTERING AND INVERSE FILTERING OF SEISMIC DATA

In addition to the data *reduction* procedures of static and dynamic time correction and CDP stacking, several digital data *processing* techniques are available for the enhancement of seismic sections. In general, the aim of reflection data processing is to increase further the SNR and improve the vertical resolution of the individual seismic traces by wave-

form manipulation, in contrast to the simple adjustments of reflection times that characterize data reduction. The two main types of waveform manipulation are frequency filtering and inverse filtering (deconvolution).

4.7.1 Frequency filtering

Any coherent or incoherent noise event whose dominant frequency is different from that of reflected arrivals may be suppressed by frequency filtering (see Chapter 2). Thus, for example, ground roll in land surveys and several types of ship-generated noise in marine seismic surveying can often be markedly attenuated by low-cut filtering, and wind noise by high-cut filtering.

Since the dominant frequency of reflected arrivals reduces with increasing length of travel path, due to the more rapid absorption of the higher frequencies, the characteristics of frequency filters are normally varied as a function of reflection time. For example, the first second of a 3-second seismic trace might typically be band-pass filtered between limits of 15 and 75 Hz, whereas the frequency limits for the third second might be 10 and 45 Hz. As the frequency characteristics of reflected arrivals are also influenced by the prevailing geology, the appropriate time-variant frequency filtering may also vary as a function

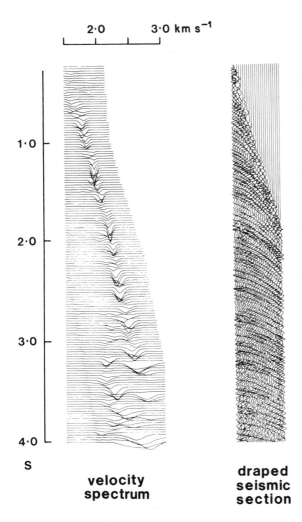

2·0   3·0 km s⁻¹

**velocity
spectrum**

**draped
seismic
section**

Fig. 4.22 The velocity spectrum defines the stacking velocity as a function of reflection time. The cross-power function used to define the stacking velocity is calculated for a large number of narrow time windows along a seismic section. The velocity spectrum is typically displayed alongside the draped section. (From Taner & Koehler 1969.)

of distance along a seismic profile. The filtering may be carried out by computer in the time domain or the frequency domain (see Chapter 2).

4.7.2 Inverse filtering (deconvolution)

Many components of seismic noise lie within the frequency spectrum of a reflected pulse and therefore cannot be removed by frequency filtering. The sup-

pression of these noise components can, however, be effected by various types of inverse filtering. Inverse filters discriminate against noise and improve signal character using criteria other than simply frequency. They are thus able to suppress types of noise that have the same frequency characteristics as the reflected signal. A wide range of inverse filters is available for reflection data processing, each designed to remove some specific adverse effect of filtering in the ground along the transmission path, such as absorption or multiple reflection.

Deconvolution is the analytical process of removing the effect of some previous filtering operation (convolution). Inverse filters are designed to deconvolve seismic traces by removing the adverse filtering effects associated with the propagation of seismic pulses through a layered ground or through a recording system. In general such effects lengthen the seismic pulse; for example, by the generation of multiple wave trains and by progressive absorption of the higher frequencies. Mutual interference of extended reflection wave trains from individual interfaces seriously degrades seismic records since onsets of reflections from deeper interfaces are totally or partially concealed by the wave trains of reflections from shallower interfaces.

Examples of inverse filtering to remove particular filtering effects include: *dereverberation* to remove ringing associated with multiple reflections in a water layer; *deghosting* to remove the short-path multiple associated with energy travelling upwards from the source and reflected back from the base of the weathered layer or the surface; and *whitening* to equalize the amplitude of all frequency components within the recorded frequency band (see below). All these deconvolution operations have the effect of shortening the pulse length on processed seismic sections and, thus, improve the vertical resolution.

Consider a composite waveform w_k resulting from an initial spike source extended by the presence of short-path multiples near source such as, especially, water layer reverberations. The resultant seismic trace x_k will be given by the convolution of the reflectivity function r_k with the composite input waveform w_k (neglecting the effects of attenuation and absorption)

$$x_k = r_k * w_k \text{ (plus noise)}$$

Reflected waveforms from closely-spaced reflectors will overlap in time on the seismic trace and, hence, will interfere. Deeper reflections may thus be concealed by the reverberation wave train associated with reflections from shallower interfaces, so that only by the elimination of the multiples will all the primary reflections be revealed. Note that short-path multiples have effectively the same normal moveout as the related primary reflection and are therefore not suppressed by CDP stacking, and they have a similar frequency content to the primary reflection so that they cannot be removed by frequency filtering.

Deconvolution has the general aim, not fully realizable, of compressing every occurrence of a composite waveform w_k on a seismic trace into a spike output, in order to reproduce the reflectivity function r_k that would fully define the subsurface layering. This is equivalent to the elimination of the multiple wave train. The required deconvolution operator is an inverse filter i_k which, when convolved with the composite waveform w_k, yields a spike function δ_k

$$i_k * w_k = \delta_k$$

Convolution of the same operator with the entire seismic trace yields the reflectivity function

$$i_k * x_k = r_k$$

In communications systems where w_k is known, deconvolution can be achieved by the use of *matched filters* which effectively cross-correlate the output with the known input signal (as in the initial processing of Vibroseis® seismic records to compress the long source wave train; see Section 3.8.2). *Wiener filters* may also be used when the input signal is known. A Wiener filter (Fig. 4.23) converts the known input signal into an output signal that comes closest, in a least-squares sense, to a desired output signal. The filter optimizes the output signal by arranging that the sum of squares of differences between the actual output and the desired output is a minimum.

Although special attempts are sometimes made in marine surveys to measure the source signature directly, by suspending hydrophones in the vicinity of the source, both w_k and r_k are generally unknown in reflection surveying (the reflectivity function r_k is, of course, the main target of reflection interpretation). Since, normally, only the seismic time series x_k is known, a special approach is required to design suitable inverse filters. This approach utilizes statistical analysis of the seismic time series, as in *predictive deconvolution* which attempts to remove the effect of multiples by predicting their arrival times from knowledge of the arrival times of the relevant primary events. Two important assumptions under-

lying predictive deconvolution (see, e.g. Robinson & Treitel 1980) are: (1) that the reflectivity function represents a random series (i.e. that there is no systematic pattern to the distribution of reflecting interfaces in the ground); and (2) that the composite waveform w_k for an impulsive source is minimum-delay (i.e. that its contained energy is concentrated at the front end of the pulse; see Chapter 2). From assumption (1) it follows that the autocorrelation function of the seismic trace represents the autocorrelation function of the composite waveform w_k. From assumption (2) it follows that the autocorrelation function can be used to define the shape of the waveform, the necessary phase information coming from the minimum-delay assumption.

Such an approach allows prediction of the shape of the composite waveform for use in Wiener filtering. A particular case of Wiener filtering in seismic deconvolution is that for which the desired output is a spike function. This is the basis of *spiking deconvolution*, also known as *whitening deconvolution* because a spike has the amplitude spectrum of *white noise* (i.e. all frequency components have the same amplitude).

A wide variety of deconvolution operators can be designed for inverse filtering of real seismic data, facilitating the suppression of multiples (dereverberation and deghosting) and the compression of reflected pulses. The presence of short-period reverberation in a seismogram is revealed by an autocorrelation function with a series of decaying waveforms (Fig. 4.24(a)). Long-period reverberations appear in the autocorrelation function as a series of separate side lobes (Fig. 4.24(b)), the lobes occurring at lag values for which the primary reflection aligns with a multiple reflection. Thus the spacing of the side lobes represents the periodicity of the reverberation pattern. The first multiple is phase-reversed with respect to the primary reflection, due to reflection at the ground surface or the base of the weathered layer. Thus the first side lobe has a negative peak resulting from cross-correlation of the out-of-phase signals. The second multiple undergoes a further phase reversal so that it is in phase with the primary reflection and therefore gives rise to a second side lobe with a positive peak (see Fig. 4.24(b)). Autocorrelation functions such as those shown in Fig. 4.24 form the basis of predictive deconvolution operators for removing reverberation events from seismograms.

Practically achievable inverse filters are always approximations to the ideal filter that would produce

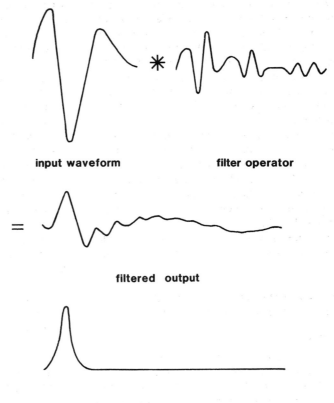

input waveform   filter operator

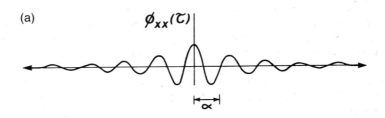

filtered output

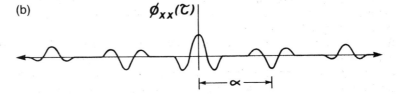

desired output

Fig. 4.23 The principle of Wiener filtering.

(a) $\phi_{xx}(\tau)$

Fig. 4.24 Autocorrelation functions of seismic traces containing reverberations. (a) A gradually decaying function indicative of short-period reverberation. (b) A function with separate side lobes indicative of long-period reverberation.

(b) $\phi_{xx}(\tau)$

a reflectivity function from a seismic trace: firstly, the ideal filter operator would have to be infinitely long; secondly, predictive deconvolution makes assumptions about the statistical nature of the seismic time series that are only approximately true. Never-

theless, dramatic improvements to seismic sections, in the way of multiple suppression and associated enhancement of vertical resolution, are routinely achieved by predictive deconvolution. An example of the effectiveness of predictive deconvolution in improving the quality of a seismic section is shown in Fig. 4.25. Deconvolution may be carried out on individual seismic traces before stacking (*deconvolution before stacking*: DBS) or on CDP gathers after stacking (*deconvolution after stacking*: DAS), and is commonly employed at both these stages of data processing.

4.7.3 Velocity filtering

The main use of *velocity filtering* (also known as *fan filtering* or *pie slice filtering*) is to remove coherent noise events from seismic records on the basis of the particular angles at which the events dip (March & Bailey 1983). The angle of dip of an event is determined from the apparent velocity with which it propagates across a spread of detectors.

A seismic pulse travelling with velocity v at an angle α to the vertical will propagate across the spread with an apparent velocity $v_a = v/\sin \alpha$ (Fig. 4.26). Along the spread direction, each individual sinusoidal component of the pulse will have an apparent wavenumber k_a related to its individual frequency f, where:

$$f = v_a k_a$$

Hence, a plot of frequency f against apparent wavenumber k_a for the pulse will yield a straight line curve with a gradient of v_a (Fig. 4.27). Any seismic event propagating across a surface spread will be characterized by an $f-k$ curve radiating from the origin at a particular gradient determined by the apparent velocity with which the event passes across the spread. The overall set of curves for a typical shot gather containing reflected and surface propagating seismic events is shown in Fig. 4.28. Events that appear to travel across the spread away from the source will plot in the positive wavenumber field; events travelling towards the source, such as backscattered rays, will plot in the negative wavenumber field.

It is apparent that different types of seismic event fall within different zones of the $f-k$ plot and this fact provides a means of filtering to suppress unwanted events on the basis of their apparent velocity. The normal means by which this is achieved, known as $f-k$ *filtering*, is to enact a two-dimensional Fourier transformation of the seismic data from the $t-x$ domain to the $f-k$ domain, then to filter the $f-k$ plot by removing a wedge-shaped zone or zones containing the unwanted noise events (March &

Bailey 1983), and finally to transform back into the $t-x$ domain.

An important application of velocity filtering is the removal of ground roll from shot gathers. This leads to marked improvement in the subsequent stacking process, facilitating better estimation of stacking velocities and better suppression of multiples. Velocity filtering can also be applied to portions of seismic record sections, rather than individual shot gathers, in order to suppress coherent noise events evident because of their anomalous dip, such as diffraction patterns. An example of such velocity filtering is shown in Fig. 4.29.

It may be noted that individual detector arrays operate selectively on seismic arrivals according to their apparent velocity across the array (Section 4.5.2), and therefore function as simple velocity filters at the data acquisition stage.

4.8 MIGRATION OF REFLECTION DATA

On seismic sections such as that illustrated in Fig. 4.25 each reflection event is mapped directly beneath the midpoint of the appropriate CDP gather. However, the reflection point is located beneath the midpoint only if the reflector is horizontal. In the presence of a component of dip along the survey line the actual reflection point is displaced in the updip direction; in the presence of a component of dip across the survey line (cross-dip) the reflection point is displaced out of the plane of the section. *Migration* is the process of reconstructing a seismic section so that reflection events are repositioned under their correct surface location and at a corrected vertical reflection time. Migration also improves the resolution of seismic sections by focusing energy spread over a Fresnel zone and by collapsing diffraction patterns produced by point reflectors and faulted beds. In *time migration*, the migrated seismic sections still have time as the vertical dimension. In *depth migration*, the migrated reflection times are converted into reflector depths using appropriate velocity information.

Two-dimensional survey data provide no information on cross-dip and, hence, in the migration of two-dimensional data the migrated reflection points are constrained to lie within the plane of the section. In the presence of cross-dip, this *two-dimensional migration* is clearly an imperfect process that will fail to remove all structural distortion. The areal reflector coverage obtained in three-dimensional surveying

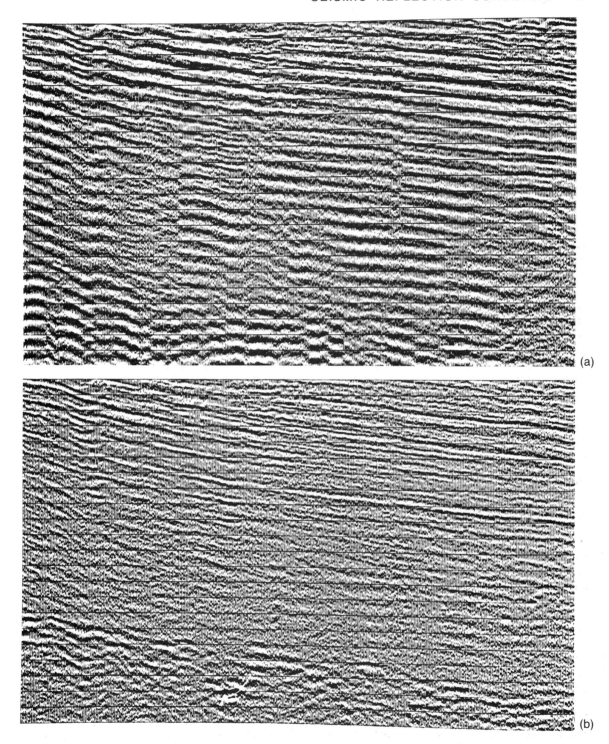

Fig. 4.25 Removal of reverberations by predictive deconvolution. (a) Seismic record dominated by strong reverberations. (b) Same section after spiking deconvolution. (Courtesy Prakla Seismos GMBH.)

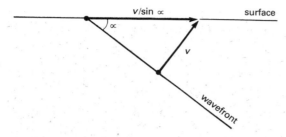

Fig. 4.26 A wave travelling at an angle α to the vertical will pass across an in-line spread of surface detectors at a velocity of $v/\sin \alpha$.

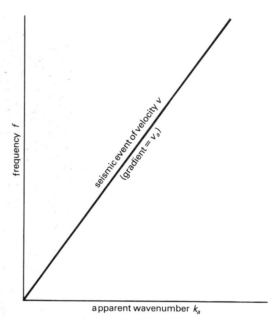

Fig. 4.27 An $f-k$ plot for a seismic pulse passing across a surface spread of detectors.

provides the additional information necessary to permit full *three-dimensional migration* in which reflection points can be migrated in any azimuthal direction. This ability fully to migrate three-dimensional survey data further enhances the value of such surveys over two-dimensional surveys in areas of complex structure.

The conversion of reflection times recorded on non-migrated sections into reflector depths, using one-way reflection times multiplied by the appropriate velocity, yields a reflector geometry known as the *record surface*. This coincides with the actual *reflector surface* only when the latter is horizontal. In the case of dipping reflectors the record surface departs from the reflector surface, i.e. it gives a distorted picture of the reflector geometry. Migration removes the distorting effects of dipping reflectors from seismic sections and their associated record

surfaces. Migration also removes the diffracted arrivals resulting from point sources since every diffracted arrival is migrated back to the position of the point source. A variety of geological structures and sources of diffraction are illustrated in Fig. 4.30(a) and the resultant non-migrated seismic section is shown in Fig. 4.30(b). Structural distortion in the

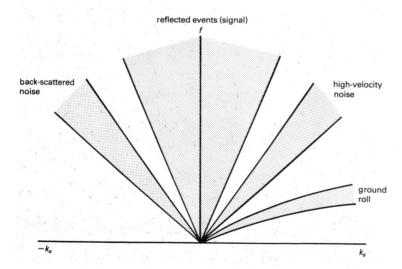

Fig. 4.28 An $f-k$ plot for a typical shot gather (such as that illustrated in Fig. 4.10) containing reflection events and different types of noise.

(a)

(b)

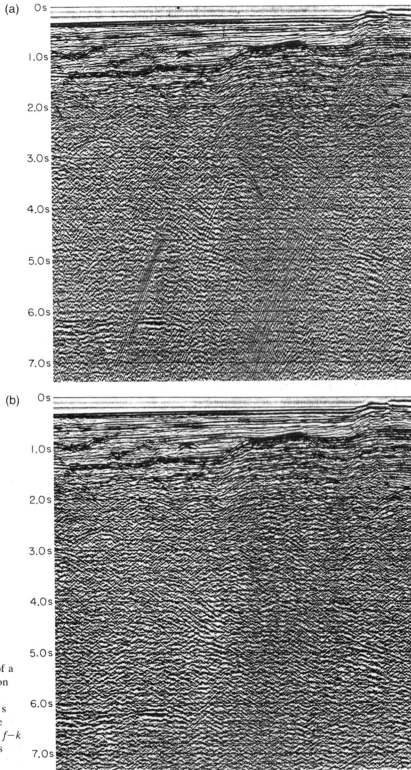

Fig. 4.29 Effect of $f-k$ filtering of a
seismic section. (a) Stacked section
showing steeply-dipping coherent
noise events, especially below 4.5 s
two-way reflection time. (b) Same
section after rejection of noise by $f-k$
filtering (Courtesy Prakla-Seismos
GMBH).

(a)

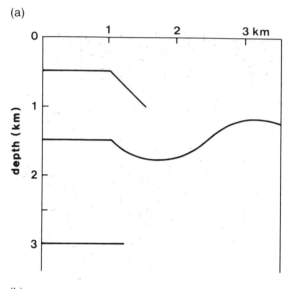

(b)

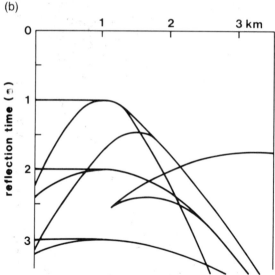

Fig. 4.30 (a) A structural model of the subsurface and (b) the resultant reflection events that would be observed in a non-migrated seismic section, containing numerous diffraction events. (After Sheriff 1978.)

non-migrated section (and record surfaces derived from it) includes a broadening of anticlines and a narrowing of synclines. The edges of fault blocks act as point sources and typically give rise to strong diffracted phases, represented by hyperbolic patterns of events in the seismic section. Synclines within which the reflector curvature exceeds the curvature

of the incident wavefront are represented on non-migrated seismic sections by a 'bow-tie' event resulting from the existence of three discrete reflection points for any surface location (see Fig. 4.31).

Various aspects of migration are discussed below using the simplifying assumption that the source and detector have a common surface position (i.e. the detector has a zero offset, which is approximately the situation involved in CDP stacks). In such a case, the incident and reflected rays follow the same path and the rays are normally incident on the reflector surface. Consider a source-detector on the surface of a medium of constant seismic velocity (Fig. 4.32). Any reflection event is conventionally mapped to lie directly beneath the source-detector but in fact it may lie anywhere on the locus of equal two-way reflection times, which is a semi-circle centred on the source-detector position.

Now consider a series of source-detector positions overlying a planar dipping reflector beneath a medium of uniform velocity (Fig. 4.33). The reflection events are mapped to lie below each source-detector location but the actual reflection points are offset in the updip direction. The construction of arcs of circles (wavefront segments) through all the mapped reflection points enables the actual reflector geometry to be mapped. This represents a simple example of migration. The migrated section indicates a steeper reflector dip than the record surface derived from the non-migrated section. In general, if α_s is the dip of the record surface and α_t is the true dip of the reflector, $\sin \alpha_t = \tan \alpha_s$. Hence the maximum dip of a record surface is 45° and represents the case of horizontal reflection paths from a vertical reflector. This *wavefront common-envelope* method of migration can be extended to deal with reflectors of irregular geometry. If there is a variable velocity above the reflecting surface to be migrated, the reflected ray paths are not straight and the associated wavefronts are not circular. In such a case, a *wavefront chart* is constructed for the prevailing velocity–depth relationship and this is used to construct the wavefront segments passing through each reflection event to be migrated.

An alternative approach to migration is to assume that any continuous reflector is composed of a series of closely-spaced point reflectors, each of which is a source of diffractions, and that the continuity of any reflection event results from the constructive and destructive interference of these individual diffraction events. A set of diffracted arrivals from a single point reflector embedded in a uniform vel-

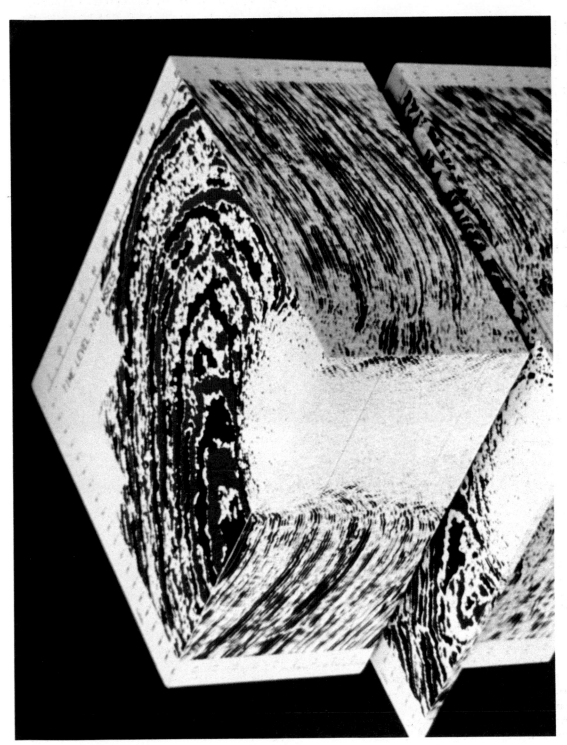

Plate 1. Three-dimensional data volume showing a Gulf of Mexico salt dome with an associated rim syncline. (Reproduced from *AAPG Memoir* No. 42, with the permission of the publishers.)

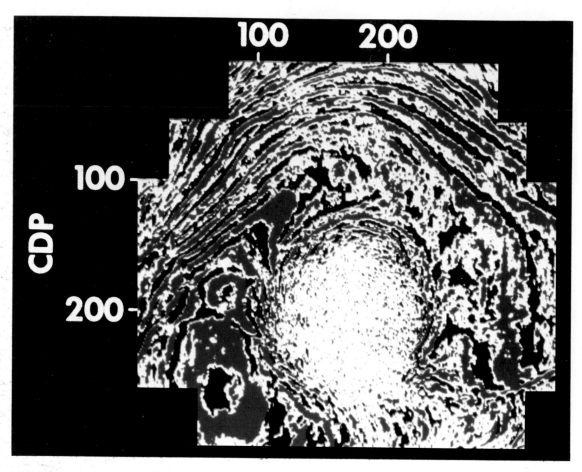

Plate 2. Seiscrop section at 3760 ms from a three-dimensional survey in the Eugene Island area of the Gulf of Mexico. (Reproduced from *AAPG Memoir* No. 42, with the permission of the publishers.)

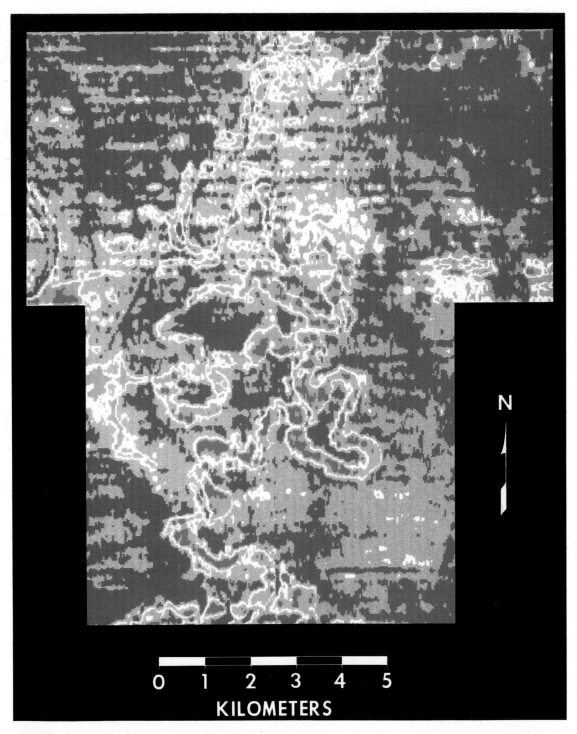

N

0 1 2 3 4 5
KILOMETERS

Plate 4 (a). Seiscrop section at 196 ms from a three-dimensional survey in the Gulf of Thailand area, showing a meandering stream channel. (Both illustrations reproduced from *AAPG Memoir* No. 42, with the permission of the publishers.)

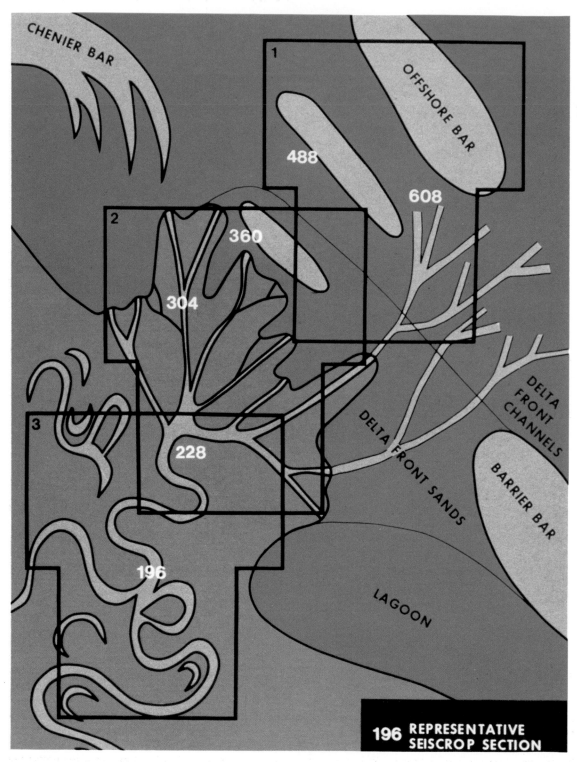

Plate 4 (b). Diagrammatic map of a former prograding delta system within the Gulf of Thailand survey area, based on interpretation of seiscrop sections 1, 2 and 3 shown on map.

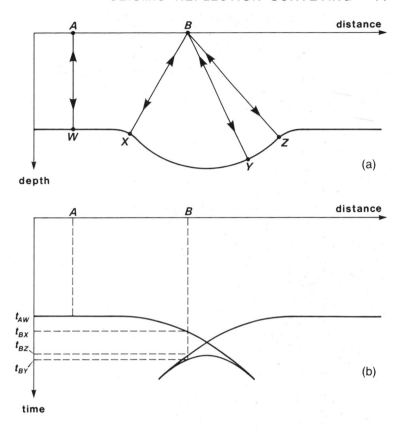

Fig. 4.31 (a) A sharp synclinal feature in a reflecting interface, and (b) the resultant 'bow-tie' shape of the reflection event on the non-migrated seismic section.

Fig. 4.32 For a given reflection time, the reflection point may lie anywhere on the arc of a circle centred on the source-detection position. On a non-migrated seismic section the point is mapped to lie immediately below the source-detector.

ocity medium is shown in Fig. 4.34. The two-way reflection times to different surface locations define a hyperbola. If arcs of circles (wavefront segments) are drawn through each reflection event they intersect at the actual point of diffraction (Fig. 4.34). In the case of a variable velocity above the point reflector the diffraction event will not be a hyperbola but a curve of similar convex shape. No reflection event on a seismic section can have a greater convexity than a diffraction event, hence the latter is referred to as a *curve of maximum convexity*. In *diffraction migration* all dipping reflection events are assumed to be tangential to some curve of maximum convexity. By the use of a wavefront chart appropriate to the prevailing velocity−depth relationship, wavefront segments can be drawn through dipping reflection events on seismic sections and the events migrated back to their diffraction points (Fig. 4.34). Events so migrated will, overall, map the prevailing reflector geometry.

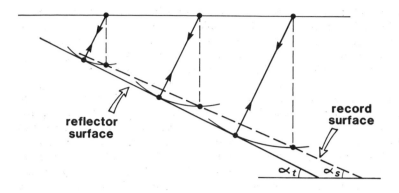

Fig. 4.33 A planar-dipping reflector surface and its associated record surface derived from a non-migrated seismic section.

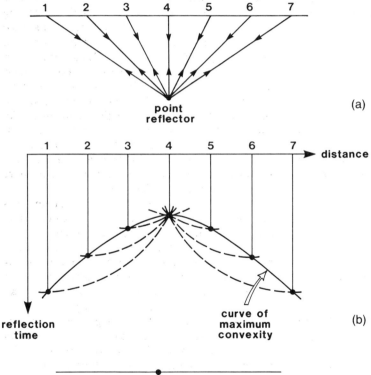

(a)

(b)

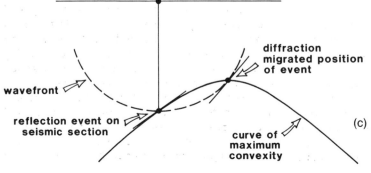

(c)

Fig. 4.34 Principles of diffraction migration. (a) Reflection paths from a point reflector. (b) Migration of individual reflection events back to position of point reflector. (c) Use of wavefront chart and curve of maximum convexity to migrate a specific reflection event: the event is tangential to the appropriate curve of maximum convexity, and the migrated position of the event is at the intersection of the wavefront with the apex of the curve.

All modern approaches to migration utilize the wave equation which is a partial differential equation describing the motion of waves, within a medium, that have been generated by a wave source: in the case under consideration, the motion of seismic waves in the ground generated by a seismic source. The migration problem can be considered in terms of wave propagation through the ground in the following way: for any reflection event, the form of the seismic wavefield at the surface can be reconstructed from the travel times of reflected arrivals to different source-detector locations, and for the purpose of migration it is required to reconstruct the form of the wavefield within the ground, in the vicinity of a reflecting interface. This reconstruction can be achieved by solution of the wave equation to effectively propagate the wave backwards in time. Propagation of the wavefield of a reflection event halfway back to its origin time should place the wave on the reflecting interface, hence, the form of the wavefield at that time should define the reflector geometry.

Migration using the wave equation is known as *wave equation migration* (Robinson & Treitel 1980). There are several approaches to the problem of solving the wave equation and these give rise to specific types of wave equation migration such as *finite difference migration*, in which the wave equation is approximated by a finite difference equation suitable for solution by computer, and *frequency-domain migration*, in which the wave equation is solved by means of Fourier transformations, the necessary spatial transformations to achieve migration being enacted in the frequency domain and recovered by an inverse Fourier transformation.

Migration by computer can also be carried out by direct modelling of ray paths through hypothetical models of the ground, the geometry of the reflecting interfaces being adjusted iteratively to remove discrepancies between observed and calculated reflection times. Particularly in the case of seismic surveys over highly complex subsurface structures, e.g. those encountered in the vicinity of salt domes and salt walls, this *ray trace migration* method may be the only method capable of successfully migrating the seismic sections.

In order to migrate a seismic section accurately it would be necessary to define fully the velocity field of the ground, i.e. to specify the value of velocity at all points. In practice, for the purposes of migration, an estimate of the velocity field is made from prior analysis of the non-migrated seismic section, together with information from borehole logs where available. In spite of this approximation, migration almost invariably leads to major improvement in the seismic imaging of reflector geometry.

Migration of seismic profile data is normally carried out on CDP stacks, thus reducing the number of traces to be migrated by a factor equal to the fold of the survey and thereby reducing the computing time and associated costs. Migration of stacked traces is based on the assumption that the stacks closely resemble the form of individual traces recorded at zero offset and containing only normal-incidence reflection events. This assumption is clearly invalid in the case of recordings over a wide range of offsets in areas of structural complexity. A better approach is to migrate the individual seismic traces (assembled into a series of profiles containing all traces with a common offset), then to assemble the migrated traces into CDP gathers and stacks. Such an approach is not necessarily cost-effective in the case of high-fold CDP surveys, and a compromise is to migrate subsets of CDP stacks recorded over a narrow range of offset distances, and then produce a full CDP stack by summing the migrated partial stacks after correction for normal moveout. Procedures involving migration before final stacking involve extra cost but can lead to significant improvements in the migrated sections and to more reliable stacking velocities.

Any system of migration represents an approximate solution to the problem of mapping reflecting surfaces into their correct spatial positions and the various methods have different performances with real data. For example, the diffraction method performs well in the presence of steep reflector dips but is poor in the presence of a low SNR. The best all round performance is given by frequency-domain migration. Examples of the migration of seismic sections are illustrated in Figs 4.35 and 4.36. Note in particular the clarification of structural detail, including the removal of bow-tie effects, and the repositioning of structural features in the migrated sections. Clearly, when planning to test hydrocarbon prospects in areas of structural complexity (as on the flank of a salt dome) it is important that drilling locations are based on interpretation of migrated rather than non-migrated seismic sections.

The essential difference between two-dimensional and three-dimensional migration may be illustrated with reference to a point reflector embedded in an homogeneous medium. On a seismic section derived from a two-dimensional survey the point reflector is

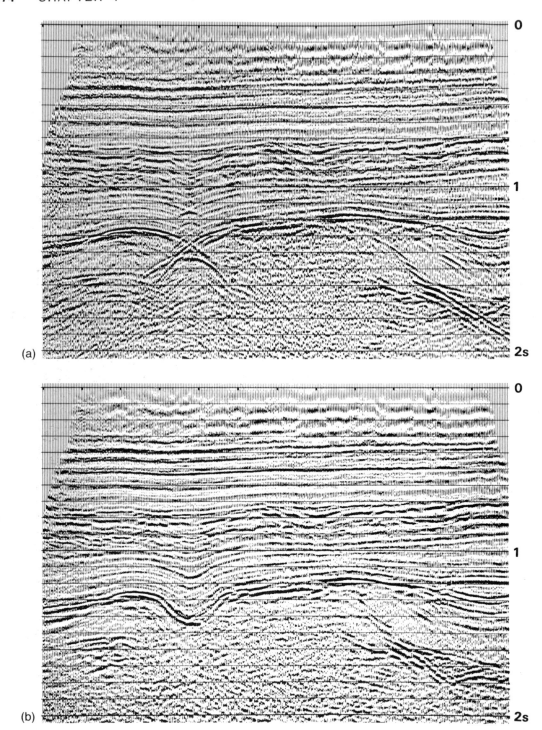

Fig. 4.35 (a) A non-migrated seismic section. (b) The same seismic section after wave equation migration. (Courtesy Prakla-Seismos GMBH.)

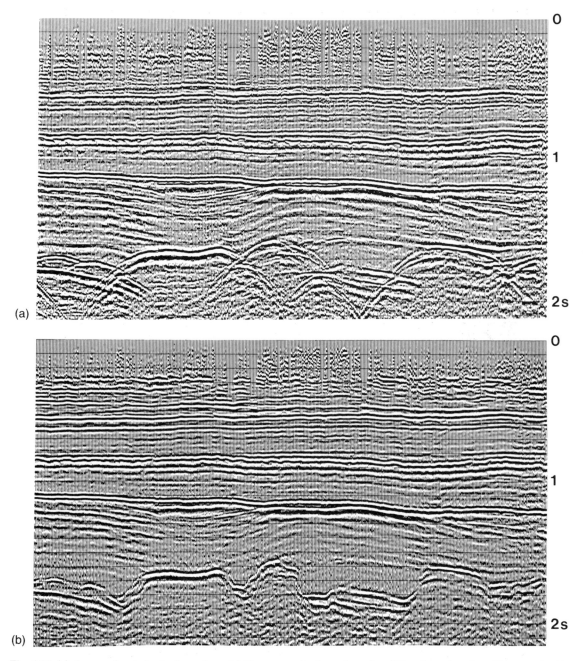

Fig. 4.36 (a) A non-migrated seismic section. (b) The same seismic section after diffraction migration. (Courtesy Prakla-Seismos GMBH.)

imaged as a diffraction hyperbola, and migration involves summing amplitudes along the hyperbolic curve and plotting the resultant event at the apex of the hyperbola (see Fig. 4.25). The actual three-

dimensional pattern associated with a point reflector is a hyperboloid of rotation, the diffraction hyperbola recorded in a two-dimensional survey representing a vertical slice through this hyperboloid. In a three-dimensional survey, reflections are recorded from a surface area of the hyperboloid and three-dimensional migration involves summing amplitudes over the surface area to define the apex of the hyperboloid.

A practical way of achieving this aim with crossed-array data from a three-dimensional land survey is the *two-pass* method (Fig. 4.37). The first pass involves collapsing diffraction hyperbolas recorded in vertical sections along one of the orthogonal line directions. The series of local apexes in these sections together define a hyperbola in a vertical section along the perpendicular direction. This hyperbola can then be collapsed to define the apex of the hyperboloid.

4.9 INTERPRETATION

Differing procedures are adopted for the interpretation of two- and three-dimensional seismic data. The results of two-dimensional surveys are presented to the seismic interpreter as non-migrated and migrated seismic sections, from which the geological information is extracted by suitable analysis of the pattern of reflection events. Interpretations are correlated from line to line, and the reflection times of picked events are compared directly at profile intersections. There are two main approaches to the interpretation of seismic sections: *structural analysis*, which is the study of reflector geometry on the basis of reflection times, and *stratigraphical analysis* (or *seismic stratigraphy*), which is the analysis of reflection sequences as the seismic expression of lithologically-distinct depositional sequences. Both structural and stratigraphical analyses are greatly assisted by *seismic modelling* in which theoretical (synthetic) seismograms are constructed for layered models in order to derive insight into the physical significance of reflection events contained in seismic sections.

In the interpretation of three-dimensional survey data, the interpreter has direct access at a computer work station to all the reflection data contained within the seismic data volume (see Section 4.5.4), and is able to select various types of data for colour display, for example, vertical sections or horizontal sections (time slices) through the data volume. The two most important shortcomings of two-dimensional interpretation are the problem of correlation between adjacent profile lines and the inaccuracy of reflector positioning due to the limitations of two-dimensional migration. The improved cover-

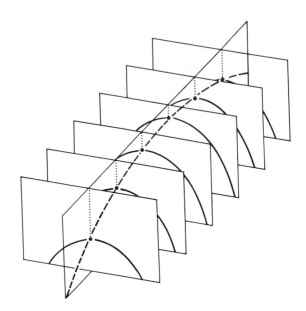

Fig. 4.37 The two-pass method of three-dimensional migration for the case of a point reflector. The apices of diffraction hyperbolae in one line direction may be used to construct a diffraction hyperbola in the orthogonal line direction. The apex of the latter hyperbola defines the position of the point reflector.

age and resolution of three-dimensional data often lead to substantial improvements in interpretation as compared with pre-existing two-dimensional interpretation. As with two-dimensional interpretation, both structural and stratigraphic analysis may be carried out, and in the following sections examples are taken from both two- and three-dimensional survey applications.

4.9.1 Structural analysis

The main application of structural analysis of seismic sections is in the search for structural traps containing hydrocarbons. Interpretation usually takes place against a background of continuing exploration activity and an associated increase in the amount of information related to the subsurface geology. Reflection events of interest are usually colour-coded initially and labelled as, e.g. 'red reflector' 'blue reflector', until their geological significance is established. Whereas an initial interpretation of reflections displayed on seismic sections may lack geological control, at some point the geological nature of the reflectors is likely to become established by tracing reflection events back either to outcrop or to an existing borehole for stratigraphic control. Subsurface reflectors may then be referred to by an appropriate stratigrapical indicator such as 'base Tertiary', 'top Lias'.

Most structural interpretation is carried out in units of two-way reflection time rather than depth, and *time-structure maps* are constructed to display the geometry of selected reflection events by means of contours of equal reflection time (Fig. 4.38). *Structural contour maps* can be produced from time-structure maps by conversion of reflection times into depths using appropriate velocity information (e.g. local stacking velocities derived from the reflection survey or sonic log data from boreholes). Time-structure maps obviously bear a close similarity to structural contour maps but are subject to distortion associated with lateral or vertical changes of velocity in the subsurface interval overlying the reflector. Other aspects of structure may be revealed by contouring variations in the reflection time interval between two reflectors, sometimes referred to as *isochron maps*, and these can be converted into

isopach maps by the conversion of reflection time intervals into thicknesses using the appropriate interval velocity.

Problems often occur in the production of time-structure or isochron maps. The difficulty of correlating reflection events across areas of poor signal-to-noise ratio, structural complexity or rapid stratigraphic transition often leaves the disposition of a reflector poorly resolved. Intersecting survey lines facilitate the checking of an interpretation by comparison of reflection times at intersection points. Mapping reflection times around a closed loop of survey lines reveals any errors in the identification or correlation of a reflection event across the area of a seismic survey.

Reprocessing of data, or migration, may be employed to help resolve uncertainties of interpretation, but additional seismic lines are often needed to resolve problems associated with an initial phase of interpretation. It is common for several rounds of seismic exploration to be necessary before a prospective structure is sufficiently well defined to locate the optimal position of an exploration borehole.

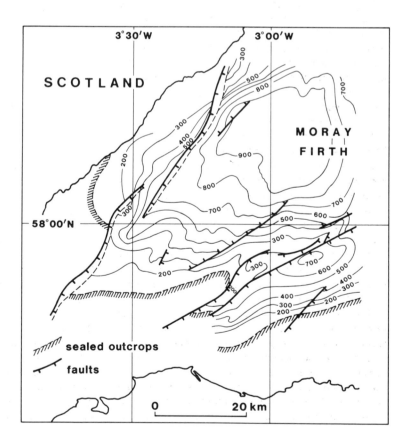

Fig. 4.38 Time-structure map of reflector at the base of the Lower Cretaceous in the Moray Firth off northeast Scotland, UK. Contour values represent two-way travel times of reflection event in milliseconds. (Courtesy British Geological Survey, Edinburgh, UK.)

Structural interpretation of three-dimensional data is able to take advantage of the areal coverage of reflection points, the improved resolution associated with three-dimensional migration and the improved methods of data access, analysis and display provided by dedicated seismic work stations. Examples of the display of geological structures using three-dimensional data volumes are illustrated in Plates 1 and 2. Interpretation of three-dimensional data is often crucial to the successful development of oilfields with a complex geological structure. An example is the North Cormorant oilfield in the UK Sector of the North Sea, where three-dimensional seismics enabled the mapping of far more fault structures than had been possible using pre-existing two-dimensional data, and revealed a set of NW-SE trending faults that had previously been unsuspected.

4.9.2 **Stratigraphical analysis (seismic stratigraphy)**

Seismic stratigraphy involves the subdivision of seismic sections into sequences of reflections that are interpreted as the seismic expression of genetically-related sedimentary sequences. The principles behind this *seismic sequence analysis* are two-fold. Firstly, reflections are taken to define chronostratigraphical units, since the types of rock interface that produce reflections are stratal surfaces and unconformities; by contrast, the boundaries of diachronous lithological units tend to be transitional and not to produce reflections. Secondly, genetically related sedimentary sequences normally comprise a set of concordant strata that exhibit discordance with underlying and overlying sequences, i.e. they are typically bounded by angular unconformities variously representing onlap, downlap, toplap or erosion (Fig. 4.39). A seismic sequence is the representation on a seismic section of a depositional sequence; as such, it is a group of concordant or near-concordant reflection events that terminate against the discordant reflections of adjacent seismic sequences. An example of a seismic sequence identified on a seismic section is illustrated in Plate 3.

Having subdivided a seismic section into its constituent sequences, each sequence may be analysed in terms of the internal disposition of reflection events and their character, to obtain insight into the depositional environments responsible for the sequence and into the range of lithofacies that may be represented within it. This use of reflection geometry and character to interpret sedimentary facies is

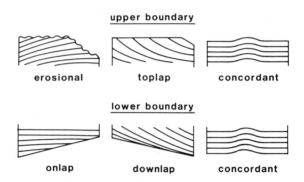

Fig. 4.39 Different types of geological boundary defining seismic sequences. (After Sheriff 1980.)

known as *seismic facies analysis*. Individual seismic facies are identified within the seismic sequence illustrated in Plate 3.

Different types of reflection configuration (see Fig. 4.40) are diagnostic of different sedimentary environments. On a regional scale, for example, parallel reflections characterize some shallow water shelf environments whilst the deeper water shelf edge and slope environments are often marked by the development of major sigmoidal or oblique cross-bedded units. The ability to identify particular sedimentary environments and predict lithofacies from analysis of seismic sections can be of great value to exploration programmes, providing a pointer to the location of potential source, reservoir and/or seal rocks. Thus, organic-rich basinal muds represent potential source rocks; discrete sand bodies developed in shelf environments represent potential reservoir rocks; and coastal mud and evaporite sequences represent potential seals (Fig. 4.41): the identification of these components in seismic sequences can thus help to focus an exploration programme by identifying areas of high potential.

An example of seismic stratigraphy based on three-dimensional data is illustrated in Plate 4. The seiscrop of Plate 4(a) shows a meandering stream channel preserved in a Neogene sedimentary sequence in the Gulf of Thailand. The channel geometry and the distinctive lithofacies of the channel fill lead to its clear identification as a distinctive seismic facies. Use of such seiscrops over a wider area enables the regional mapping of a Neogene deltaic environment (Plate 4(b)).

Major seismic sequences can often be correlated across broad regions of continental margins and clearly give evidence of being associated with major

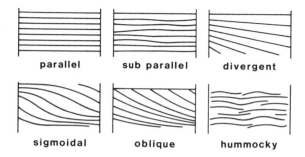

Fig. 4.40 Various internal bedforms that give rise to different seismic facies within sedimentary sequences identified on seismic sections. (After Sheriff 1980.)

sea-level changes. The widespread application of seismic stratigraphy in areas of good chrono-stratigraphical control has led to the development of a model of global cycles of major sea-level change and associated transgressive and regressive de-positional sequences throughout the Mesozoic and Cenozoic (Payton 1977). Application of the methods of seismic stratigraphy in offshore sedimentary basins with little or no geological control often enables correlation of locally recognized depositional se-quences with the worldwide pattern of sea-level changes (Payton 1977). It also facilitates identi-fication of the major progradational sedimentary sequences which offer the main potential for hydro-carbon generation and accumulation. Stratigraphic analysis therefore greatly enhances the chances of successfully locating hydrocarbon traps in sed-imentary basin environments.

Hydrocarbon accumulations are sometimes re-vealed directly on true-amplitude seismic sections (see below) by localized zones of anomalously strong reflections known as *bright spots*. These high-amplitude reflection events (Fig. 4.42) are attribu-table to the large reflection coefficients at the top and bottom of gas zones (typically, gas-filled sands) within a hydrocarbon reservoir. In the absence of bright spots, fluid interfaces may nevertheless be directly recognizable by *flat spots* which are hor-izontal or near horizontal reflection events dis-cordant to the local geological dip.

4.9.3 Seismic modelling

Conventionally, reflection amplitudes are nor-malized prior to their presentation on seismic sec-tions so that original distinctions between weak and strong reflections are suppressed. This practice tends to increase the continuity of reflection events across a section and therefore aids their identification and structural mapping. However, much valuable geo-logical information is contained in the true am-plitude of a reflection event, which can be recovered from suitably calibrated field recordings. Any lateral variation of reflection amplitude is due to lateral change in the lithology of a rock layer or in its pore fluid content. Thus, whilst the production of normalized-amplitude sections may assist structural mapping of reflectors, it suppresses information that is vital to a full stratigraphic interpretation of the data. With increasing interest centering on stra-tigraphic interpretation, true-amplitude seismic sec-tions are becoming increasingly important.

In addition to amplitude, the shape and polarity of a reflection event also contain important geo-logical information (Meckel & Nath 1977). Analysis of the significance of lateral changes of shape,

Dark grey marls, black organic-rich mudstones	Siltstones, shales, reef limestones	Calcareous sandstones, oolites, bioclastic limestones	Sandstones, mudstones, dolomitic mudstones, evaporites	TYPICAL LITHOLOGIES
Very thin and continuous units	Thin to intermediate tabular bodies with len-soid reef limestones	Intermediate continuous to lensoid bodies	Irregular to discontinuous units	BED GEOMETRY
25 - 50	150 - 450	100 - 50	50 - 25	THICKNESS (m)
BASINAL	OUTER SHELF	INNER SHELF	COASTAL	ENVIRONMENT

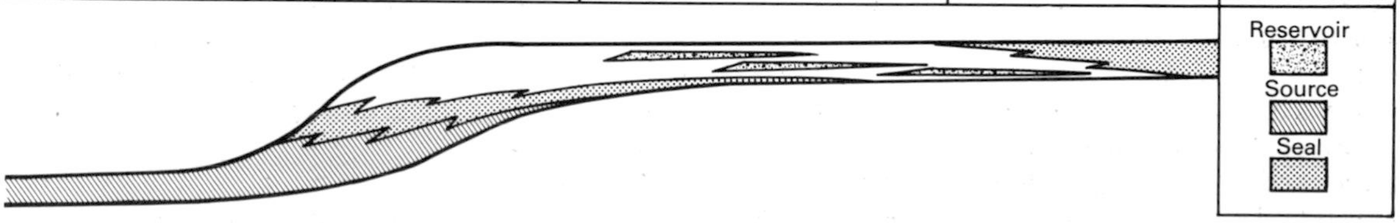

Fig. 4.41 The overall geometry of a typical depositional sequence and its contained sedimentary facies.

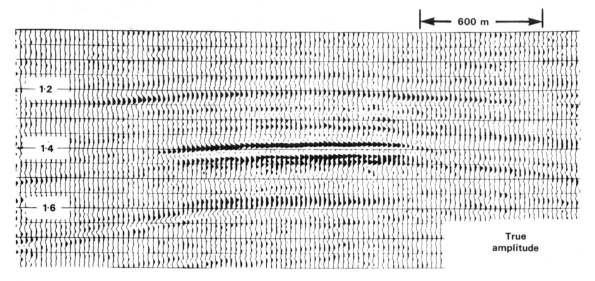

Fig. 4.42 Part of a true-amplitude seismic section containing a seismic bright spot associated with a local hydrocarbon accumulation. (From Sheriff 1980, after Schramm *et al*. 1977.)

polarity and amplitude observed in true-amplitude seismic sections is carried out by *seismic modelling*, often referred to in this context as *stratigraphic modelling*. Seismic modelling involves the production of synthetic seismograms for layered sequences to investigate the effects of varying the model parameters on the form of the resulting seismograms. Synthetic seismograms and synthetic seismic sections can be compared with real data, and models can be manipulated in order to simulate the real data. By this means, valuable insights can be obtained into the subsurface geology responsible for a particular seismic section. The standard type of synthetic seismogram represents the seismic response to vertical propagation of an assumed source wavelet through a model of the subsurface composed of a series of horizontal layers of differing acoustic impedance. Each layer boundary reflects some energy back to the surface, the amplitude and polarity of the reflection being determined by the acoustic impedance contrast. The synthetic seismogram comprises the sum of the individual reflections in their correct travel time relationships. (see Fig. 4.43).

In its simplest form, a synthetic seismogram $x(t)$ may be considered as the convolution of the assumed source function $s(t)$ with a reflectivity function $r(t)$ representing the acoustic impedance contrasts in the layered model.

$$x(t) = s(t) * r(t)$$

However, filtering effects along the downgoing and upgoing ray paths and the overall response of the recording system need to be taken into account. Multiples may or may not be incorporated into the synthetic seismogram.

The acoustic impedance values necessary to compute the reflectivity function may be derived directly from sonic log data. This is normally achieved assuming density to be constant throughout the model, but it may be important to derive estimates of layer densities in order to compute more accurate impedance values.

Synthetic seismograms can be derived for more complex models using ray tracing techniques.

Particular stratigraphic features that have been investigated by seismic modelling, to determine the nature of their representation on seismic sections, include thin layers, discontinuous layers, wedge-shaped layers, transitional layer boundaries, variable porosity and type of pore fluid. Fig. 4.44 illustrates synthetic seismograms computed across a section of stratigraphic change. These show how the varying pattern of interference between reflection events expresses itself in lateral changes of pulse shape and peak amplitude.

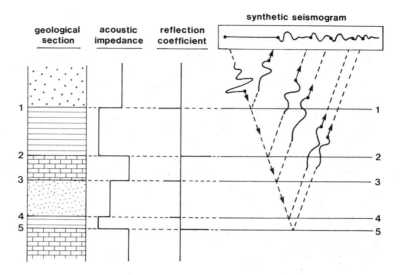

Fig. 4.43 The synthetic seismogram.

4.10 SINGLE-CHANNEL MARINE REFLECTION PROFILING

Single-channel reflection profiling is a simple but highly effective method of seismic surveying at sea that finds wide use in a variety of offshore applications. It represents reflection surveying reduced to its bare essentials: a marine seismic/acoustic source is towed behind a survey vessel and triggered at a fixed firing rate, and signals reflected from the sea bed and from sub-bottom reflectors are detected by a hydrophone streamer towed in the vicinity of the source (Fig. 4.45). The outputs of the individual hydrophone elements are summed and fed to a single channel amplifier/processor unit and thence to a chart recorder. This survey procedure is not possible on land because only at sea can the source and detectors be moved forward continuously, and a sufficiently high firing rate achieved, to enable surveys to be carried out continuously from a moving vehicle.

The source and hydrophone array are normally towed at shallow depth but some deep water applications utilize deep-tow systems in which the source and receiver are towed close to the sea bed. Deep-tow systems overcome the transmission losses associated with a long water path, thus giving improved penetration of seismic/acoustic energy into the sea bed. Moreover, in areas of rugged bathymetry they produce records that are much simpler to interpret: there is commonly a multiplicity of reflection paths

from a rugged sea bed to a surface source-detector location, so that records obtained in deep water using shallow-tow systems commonly exhibit hyperbolic diffraction patterns, bow-tie effects and other undesirable features of non-migrated seismic sections.

In place of the oscillographic recorder used in multichannel seismic surveying, single-channel profiling typically utilizes an *oceanographic recorder* in which a stylus repeatedly sweeps across the surface of an electrically-conducting recording paper that is continuously moving forward at a slow speed past a strip electrode in contact with the paper. A mark is burnt into the paper whenever an electrical signal is fed to the stylus and passes through the paper to the strip electrode. The seismic/acoustic source is triggered at the commencement of a stylus sweep and all seismic pulses returned during the sweep interval are recorded as a series of dark bands on the recording paper (Fig. 4.46). The triggering rate and sweep speed are variable over a wide range: for a shallow penetration survey the source may be triggered every 500 ms and the recording interval may be 0–250 ms, whereas for a deep penetration survey in deep water the source may be triggered every 8 s and the recording interval may be 2–6 s.

The analogue recording systems used in single-channel profiling are relatively cheap to operate. There are no processing costs and seismic records are produced in real time by the continuous chart recording of band-pass filtered and amplified signals,

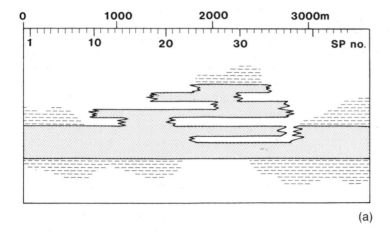

(a)

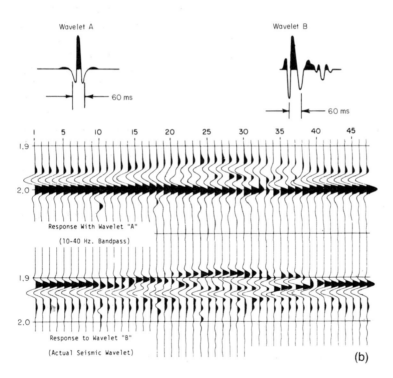

(b)

Fig. 4.44 A set of synthetic seismograms simulating a seismic section across a zone of irregular sandstone geometry. (From Neidell & Poggiagliolmi 1977.)

sometimes with time variable gain (TVG). When careful consideration is given to source and hydrophone array design and deployment, good basic reflection records may be obtained from a single-channel system, but they cannot compare in quality with the type of seismic record produced by computer processing of multichannel data. Moreover, single-channel recordings cannot provide velocity information so that the conversion of reflection times into reflector depths has to utilize independent estimates of seismic velocity. Nonetheless, single-channel profiling often provides good imaging of subsurface geology and permits estimates of reflector depth and geometry that are sufficiently accurate for many purposes.

The record sections suffer from the presence of multiple reflections, especially multiples of the sea bed reflection, which may obliterate primary reflec-

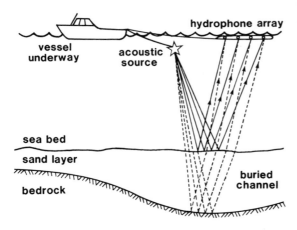

Fig. 4.45 The survey set-up for single-channel seismic reflection profiling.

tion events in the later parts of the records. Multiples are a particular problem when surveying in very shallow water, since they then occur at a short time interval after the primary events (e.g. see Fig. 4.49). Record sections are often difficult to interpret in areas of complex reflector geometry because of the presence of bow-tie effects, diffraction events and other features of non-migrated seismic sections.

As discussed in Chapter 3 there is a variety of marine seismic/acoustic sources, operating at differing energy levels and characterized by different dominant frequencies. Consequently by selection of a suitable source, single-channel profiling can be applied to a wide range of offshore investigations from high-resolution surveys of near-surface sedimentary layers to surveys of deep geological structure. In general there is a trade-off between depth of penetration and degree of vertical resolution, since the higher energy sources required to transmit signals to greater depths are characterized by lower dominant frequencies and longer pulse lengths that adversely affect the resolution of the resultant seismic records.

Pingers are low energy (typically about 5 J), tunable sources that can be operated within the frequency range from 3 to 12 kHz. The piezoelectric transducers used to generate the pinger signal also serve as receivers for reflected acoustic energy and, hence, a separate hydrophone streamer is not required in pinger surveying. Vertical resolution can be as good as 10−20 cm but depth penetration is limited to a few tens of metres in muddy sediments or several metres in coarse sediments, with virtually no penetration into solid rock. Pinger surveys are commonly used in offshore engineering site investigation and are of particular value in submarine pipeline route surveys. Repeated pinger surveying along a pipeline route enables monitoring of local sediment movement and facilitates location of the pipeline where it has become buried under recent sediments. A typical pinger record is shown in Fig. 4.47.

Boomer sources provide a higher energy output

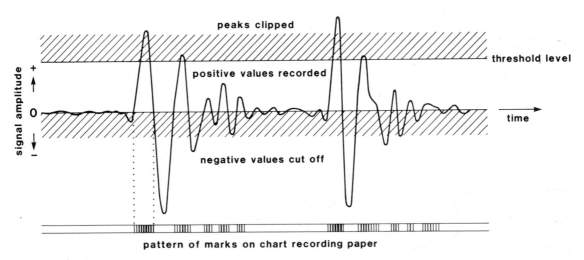

Fig. 4.46 Seismic signals and their representation on the chart recording paper of an oceanographic recorder. (From Le Tirant 1979.)

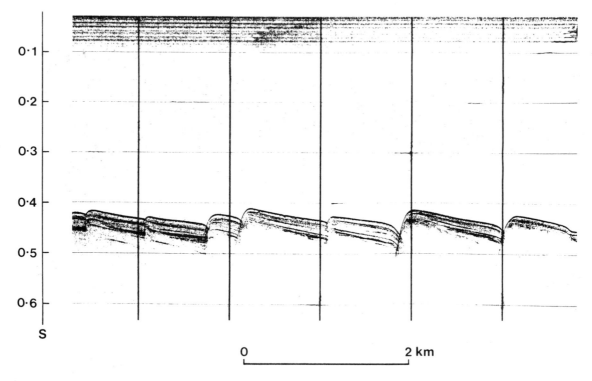

Fig. 4.47 Pinger record from the northern Aegean Sea, Greece, across a zone of active growth faults extending up to the sea bed. The sea floor is underlain by a layered sequence of Holocene muds and silts that can be traced to a depth of about 50 m. Note the diffraction patterns associated with the edges of the individual fault blocks.

(typically 300–500 J) and operate at lower dominant frequencies (1–5 kHz) than pingers and therefore provide greater penetration (up to 100 m in bedrock) with good resolution (0.5–1 m). Boomer surveys are useful for mapping thick sedimentary sequences, in connection with channel dredging or sand and gravel extraction, or for high resolution surveys of shallow geological structures. A boomer record section is illustrated in Fig. 4.48.

Sparker sources can be operated over a wide range of energy levels (300–30 000 J), though the production of spark discharges of several thousand joules every few seconds requires a large power supply and a large bank of capacitors. Sparker surveying therefore represents a versatile tool for a wide range of applications, from shallow penetration surveys (100 m) with moderate resolution (2 m) to deep penetration surveys (> 1 km) where resolution is not important. However, sparker surveying cannot match the resolution of precision boomer surveying, and sparkers do not offer as good a source signature as air guns for deeper penetration surveys.

By suitable selection of chamber size and rate of release of compressed air, air gun sources can be tailored to high resolution or deep penetration profiling applications and therefore represent the most versatile source for singe-channel profiling. A reflection record obtained in a shallow water area with a small air gun (40 in³) is shown in Fig. 4.49.

Single-channel reflection profiling systems (sometimes referred to as *sub-bottom profiling systems*) are commonly operated in conjunction with a precision echo-sounder, for high-quality bathymetric information, and/or with a sidescan sonar system. *Sidescan sonar* is a sideway-scanning acoustic survey method in which the sea floor to one or both sides of the survey vessel is insonified by beams of high-frequency sound (30–110 kHz) transmitted by hull-mounted or fish-mounted transceiving transducers (Fig. 4.50). Sea bed features facing towards the survey vessel, such as rock outcrops or sedimentary bedforms, reflect acoustic energy back towards the transducers whilst in the case of features facing away from the vessel, or a featureless sea floor, the

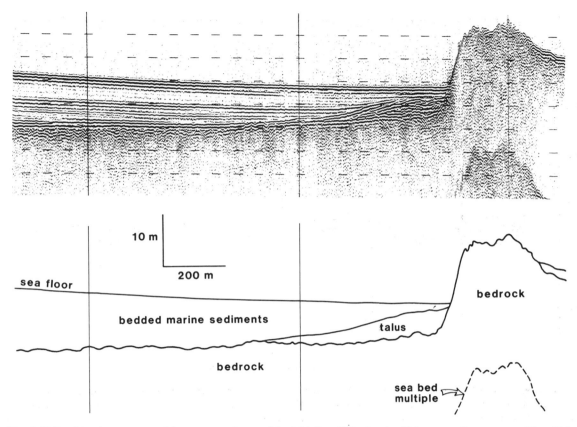

Fig. 4.48 Precision boomer record from a coastal area of the Irish Sea, UK, showing Holocene sediments up to 10 m thick banked against a reef of Lower Palaeozoic rocks. (Courtesy C.R. Price.)

acoustic energy is reflected away from the transducers. Signals reflected back to the transducers are fed to the same type of recorder that is used to produce seismic profiling records, and the resulting pattern of returned acoustic energy is known as a *sonograph*. The oblique insonification produces scale distortion resulting from the varying path lengths and angles of incidence of returning rays (Fig. 4.50(b)). This distortion can be automatically corrected for prior to display so that the sonograph provides an isometric plan view of sea bed features. A sonograph is shown in Fig. 4.51.

Although not strictly a seismic surveying tool, sidescan sonar provides valuable information on, for example, the configuration and orientation of sedimentary bedforms or on the pattern of rock outcrops. This information is often very useful in complementing the subsurface information derived from shallow seismic reflection surveys. Sidescan sonar is also useful for locating artifacts on the sea floor such

as wrecks, cables or pipes. As with sub-bottom profiling systems, results in deep water are much improved by the use of deep-tow systems.

4.11 VERTICAL SEISMIC PROFILING

Vertical seismic profiling (VSP) is a form of seismic reflection surveying that utilizes boreholes. Shots are normally fired at surface, at the wellhead or offset laterally from it, and recorded at different depths within the borehole using special detectors clamped to the borehole wall. Alternatively, small shots may be fired at different depths within the borehole and recorded at surface using conventional geophones, but in the following account the former configuration is assumed throughout. Typically, for a borehole a kilometre or more deep, seismic data are recorded at more than 100 different levels down the borehole. If the surface shot location lies at the

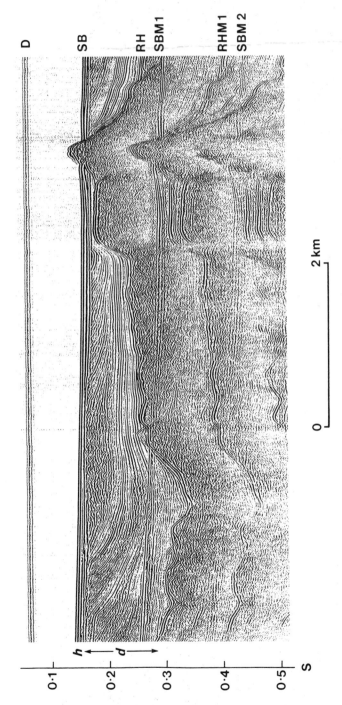

Fig. 4.49 Air gun record from the Gulf of Patras, Greece, showing Holocene hemipelagic (h) and deltaic (d) sediments overlying an irregular erosion surface (rockhead, RH) cut into tectonized Mesozoic and Tertiary rocks of the Hellenide (Alpine) orogenic belt. SB: sea bed reflection; SBM1 and SBM2: first and second multiples of sea bed reflection; RHM1: first multiple of rockhead reflection.

Plate 3. A seismic section from the northern Amadeus basin, central Australia, illustrating a depositional sequence bounded by major unconformities. (Reproduced from *AAPG Memoir* No. 39 with the permission of the publishers.)

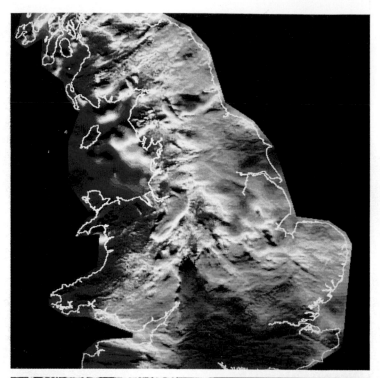

Plate 5 (a). Colour shaded-relief image of the gravity field of Central Britain illuminated from the north. Blue represents low values, red high values. (From Lee *et al*. 1990, with permission.)

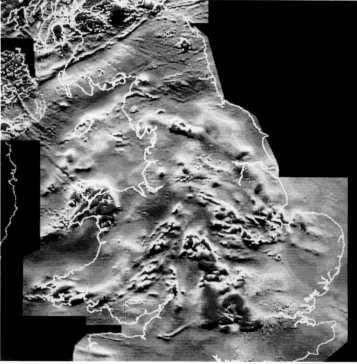

Plate 5 (b). Colour shaded-relief image of the magnetic field of Central Britain illuminated from the north. Blue represents low values, red high values. (From Lee *et al*. 1990, with permission.)

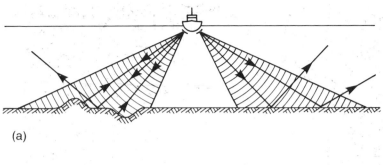

(a)

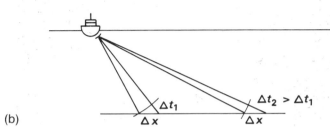

(b)

Fig. 4.50 Principles of sidescan sonar. (a) Individual reflected ray paths within the transmitted lobes, showing signal return from topographic features on the sea bed. (b) Scale distortion resulting from oblique incidence: the same widths of sea floor Δx are represented by different time intervals Δt_1 and Δt_2 at the inner and outer edges of the sonograph, respectively.

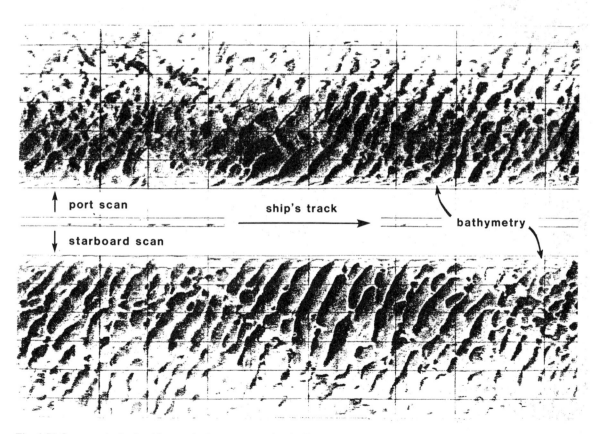

Fig. 4.51 Sonograph obtained from a dualscan survey of a pipeline route across an area of linear sand waves in the southern North Sea. The inner edges of the two swathes define the bathymetry beneath the survey vessel. (Scanning range: 100 m).

wellhead vertically above the borehole detector locations, so that the recorded rays have travelled along vertical ray paths, the method is known as *zero-offset VSP*; if the surface shot locations are offset laterally, so that the recorded rays have travelled along inclined ray paths, the method is known as *offset VSP* (Fig. 4.52).

VSP has several major applications in seismic exploration (Balch *et al.* 1982; Cassell 1984). Perhaps most importantly, reflection events recorded on seismic sections obtained at surface from conventional reflection surveys can be traced by VSP to their point of origin in the subsurface, thus calibrating the seismic sections geologically. Ambiguity as to whether particular events observed on conventional seismic sections represent primary or multiple reflections can be removed by direct comparison of the sections with VSP data. The reflection properties of particular horizons identified in the borehole section can be investigated directly using VSP and it can therefore be determined, for example, whether or not an horizon returns a detectable reflection to the surface.

Uncertainty in interpreting subsurface geology using conventional seismic data is in part due to the surface location of shot points and detectors. VSP recording in a borehole enables the detector to be located in the immediate vicinity of the target zone, thus shortening the overall path length of reflected rays, reducing the effects of attenuation, and reducing the dimensions of the Fresnel zone (Section 4.5.1). By these various means, the overall accuracy of a seismic interpretation may be markedly increased. A particular uncertainty in conventional seismics is the nature of the downgoing pulse that is reflected back to surface from layer boundaries. This uncertainty often reduces the effectiveness of deconvolution of conventional seismic data. By contrast, an intrinsic feature of VSP surveys is that both downgoing and upgoing rays are recorded, and the waveform of the downgoing pulse may be used to optimize the design of a deconvolution operator for inverse filtering of VSP data to enhance resolution. Direct comparison with such VSP data leads to much improved reliability in the geological interpretation of seismic sections recorded at the surface in the vicinity of the borehole.

The nature of VSP data may be considered by reference to Fig. 4.53 which illustrates a synthetic zero-offset VSP data set for the velocity–depth model shown, each trace being recorded at a different depth. Two sets of events are recorded which

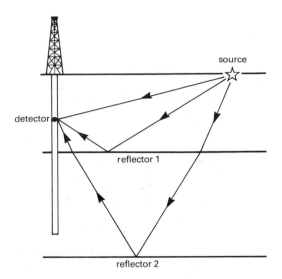

Fig. 4.52 An offset VSP survey configuration.

have opposite directions of dip in the VSP section. Events whose travel time increases as a function of detector depth represent downgoing rays; the weaker events, whose travel time reduces as a function of detector depth, represent upgoing, reflected rays. Note that the direct downgoing pulse (the first arrival, DO) is followed by other events (DS1, DS2, DS3) with the same dip, representing downgoing near-surface and peg-leg multiples. Each reflected event (U1, U2, U3) terminates at the relevant reflector depth, where it intersects the direct downgoing event.

For most purposes, it is desirable to separate downgoing and upgoing events to produce a VSP section retaining only upgoing, reflected arrivals. The opposite dip of the two types of event in the original VSP section enables this separation to be carried out by $f - k$ filtering (see Section 4.7.3). Fig. 4.54(a) illustrates a synthetic VSP section after removal of downgoing events. The removal of the stronger downgoing events has enabled representation of the upgoing events at enhanced amplitude, and weak multiple reflection events are now revealed. Note that these terminate at the same depth as the relevant primary event, and therefore do not extend to the point of intersection with the direct downgoing event. It is now possible to apply a time correction to each trace in the VSP section, based on the travel time of the downgoing direct event, in order to predict the form of seismic trace that would be obtained at surface (Fig. 4.54(b)). By stacking

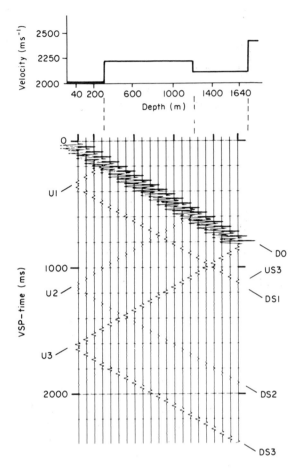

Fig. 4.53 A synthetic zero-offset VSP record section for the velocity–depth model shown. The individual traces are recorded at the different depths shown. DO is the direct downgoing wave; DS1, DS2 and DS3 are downgoing waves with multiple reflection between the surface and interfaces 1, 2 and 3 respectively. U1, U2 and U3 are primary reflections from the three interfaces; US3 is a reflection from the third interface with multiple reflection in the top layer. (From Cassell 1984.)

these traces within a time corridor that avoids the multiple events, it is possible to produce a stacked trace containing only primary reflection events. Comparison of this stacked trace with a conventional seismic section from the vicinity of the borehole (Fig. 4.55) enables the geological content of the latter to be identified reliably.

4.12 APPLICATIONS OF SEISMIC REFLECTION SURVEYING

The 1980s saw major developments in reflection seismic surveying. Over that period the general quality of seismic record sections improved markedly due to the move to digital data acquisition systems and the use of increasingly powerful processing techniques. At the same time, the range of applications of the method increased considerably. Previously, reflection surveying was concerned almost exclusively with the search for hydrocarbons and coal, down to depths of a few kilometres; now, the method is being used increasingly for studies of the entire continental crust and the uppermost mantle to depths of several tens of kilometres, and, at the other end of the spectrum of target depths, for high-resolution onshore mapping of shallow geology to depths of a few tens or hundreds of metres.

The search for hydrocarbons, onshore and off-shore, nevertheless remains by far the largest single application of reflection surveying. This reflects the particular strength of the method in producing well-resolved images of sedimentary sequences down to a depth of several kilometres. The method is used at all stages of an exploration programme for hydrocarbons, from the early reconnaissance stage through to the detailed mapping of specific structural targets in preparation for exploration drilling, and on into the field development stage when the overall reservoir geometry requires further detailing.

Because of its relatively high cost, three-dimensional seismic surveying still does not find routine application in hydrocarbon exploration programmes. However, whereas it was originally used only at the field development stage, it now finds widespread application also at the exploration stage in some oilfields. Vertical seismic profiling is another important new technique that is being applied increasingly at the stage of oilfield development because of its ability to reveal subsurface detail that is generally unobtainable from surface seismic data alone.

The initial round of seismic exploration for hydrocarbons normally involves speculative surveys along widely-spaced profile lines covering large areas. In this way the major structural or stratigraphic elements of the regional geology are delineated, so enabling the planning of detailed, follow-up reflection surveys in more restricted areas containing the main prospective targets. Where good geological mapping of known sedimentary sequences

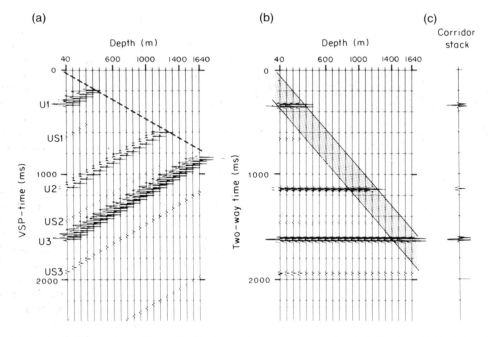

Fig. 4.54 (a) Synthetic VSP section of Fig. 4.53 with downgoing waves removed by f−k filtering. (b) Each trace has been time shifted by the relevant uphole time to simulate a surface recording. (c) Stacked seismogram produced by stacking in the shaded corridor zone of Fig. 4.54(b) to avoid multiple events. (From Cassell 1984.)

exists, the need for expenditure on initial speculative seismic surveys is often much reduced and effort can be concentrated from an early stage on the seismic investigation of areas of particular interest.

Detailed reflection surveys involve closely spaced profile lines and a high density of profile intersection points in order that reflection events can be traced reliably from profile to profile and used to define the prevailing structure. Initial seismic interpretation is likely to involve structural mapping, using time-structure and/or isochron maps (Section 4.9.1) in the search for the structural closures that may contain oil or gas. Any closures that are identified may need further delineation by a second round of detailed seismic surveying before the geophysicist is sufficiently confident to select the location of an exploration borehole from a time-structure map. Three-dimensional seismics may need to be employed when critical structural details are unresolved by interpretation of the two-dimensional survey data.

Exploration boreholes are normally sited on seismic profile lines so that the borehole logs can be correlated directly with the local seismic section. This facilitates precise geological identification of specific seismic reflectors, especially if vertical seismic profiling surveys (Section 4.11) are carried out at the site of the borehole.

Particularly in offshore areas, where the best quality seismic data are generally obtained, the methods of seismic stratigraphy (Section 4.9.2) are increasingly employed on sections displaying seismic sequences, to obtain insight into the associated sedimentary lithologies and depositional environments. Such stratigraphic information, derived from seismic facies analysis of the individual sequences, is often of great value to an exploration programme in highlighting the location of potential source rocks (e.g. organic-rich mudstones) and potential reservoir rocks (e.g. a deltaic or reef facies).

The contribution of reflection surveying to the development of hydrocarbon reserves does not end with the discovery of an oil or gas field. Refinement of the seismic interpretation using information from, variously, additional seismic profiles, three-dimensional seismics and vertical seismic profiling data will assist in optimizing the location of production boreholes. In addition, seismic modelling (Section 4.9.3) of amplitude variations and other

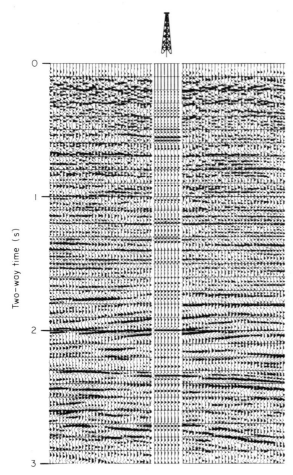

Fig. 4.55 Corridor stack of the zero-offset VSP section (Fig. 4.54(c)) reproduced eight times and spliced into a conventional seismic section based on surface profiling data from the vicinity of the borehole site. Comparison of the VSP stack with the surface recorded data enables the primary events in the seismic section to be reliably distinguished from multiple events. (From Cassell 1984.)

aspects of reflection character displayed on seismic sections across the producing zone can be used to obtain detailed information on the geometry of the reservoir and on internal lithological variations that may affect the hydrocarbon yield.

Examples of seismic sections from hydrocarbon fields in the North Sea area are shown in Figs 4.56 and 4.57. Fig. 4.56 represents a seismic section across the North Viking gas field in the southern North Sea. The gas is trapped in the core of a NW–SE trending anticlinal structure that is extensively faulted at the level of the Lower Permian. A typical

combined structural/stratigraphic trap in the northern North Sea is represented by the Brent oilfield structure, and Fig. 4.57 illustrates a seismic section across the field. A tilted fault block containing Upper Palaeozoic, Triassic and Jurassic strata is overlain unconformably by Upper Jurassic, Cretaceous and Tertiary sediments. Two Jurassic sands in the tilted fault block constitute the main reservoirs, the oil and gas being trapped beneath a capping of unconformably overlying shales of Late Jurassic and Cretaceous age.

Reflection profiling at crustal and lithospheric scale is now being carried out by many developed countries. Following on from the extensive use of multichannel reflection profiling to investigate the crustal structure of oceanic areas, national programmes such as the US COCORP project (Consortium for Continental Reflection Profiling; Brewer & Oliver 1980) and the British BIRPS project (British Institutions Reflection Profiling Syndicate; Brewer 1983) are now producing seismic sections through the entire continental crust and the uppermost part of the underlying mantle. These national programmes utilize essentially the same data acquisition systems and processing techniques as the oil industry, whilst increasing the size of source arrays and detector spread lengths; recording times of 15 s are commonly employed, as compared with a standard oil industry recording time of about 4 s. A typical BIRPS section is illustrated in Fig. 4.58.

Crustal reflection profiling results from several different continental areas (see, for example, Barazangi & Brown (1986) and the special issue of *Tectonophysics* **173** (1990) for a wide range of relevant papers) reveal that the upper part of the continental crust typically has a rather transparent seismic character with localized bands of dipping reflectors, interpreted as fault zones, passing down into the lower crust. By contrast, the lower crust is often found to be highly reflective with discontinuous horizontal or gently dipping events giving an overall layered appearance (Fig. 4.58). The origin of this layering is uncertain, but the main possibilities appear to be primary igneous layering, horizontal shear zones and zones of fluid concentration (e.g. Klemperer *et al.* 1987). All may contribute in some measure to the observed reflectivity. Where refraction and reflection data both exist, the base of the zone of reflectivity is found to coincide with the Mohorovičić discontinuity as defined by refraction interpretation of head wave arrivals from the uppermost mantle (Barton 1986).

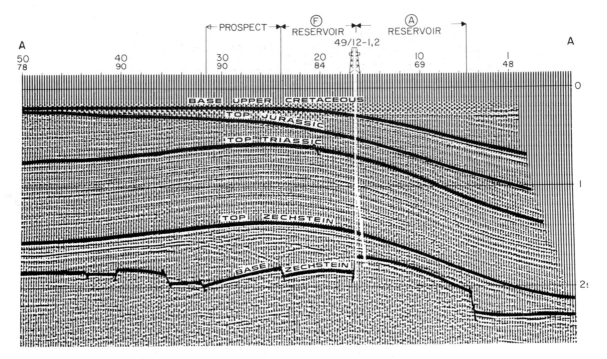

Fig. 4.56 Interpreted seismic section across the North Viking gas field, North Sea. (Courtesy Conoco UK Ltd.)

The use of reflection seismics for high-resolution studies of shallow geology is a field of growing importance in which developments are linked directly to recent technical advances. Highly portable digital multichannel data acquisition systems, backed up by PC-based processing packages, make it possible to produce seismic sections of shallow sub-surface geology at reasonable cost. High-resolution reflection seismology is particularly well suited to the investigation of Quaternary sedimentary sequences and for the detailed mapping of concealed bedrock surfaces of irregular geometry.

4.13 PROBLEMS

1 A seismic wave is incident normally on a reflector with a reflection coefficient R of 0.01. What proportion of the incident *energy* is transmitted?

2 What is the root-mean-square velocity in reflection surveying, and how is it related to interval velocity and to stacking velocity?

3 A zero-offset reflection event at 1.000 s has a normal moveout (NMO) of 0.005 s at 200 m offset. What is the stacking velocity?

4 (a) Reflection profiling is used to investigate lower crustal structure at a depth of about 30 km. The domi-

nant frequency of the reflected pulse is found to be 10 Hz. Using a typical average crustal velocity of $6.5 \, \mathrm{km \, s^{-1}}$ calculate the approximate dimensions of the Fresnel zone.

(b) A high resolution reflection survey is used to map rockhead beneath a Quaternary sediment cover about 100 m thick using a high frequency source. The dominant frequency of the reflected pulse is found to be 150 Hz. Using a sediment velocity of $2 \, \mathrm{km \, s^{-1}}$, calculate the approximate dimensions of the Fresnel zone.

(c) Discuss the importance of the above Fresnel zone dimensions as indications of the inherent limits on horizontal resolution achievable in different types of reflection survey.

(d) Use the frequency and velocity information to calculate the vertical resolution of the two types of survey (Section 4.5) and again discuss the general importance of the results obtained to the vertical resolution that is achievable in reflection seismics.

5 In the initial stages of a seismic reflection survey, a noise test indicates a direct wave with a velocity of $3.00 \, \mathrm{km \, s^{-1}}$ and a dominant frequency of 100 Hz, and ground roll with a velocity of $1.80 \, \mathrm{km \, s^{-1}}$ and a dominant frequency of 30 Hz. What is the optimum spacing of individual geophones in five-element linear arrays in order to suppress these horizontally-travelling phases?

6 In CDP stacking, the method of applying a NMO

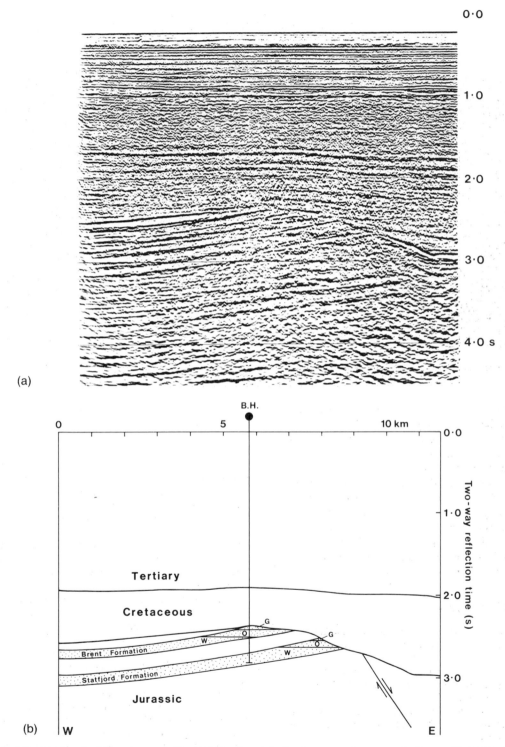

Fig. 4.57 (a) Seismic section (courtesy Shell UK Ltd.) and (b) line interpretation across the Brent oilfield, North Sea. G: gas, O: oil, W: water.

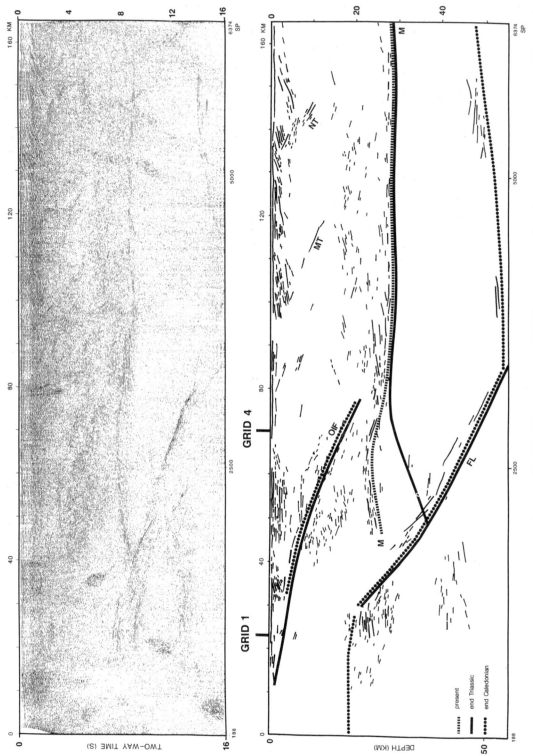

Fig. 4.58 A non-migrated crustal reflection section from the 1986/87 GRID survey of the BIRPS programme, collected along a west–east line about 30 km north of Scotland, UK, and a migrated line drawing of the main reflection events. The main structures are interpreted to be of Caledonian age with later re-activation (FL: Flannan reflection; OIF: Outer Isles fault; MT: Moine thrust; NT: Naver thrust; M: Moho). (From Snyder & Flack 1990.)

correction to individual seismic traces creates distortion in seismic pulses recorded at large offset that can degrade the stacking process. Why?

7 Along a two-dimensional marine survey line involving a 48-channel streamer with a hydrophone array interval of 50 m, shots are fired every 25 m.
(a) What is the fold of CDP cover?
(b) If the cover is to be increased to 96-fold, what must the new shot interval be?

8 In single-channel seismic profiling, what is the optimal depth for towing an air gun source with a dominant frequency of 100 Hz such that the reflected ray from the sea surface will interfere constructively with the downgoing primary pulse? (The compressional wave velocity in sea water is $1.505\,\mathrm{km\,s^{-1}}$.

9 What is the significance of the curved boundary lines to the typical ground roll sector of the $f-k$ plot illustrated in Fig. 4.28, and how may it be explained?

10 How may three-dimensional seismic survey data be used to study velocity anisotopy?)

FURTHER READING

Al-Sadi, H.N. (1980) *Seismic Exploration*. Birhauser Verlag, Basel.

Anstey, N.A. (1982) *Simple Seismics*. IHRDC, Boston.

Bally, A.W. (ed) (1983) *Seismic Expression of Structural Styles (a picture and work atlas): Vol 1 — The layered Earth; Vol 2 — Tectonics of extensional provinces; Vol 3 — Tectonics of compressional provinces/Strike-slip tectonics*. AAPG Studies in Geology No. **15**, American Association of Petroleum Geologists, Tulsa.

Bally, A.W. (ed) (1987) *Atlas of Seismic Stratigraphy* (3 vols). AAPG Studies in Geology No **27**, American Association of Petroleum Geologists, Tulsa.

Barazangi, M. & Brown, L. (eds) (1986) *Reflection Seismology: The Continental Crust*. AGU Geodynamics Series, No. 14. American Geophysical Union, Washington.

Berg, O.R. & Woolverton, D.G. (eds) (1985) *Seismic Stratigraphy II: An Integrated Approach to Hydrocarbon Exploration*. AAPG Memoir 39, American Association of Petroleum Geologists, Tulsa.

Brown, A.R. (1986) *Interpretation of Three-dimensional Seismic Data*. AAPG Memoir 42, American Association of Petroleum Geologists, Tulsa.

Camina, A.R. & Janacek, G.J. (1984) *Mathematics for Seismic Data Processing and Interpretation*. Graham & Trotman, London.

Cassell, B. (1984) Vertical seismic profiles — an introduction. *First Break*, **2**(11), 9–19.

Claerbout, J.F. (1985) *Fundamentals of Geophysical Data Processing*. McGraw Hill, New York.

Dobrin, M.B. & Savit, C.H. (1988) *Introduction to Geophysical Prospecting* (4th edn). McGraw Hill, New York.

Hatton, L., Worthington, M.H. & Makin, J. (1986) *Seismic Data Processing*. Blackwell Scientific Publications, Oxford.

Hubral, P. & Krey, T. (1980) *Interval Velocities from Seismic Reflection Time Measurements*. Society of Exploration Geophysicists, Tulsa.

Kleyn, A.H. (1983) *Seismic Reflection Interpretation*. Applied Science Publishers, London.

Lavergne, M. (1989) *Seismic Methods*. Editions Technip, Paris.

McQuillin, R., Bacon, M. & Barclay, W. (1979) *An Introduction To Seismic Interpretation*. Graham and Trotman, London.

Payton, C.E. (ed.) (1977) *Seismic Stratigraphy — Application to Hydrocarbon Exploration*. Memoir 26, American Association of Petroleum Geologists, Tulsa.

Robinson, E.A. (1983) *Migration of Geophysical Data*. IHRDC, Boston.

Robinson, E.A. (1983) *Seismic Velocity Analysis and the Convolutional Model*. IHRDC, Boston.

Robinson, E.S. & Çoruh, C. (1988) *Basic Exploration Geophysics*. Wiley, New York.

Sengbush, R.L. (1983) *Seismic Exploration Methods*. IHRDC, Boston.

Sheriff, R.E. (1980) *Seismic Stratigraphy*. IHRDC, Boston.

Sheriff, R.E. (1982) *Structural Interpretation of Seismic Data*. AAPG Continuing Education Course Note Series No. 23.

Sheriff, R.E. & Geldart, L.P. (1983) *Exploration Seismology, Vol. 2: Data-processing and Interpretation*. Cambridge University Press, Cambridge.

Waters, K.H. (1978) *Reflection Seismology — A Tool For Energy Resource Exploration*. Wiley, New York.

Ziolkowski, A. (1983) *Deconvolution*. IHRDC, Boston.

5 / Seismic refraction surveying

5.1 INTRODUCTION

The seismic refraction surveying method utilizes seismic energy that returns to the surface after travelling through the ground along refracted ray paths. The method is normally used to locate refracting interfaces (refractors) separating layers of different seismic velocity, but the method is also applicable in cases where velocity varies smoothly as a function of depth or laterally.

Refraction seismology is applied to a very wide range of scientific and technical problems, from engineering site investigation surveys to large-scale experiments designed to study the structure of the entire crust or lithosphere. Refraction measurements can provide valuable velocity information for use in the interpretation of reflection surveys, and refracted arrivals recorded during land reflection surveys are used to map the weathered layer, as discussed in Chapter 4. This wide variety of applications leads to an equally wide variety of field survey methods and associated interpretation techniques.

As briefly discussed in Chapter 3, the first arrival of seismic energy at a detector offset from a seismic source always represents either a direct ray or a refracted ray. This fact allows simple refraction surveys to be performed in which attention is concentrated solely on the first arrival (or *onset*) of seismic energy, and time−distance curves of these first arrivals are interpreted to derive information on the depth to refracting interfaces. As will be seen later in the chapter, this simple approach does not always yield a full or accurate picture of the subsurface.

Refraction seismograms may, of course, contain reflection events as subsequent arrivals, though generally no special attempt is made to enhance reflected arrivals in refraction surveys. Nevertheless, the relatively high reflection coefficients associated with rays incident on an interface at angles near to the critical angle often lead to strong *wide-angle reflections* which are quite commonly detected at the greater recording ranges that characterize large-scale refraction surveys. These wide-angle reflections often provide valuable additional information on

subsurface structure such as, for example, indicating the presence of a low velocity layer which would not be revealed by refracted arrivals alone.

The vast majority of refraction surveying is carried out along profile lines which are normally arranged to be sufficiently long to ensure that refracted arrivals from target layers are recorded as first arrivals. To ensure that the relevant crossover distance is well exceeded, refraction profiles typically need to be between five and ten times as long as the required depth of investigation, although the actual profile length required in a particular case depends upon the distribution of velocities with depth. The requirement in refraction surveying for an increase in profile length with increase in the depth of investigation contrasts with the situation in conventional reflection surveying, where near-normal incidence reflections from deep interfaces are recorded at small offset distances. A consequence of this requirement is that large seismic sources are needed for the detection of deep refractors in order that sufficient energy is transmitted over the long range necessary for the recording of deep refracted phases as first arrivals.

In many geological situations, subsurface refractors may approximate planar surfaces over the linear extent of a refraction line. In such cases the observed travel−time curves are commonly assumed to derive from a set of planar layers and are analysed to determine depths to, and dips of, individual planar refractors. The geometry of refracted ray paths through planar layer models of the subsurface is considered first, after which, consideration is given to methods of dealing with refraction at irregular (non-planar) interfaces.

5.2 GEOMETRY OF REFRACTED RAY PATHS: PLANAR INTERFACES

The general assumptions relating to the ray path geometries considered below are that the subsurface is composed of a series of layers separated by plane and possibly dipping interfaces, that seismic velocities are uniform within each layer, that layer velocities increase with depth, and that ray paths are restricted to a vertical plane containing the profile line (i.e. that there is no component of cross-dip).

5.2.1 Two-layer case with horizontal interface

Figure 5.1 illustrates progressive positions of the wavefront associated with energy travelling directly through an upper layer and energy critically refracted in a lower layer, from a seismic source at A. Direct and refracted ray paths to a detector at D, a distance x from the source, are also shown. The layer velocities are v_1 and v_2 ($> v_1$) and the refracting interface is at a depth z.

The direct ray travels horizontally through the top of the upper layer from A to D at velocity v_1. The refracted ray travels down to the interface and back up to the surface at velocity v_1 along slant paths AB and CD that are inclined at the critical angle θ, and travels along the interface between B and C at the higher velocity v_2.

The total travel time along the refracted ray path $ABCD$ is

$$t = t_{AB} + t_{BC} + t_{CD}$$

$$= \frac{z}{v_1 \cos \theta} + \frac{(x - 2z \tan \theta)}{v_2} + \frac{z}{v_1 \cos \theta}$$

Noting that $\sin \theta = v_1/v_2$ (Snell's Law) and $\cos \theta = (1 - v_1^2/v_2^2)^{1/2}$, the travel time equation may be expressed in a number of different forms, a useful general form being

$$t = \frac{x \sin \theta}{v_1} + \frac{2z \cos \theta}{v_1} \tag{5.1}$$

Alternatively

$$t = \frac{x}{v_2} + \frac{2z(v_2^2 - v_1^2)^{1/2}}{v_1 v_2} \tag{5.2}$$

or

$$t = \frac{x}{v_2} + t_i \tag{5.3}$$

where, plotting t against x (Fig. 5.2), t_i, the intercept on the time axis of a travel–time curve or *time–distance curve* having a gradient of $1/v_2$ .t_i, known as the *intercept time*, is given by

$$t_i = \frac{2z(v_2^2 - v_1^2)^{1/2}}{v_1 v_2} \qquad \text{(from (5.2))}$$

Solving for refractor depth

$$z = \frac{t_i v_1 v_2}{2(v_2^2 - v_1^2)^{1/2}}$$

Thus by analysis of the travel–time curves of direct and refracted arrivals, v_1 and v_2 can be derived (reciprocal of the gradient of the relevant travel–time curve, see Fig. 5.2) and from the intercept time t_i the refractor depth z can be determined.

At the crossover distance x_{cros} the travel times of direct and refracted rays are equal

$$\frac{x_{cros}}{v_1} = \frac{x_{cros}}{v_2} + \frac{2z(v_2^2 - v_1^2)^{1/2}}{v_1 v_2}$$

Thus, solving for x_{cros}

$$x_{cros} = 2z \left[\frac{v_2 + v_1}{v_2 - v_1} \right]^{1/2} \tag{5.4}$$

From this equation it may be seen that the crossover distance is always greater than twice the depth to the refractor.

5.2.2 Three-layer case with horizontal interfaces

The geometry of the ray path in the case of critical refraction at the second interface is shown in

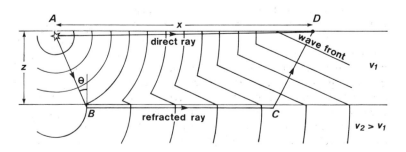

Fig. 5.1 Successive positions of the expanding wavefronts for direct and refracted waves through a two-layer model. Individual ray paths from source A to detector D are shown.

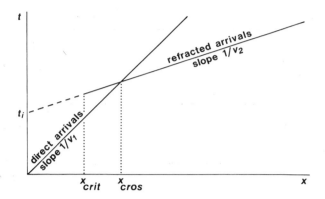

Fig. 5.2 Travel–time curves for the direct wave and the head wave from a single horizontal refractor.

Fig. 5.3. The seismic velocities of the three layers are v_1, v_2 $(> v_1)$ and v_3 $(> v_2)$. The angle of incidence of the ray on the upper interface is θ_1 and on the lower interface is θ_2 (critical angle). The thicknesses of layers 1 and 2 are z_1 and z_2 respectively.

By analogy with equation (5.1) for the two-layer case, the travel time along the refracted ray path $ABCDEF$ to an offset distance x, involving critical refraction at the second interface, can be written in the form

$$t = \frac{x \sin \theta_1}{v_1} + \frac{2z_1 \cos \theta_1}{v_1} + \frac{2z_2 \cos \theta_2}{v_2} \quad (5.5)$$

where

$$\theta_1 = \sin^{-1} (v_1/v_3)$$

and

$$\theta_2 = \sin^{-1} (v_2/v_3)$$

or

$$t = \frac{x \sin \theta_1}{v_1} + t_{i_1} + t_{i_2} \quad (5.6)$$

where t_{i_1} is the intercept on the time axis of the travel–time curve for rays critically refracted at the upper interface and t_{i_2} is the difference between t_{i_1} and the intercept of the curve for rays critically refracted at the lower interface (see Fig. 5.4).

The interpretation of travel–time curves for a three-layer case proceeds via the initial interpretation of the top two layers. Having used the travel–time curve for rays critically refracted at the upper interface to derive z_1 and v_2, the travel–time curve for rays critically refracted at the second interface can be used to derive z_2 and v_3 using equations (5.5) and (5.6) or equations derived therefrom.

5.2.3 Multilayer case with horizontal interfaces

In general the travel time t_n of a ray critically refracted along the top surface of the nth layer is given by

$$t_n = \frac{x \sin \theta_1}{v_1} + \sum_{i=1}^{n-1} \frac{2z_i \cos \theta_i}{v_i} \quad (5.7)$$

where

$$\theta_i = \sin^{-1} (v_i/v_n).$$

Equation (5.7) can be used progressively to compute layer thicknesses in a sequence of horizontal strata represented by travel–time curves of refracted arrivals.

5.2.4 Dipping-layer case with planar interfaces

In the case of a dipping refractor (Fig. 5.5(a)) the value of dip enters the travel–time equations as an additional unknown. The reciprocal of the gradient of the travel–time curve no longer represents the refractor velocity but a quantity known as the *apparent velocity* which is higher than the refractor velocity when recording along a profile line in the updip direction from the shot point and lower when recording downdip.

The conventional method of dealing with the possible presence of refractor dip is to *reverse* the refraction experiment by firing at each end of the profile line and recording seismic arrivals along the line from both shots. In the presence of a component of refractor dip along the profile direction, the *forward* and *reverse* travel–time curves for refracted rays will differ in their gradients and intercept times, as shown in Fig. 5.5(b).

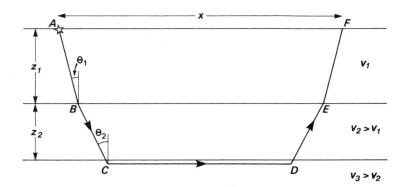

Fig. 5.3 Refracted ray path through the bottom layer of a three-layer model.

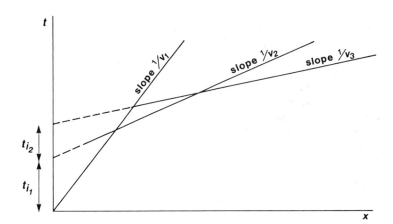

Fig. 5.4 Travel–time curves for the direct wave and the head waves from two horizontal refractors.

The general form of the equation for the travel time t_n of a ray critically refracted in the nth dipping refractor (Fig. 5.6; Johnson 1976) is given by

$$t_n = \frac{x \sin \beta_1}{v_1} + \sum_{i=1}^{n-1} \frac{h_i (\cos \alpha_i + \cos \beta_i)}{v_i} \qquad (5.8)$$

where h_i is the vertical thickness of the ith layer beneath the shot, v_i is the velocity of the ray in the ith layer, α_i is the angle with respect to the vertical made by the downgoing ray in the ith layer, β_i is the angle with respect to the vertical made by the upgoing ray in the ith layer, and x is the offset distance between source and detector.

Equation (5.8) is directly comparable with equation (5.7), the only differences being the replacement of θ by angles α and β that include a dip term. In the case of shooting downdip, for example (see Fig. 5.6), $\alpha_i = \theta_i - \gamma_i$ and $\beta_i = \theta_i + \gamma_i$, where γ_i is the dip of the ith layer and $\theta_i = \sin^{-1}(v_i/v_n)$ as before. Note that h is the vertical thickness rather than the perpendicular or true thickness of a layer.

To exemplify the use of equation (5.8) in interpreting travel–time curves, consider the two-layer case illustrated in Fig. 5.5.

Shooting downdip, along the forward profile

$$
\begin{aligned}
t_2 &= \frac{x \sin \beta}{v_1} + \frac{h(\cos \alpha + \cos \beta)}{v_1} \\
&= \frac{x \sin (\theta + \gamma)}{v_1} + \frac{h \cos (\theta - \gamma)}{v_1} + \frac{h \cos (\theta + \gamma)}{v_1} \\
&= \frac{x \sin (\theta + \gamma)}{v_1} + \frac{2h \cos \theta \cos \gamma}{v_1} \\
&= \frac{x \sin (\theta + \gamma)}{v_1} + \frac{2z \cos \theta}{v_1} \qquad (5.9)
\end{aligned}
$$

where z is the perpendicular distance to the interface beneath the shot.

Equation (5.9) defines a linear curve with a gradient of $\sin (\theta + \gamma)/v_1$ and an intercept time of $2z \cos \theta/v_1$.

Shooting updip, along the reverse profile

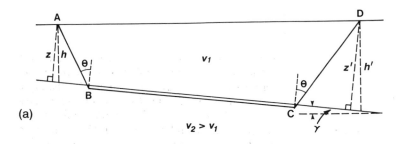

(a)

$v_2 > v_1$

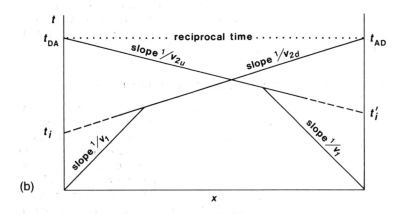

(b)

Fig. 5.5 Travel–time curves for head wave arrivals from a dipping refractor in the forward and reverse directions along a refraction profile line.

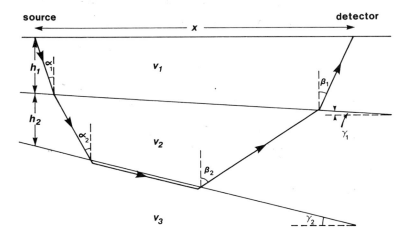

Fig. 5.6 Geometry of the refracted ray path through a multilayer, dipping model. (After Johnson 1976.)

$$t_2' = \frac{x \sin(\theta - \gamma)}{v_1} + \frac{2z' \cos \theta}{v_1} \qquad (5.10)$$

where z' is the perpendicular distance to the interface beneath the second shot.

The gradients of the travel–time curves of refracted arrivals along the forward and reverse profile lines yield the downdip and updip apparent velocities v_{2_d} and v_{2_u} respectively (Fig. 5.5(b)). From the forward direction

$$1/v_{2_d} = \sin(\theta + \gamma)/v_1 \qquad (5.11)$$

and from the reverse direction

$$1/v_{2_u} = \sin(\theta - \gamma)/v_1 \qquad (5.12)$$

Hence

$$\theta + \gamma = \sin^{-1}(v_1/v_{2_d})$$
$$\theta - \gamma = \sin^{-1}(v_1/v_{2_u})$$

Solving for θ and γ yields

$$\theta = \tfrac{1}{2}(\sin^{-1}(v_1/v_{2_d}) + \sin^{-1}(v_1/v_{2_u}))$$
$$\gamma = \tfrac{1}{2}(\sin^{-1}(v_1/v_{2_d}) - \sin^{-1}(v_1/v_{2_u}))$$

Knowing v_1, from the gradient of the direct ray travel−time curve, and θ, the true refractor velocity may be derived using Snell's Law

$$v_2 = v_1/\sin\theta$$

The perpendicular distances z and z' to the interface under the two ends of the profile are obtained from the intercept times t_i and t_i' of the travel−time curves obtained in the forward and reverse directions

$$t_i = 2z\cos\theta/v_1$$
$$\therefore z = v_1 t_i/2\cos\theta$$

And similarly

$$z' = v_1 t_i'/2\cos\theta$$

By using the computed refractor dip γ, the perpendicular depths z and z' can be converted into vertical depths h and h' using

$$h = z/\cos\gamma$$

and

$$h' = z'/\cos\gamma$$

Note that the travel time of a seismic phase from one end of a refraction profile line to the other (i.e. from shot point to shot point) should be the same whether measured in the forward or the reverse direction. Referring to Fig. 5.5(b), this means that t_{AD} should equal t_{DA}. Establishing that there is satisfactory agreement between these *reciprocal times* is a useful means of checking that travel−time curves have been drawn correctly through a set of refracted ray arrival times derived from a reversed profile.

5.2.5 Faulted planar interfaces

The effect of a fault displacing a planar refractor is to offset the segments of the travel−time curve on opposite sides of the fault (see Fig. 5.7). There are thus two intercept times t_{i_1} and t_{i_2}, one associated with each of the travel−time curve segments, and the difference between these intercept times ΔT is a measure of the throw of the fault. For example, in the case of the faulted horizontal refractor shown in Fig. 5.7 the throw of the fault Δz is given by

$$\Delta z = \Delta T v_1 v_2/(v_2^2 - v_1^2)^{1/2}$$

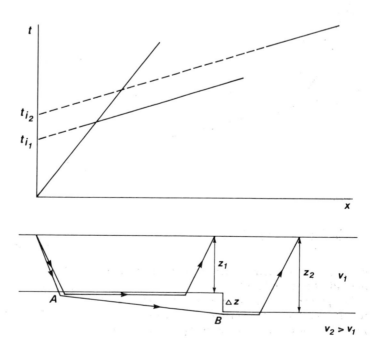

Fig. 5.7 Offset segments of the travel−time curve for refracted arrivals from opposite sides of a fault.

Note that there is some error in this formulation, since the ray travelling to the downthrown side of the fault is not the critically refracted ray at A and involves diffraction at the base B of the fault step, but the error will be negligible where the fault throw is small compared with the refractor depth.

5.3 PROFILE GEOMETRIES FOR STUDYING PLANAR LAYER PROBLEMS

As stated above, the conventional field geometry for a refraction profile involves shooting at each end of the profile line and recording seismic arrivals along the line from both shots. As will be seen with reference to Fig. 5.5(a), only the central portion of the refractor (from B to C) is in fact sampled by refracted rays detected along the line length. Interpreted depths to the refractor under the endpoints

of a profile line, using equations given above, are thus not directly measured but are inferred on the basis of the refractor geometry over the shorter length of refractor actually traversed by refracted rays. Where continuous cover of refractor geometry is required along a series of reversed profiles, therefore, individual profile lines should be arranged to overlap in order that all parts of the refractor are directly sampled by critically refracted rays.

In addition to the conventional reversed profile, illustrated schematically in Fig. 5.8(a), other methods of deriving full planar layer interpretations in the presence of dip include the *split-profile* method (Johnson 1976) and the *single-ended profile method* (Cunningham 1974).

The split-profile method (Fig. 5.8(b)) involves recording outwards in both directions from a central shot point. Although the interpretation method differs in detail from that for a conventional reversed

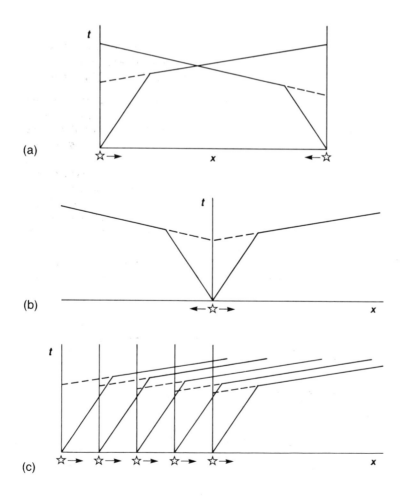

Fig. 5.8 Various types of profile geometry used in refraction surveying. (a) Conventional reversed profile with end shots. (b) Split-profile with central shot. (c) Single-ended profile with repeated shots.

profile, it is based on the same general travel–time equation (5.8).

The single-ended profile method (Fig. 5.8(c)) was developed to derive interpretations of low velocity surface layers represented by refracted arrivals in single-ended reflection spread data, for use in the calculation of static corrections. A simplified treatment is given below.

To obtain a value of refractor dip, estimates of apparent velocity are required in both the forward and reverse directions. The repeated forward shooting of the single-ended profile method enables an apparent velocity in the forward direction to be computed from the gradient of the travel time curves. For the method of computing the apparent velocity in the reverse direction, consider two refracted ray paths from surface sources S_1 and S_2 to surface detectors D_1 and D_2, respectively (Fig. 5.9). The offset distance is x in both cases, the separation Δx of S_1 and S_2 being the same as that of D_1 and D_2.

Since D_1 is on the downdip side of S_1, the travel time of a refracted ray from S_1 to D_1 is given by (equation (5.9))

$$t_1 = \frac{x \sin(\theta + \gamma)}{v_1} + \frac{2z_1 \cos \theta}{v_1} \qquad (5.13)$$

and from S_2 to D_2 the travel time is given by

$$t_2 = \frac{x \sin(\theta + \gamma)}{v_1} + \frac{2z_2 \cos \theta}{v_1} \qquad (5.14)$$

where z_1 and z_2 are the perpendicular depths to the refractor under shot points S_1 and S_2, respectively. Now,

$$z_2 - z_1 = \Delta x \sin \gamma$$

$$\therefore z_2 = z_1 + \Delta x \sin \gamma \qquad (5.15)$$

Substituting equation (5.15) in (5.14) and then subtracting equation (5.13) from (5.14) yields

$$t_2 - t_1 = \Delta t = \frac{\Delta x}{v_1} (2 \sin \gamma \cos \theta)$$

$$= \frac{\Delta x \sin(\theta + \gamma)}{v_1} - \frac{\Delta x \sin(\theta - \gamma)}{v_1}$$

Substituting equations (5.11) and (5.12) in the above equation and rearranging terms

$$\frac{\Delta t}{\Delta x} = \frac{1}{v_{2_d}} - \frac{1}{v_{2_u}}$$

where v_{2_u} and v_{2_d} are the updip and downdip apparent velocities, respectively. In the case considered v_{2_d} is derived from the single-ended travel time curves, hence v_{2_u} can be calculated from the difference in travel time of refracted rays from adjacent shots recorded at the same offset distance x. With both apparent velocities calculated, interpretation proceeds by the standard methods for conventional reversed profiles discussed in Section 5.2.4.

5.4 GEOMETRY OF REFRACTED RAY PATHS: IRREGULAR (NON-PLANAR) INTERFACES

The assumption of planar refracting interfaces would often lead to unacceptable error or imprecision in the interpretation of refraction survey data. For example, a survey may be carried out to study the form of the concealed bedrock surface beneath a valley fill of alluvium or glacial drift. Clearly such a surface could not be represented adequately by a planar refractor. It is sometimes necessary, therefore, to remove the constraint that refracting interfaces be interpreted as planar and, consequently, to employ different interpretation methods.

A test of the prevailing refractor geometry is provided by the configuration of travel–time curves

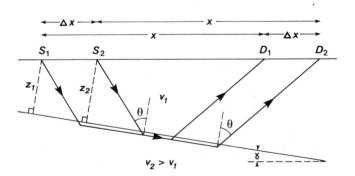

Fig. 5.9 Refraction interpretation using the single-ended profiling method. (After Cunningham 1974.)

derived from a survey. A layered sequence of planar refractors gives rise to a travel–time graph consisting of a series of straight-line segments, each segment representing a particular refracted phase and characterized by a particular gradient and intercept time. Irregular travel–time curves are an indication of irregular refractors (or, alternatively, of lateral velocity variation within individual layers – a complication not discussed here). Methods of interpreting irregular travel–time curves, to determine the non-planar refractor geometry that gives rise to them, are based on the concept of *delay time*.

Consider a horizontal refractor separating upper and lower layers of velocity v_1 and v_2 ($> v_1$), respectively (Fig. 5.1). The travel time of a head wave arriving at an offset distance x is given (see equation (5.3)) by

$$t = x/v_2 + t_i$$

The intercept time t_i can be considered to be composed of two delay times resulting from the presence of the top layer at each end of the ray path. Referring to Fig. 5.10(a), the *delay time* (or *time-term*) a is defined as the time difference between the slant path AB through the top layer and the time that would be required for a ray to travel along the projection BC of the above path through the refractor at the refractor velocity to a position vertically below the point of emergence of the ray at the surface. Thus,

$$a = t_{AB} - t_{BC}$$

$$= \frac{AB}{v_1} - \frac{BC}{v_2}$$

$$= \frac{z}{v_2 \sin\theta \cos\theta} - \frac{z}{v_2}\tan\theta$$

$$= \frac{z}{v_2 \tan\theta}$$

$$= z(v_2^2 - v_1^2)^{1/2}/v_1 v_2 \qquad (5.16)$$

Solving equation (5.16) for the depth z to the refractor yields

$$z = av_1 v_2/(v_2^2 - v_1^2)^{1/2} \qquad (5.17)$$

Thus the delay time can be converted into a refractor depth if v_1 and v_2 are known.

The intercept time t_i in equation (5.3) can be partitioned into two delay times

$$t = x/v_2 + a_s + a_d \qquad (5.18)$$

where a_s and a_d are the delay times at the shot end and detector end of the refracted ray path. Note that in this case of a horizontal refractor

$$a_s = a_d = \tfrac{1}{2}t_i = z(v_2^2 - v_1^2)^{1/2}/v_1 v_2$$

In the presence of refractor dip the delay time is similarly defined except that point C is perpendicularly, not vertically, below A (see Fig. 5.10(b)), and the delay time is again related to depth by equation (5.17), where z is now the refractor depth at A measured normal to the refractor surface. Using this definition of delay time, the travel time of a ray refracted along a dipping interface (see Fig. 5.11(a)) is given by

$$t = x'/v_2 + a_s + a_d \qquad (5.19)$$

where

$$a_s = t_{AB} - t_{BC} \text{ and } a_d = t_{DE} - t_{DF}$$

For shallow dips, x' (unknown) is closely similar to the offset distance x (known), in which case equation (5.18) can be used in place of (5.19) and methods applicable to a horizontal refractor employed. This approximation is valid also in the case of an irregular refractor if the relief on the refractor is small in amplitude compared to the average refractor depth (Fig. 5.11(b)).

Delay times cannot be measured directly but occur in pairs in the travel time equation for a refracted ray from a surface source to a surface detector. The *plus-minus method* of Hagedoorn (1959) provides a means of solving equation (5.18) to derive individual

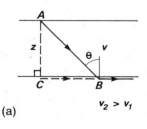

(a) $v_2 > v_1$

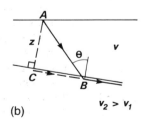

(b) $v_2 > v_1$

Fig. 5.10 The concept of delay time.

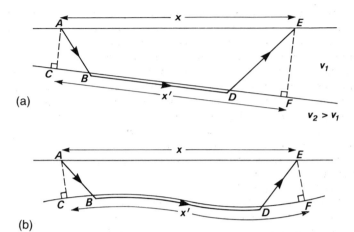

Fig. 5.11 Refracted ray paths associated with (a) a dipping and (b) an irregular refractor.

delay time values for the calculation of local depths to an irregular refractor.

Fig. 5.12(a) illustrates a two-layer ground model with an irregular refracting interface. Selected ray paths are shown associated with a reversed refraction profile line of length l between end shot points S_1 and S_2. The travel time of a refracted ray travelling from one end of the line to the other is given by

$$t_{S_1 S_2} = l/v_2 + a_{S_1} + a_{S_2} \qquad (5.20)$$

where a_{S_1} and a_{S_2} are the delay times at the shot points. Note that $t_{S_1 S_2}$ is the reciprocal time for this reversed profile (see Fig. 5.12(b)). For rays travelling to an intermediate detector position D from each end of the line, the travel times are, for the forward ray, from shot point S_1:

$$t_{S_1 D} = x/v_2 + a_{S_1} + a_D \qquad (5.21)$$

for the reverse ray, from shot point S_2:

$$t_{S_2 D} = (l - x)/v_2 + a_{S_2} + a_D \qquad (5.22)$$

where a_D is the delay time at the detector.

Adding equations (5.21) and (5.22)

$$t_{S_1 D} + t_{S_2 D} = l/v_2 + a_{S_1} + a_{S_2} + 2a_D$$

Substituting equation (5.20) in the above equation yields

$$t_{S_1 D} + t_{S_2 D} = t_{S_1 S_2} + 2a_D$$

Hence

$$a_D = \tfrac{1}{2}(t_{S_1 D} + t_{S_2 D} - t_{S_1 S_2}) \qquad (5.23)$$

This delay time is the *plus* term of Hagedoorn and may be used to compute the perpendicular depth z

to the underlying refractor at D using equation (5.17), once v_1 and v_2 have been determined. v_1 is computed from the slope of the direct ray travel–time curve (see Fig. 5.12(b)). v_2 cannot be obtained directly from the irregular travel–time curve of refracted arrivals, but it can be estimated by means of Hagedoorn's *minus* term, obtained by taking the difference of equations (5.21) and (5.22)

$$t_{S_1 D} - t_{S_2 D} = 2x/v_2 - l/v_2 + a_{S_1} - a_{S_2}$$

This subtraction eliminates the variable (site dependent) delay time a_D from the above equation and, since the last three terms on the right hand side of the equation are constant for a particular profile line, plotting the minus term $(t_{S_1 D} - t_{S_2 D})$ against the offset distance x yields a graph of slope $2/v_2$ from which v_2 may be derived. Any lateral change of refractor velocity v_2 along the profile line will show up as a change of gradient in the minus term plot.

A plus term and, hence, a local refractor depth can be computed at all detector positions at which head wave arrivals are recognized from both ends of the profile line. In practice, this normally means the portion of the profile line between the crossover distances, that is, between x_{c_1} and x_{c_2} in Fig. 5.12(b).

The plus-minus method is only applicable in the case of shallow refractor dips, generally being considered valid for dips of less than 10°. With steeper dips, x' becomes significantly different from the offset distance x. Further, there is an inherent smoothing of the interpreted refractor geometry in the plus-minus method since in computing the plus term from the travel times of forward and reverse rays arriving at any detector position, the refractor is

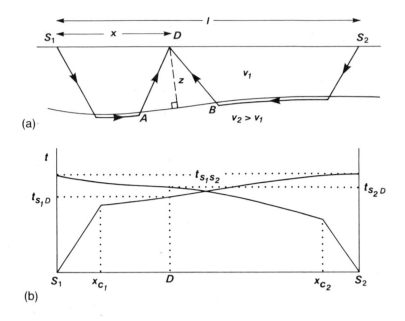

(a)

(b)

Fig. 5.12 The plus-minus method of refraction interpretation (Hagedoorn 1959). (a) Refracted ray paths from each end of a reversed seismic profile line to an intermediate detector position. (b) Travel-time curves in the forward and reverse directions.

assumed to be planar between the points of emergence from the refractor of the forward and reverse rays, e.g. between A and B in Fig. 5.12(a) for rays arriving at detector D. This problem of smoothing is solved in the *generalized reciprocal method* of refraction interpretation (Palmer 1980) by combining forward and reverse rays which, rather than arriving at the same detector, leave the refractor at approximately the same point and arrive at different detector positions separated by a distance Δx (see Fig. 5.13). The method uses a velocity analysis function t_v given by

$$t_v = (t_{S_1D_1} - t_{S_2D_2} + t_{S_1S_2})/2 \qquad (5.24)$$

the values being referred to the midpoint between each pair of detector positions D_1 and D_2. For the case where $D_1 = D_2 = D$ (i.e. $\Delta x = 0$), equation (5.24) reduces to a form similar to Hagedoorn's minus term (see above). The optimal value of Δx for a particular survey is that which produces the closest approach to a linear curve when the velocity analysis function t_v is plotted against distance along the profile line, and is derived by plotting curves for a range of possible Δx values.

Where a refractor is overlain by more than one layer, equation (5.17) cannot be used directly to derive a refractor depth from a delay time (or plus term). In such a case, either the thickness of each overlying layer is computed separately using refracted arrivals from the shallower interfaces, or an average overburden velocity is used in place of v_1 in equation (5.17) to achieve a depth conversion.

5.5 CONSTRUCTION OF WAVEFRONTS AND RAY TRACING

Given the travel−time curves in the forward and reverse directions along a profile line it is possible to reconstruct the configuration of successive wavefronts in the subsurface and thereby derive, graphically, the form of refracting interfaces. This *wavefront method* of Thornburgh (1930) represents one of the earliest refraction interpretation methods but is no longer widely used.

With the recent massive expansion in the speed and power of digital computers, and their wide availability, an increasingly important method of refraction interpretation is a modelling technique known as *ray tracing* (Červený *et al.* 1974). In this method, which is especially useful in the case of complex subsurface structures that are difficult to treat analytically, structural models are postulated and the travel times of refracted (and reflected) rays through these models are calculated by computer for comparison with observed travel times. The model is then adjusted iteratively until the calculated and observed travel times are in acceptable agreement. An example of a ray tracing interpretation is illustrated in Fig. 5.14. The ray tracing method is

Fig. 5.13 The generalized reciprocal
method of refraction interpretation.
(Palmer 1980.)

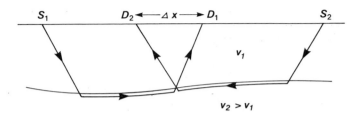

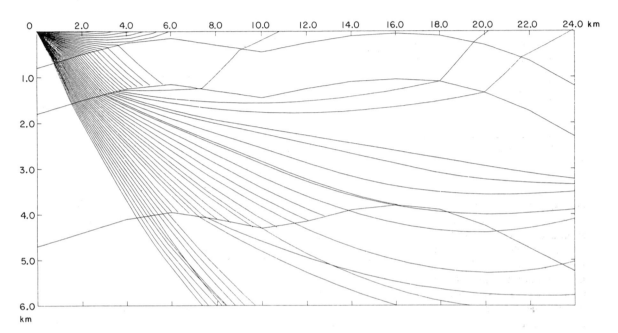

Fig. 5.14 Modelling of complex geology by ray tracing in the case of a refraction profile between quarries in south Wales, UK. Refracted ray paths from Cornelly Quarry (located in Carboniferous Limestone) are modelled through a layered Palaeozoic sedimentary sequence overlying an irregular Precambrian basement surface at a depth of about 5 km. This model accounts for the measured travel times of refracted arrivals observed along the profile. (From Bayerly & Brooks 1980.)

particularly valuable in coping with such complexities as horizontal or vertical velocity gradients within layers, highly irregular or steeply dipping refractor interfaces and discontinuous layers.

5.6 THE HIDDEN LAYER PROBLEM

A *hidden layer*, or *blind layer*, is one that is undetectable by refraction surveying. In practice, there are two different types of hidden layer problem.

Firstly, a layer may simply not give rise to first arrivals, i.e. rays travelling to deeper levels may arrive before those critically refracted at the top of

the layer in question (Fig. 5.15(a)). This may result from the thinness of the layer, or from the closeness of its velocity to that of the overlying layer. In such a case, a method of survey involving recognition of only first arrivals will fail to detect the layer.

A more insidious type of hidden layer problem is associated with a low velocity layer, as illustrated in Fig. 5.15(b). Rays cannot be critically refracted at the top of such a layer and the layer will therefore not give rise to head waves. Hence, a low velocity layer cannot be detected by refraction surveying although the top of the low velocity layer gives rise to wide angle reflections that may be detected during a refraction survey.

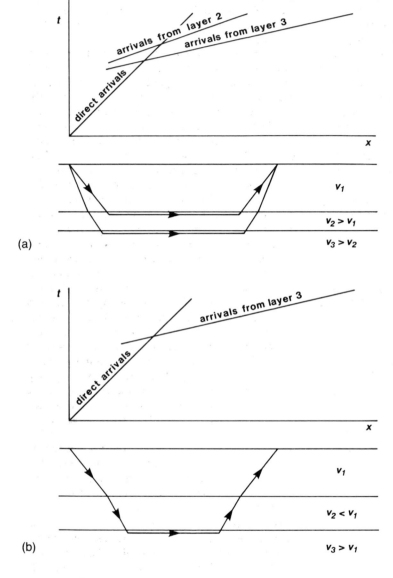

(a)

(b)

Fig. 5.15 The hidden layer problem in refraction seismology. (a) A thin layer that does not give rise to first arrivals. (b) A layer of low velocity that does not generate head waves.

In the presence of a low velocity layer, the interpretation of travel–time curves leads to an overestimation of the depth to underlying interfaces. Low velocity layers are a hazard in all types of refraction seismology. On a small scale, a peat layer in muds and sands above bedrock may escape detection, leading to a false estimation of foundation conditions and rockhead depths beneath a construction site; on a much larger scale, low velocity zones of regional extent are known to exist within the continental crust and may escape detection in crustal seismic experiments.

5.7 REFRACTION IN LAYERS OF CONTINUOUS VELOCITY CHANGE

In some geological situations, velocity varies gradually as a function of depth rather than discontinuously at discrete interfaces of lithological change. In thick clastic sequences, for example, especially clay sequences, velocity increases downwards due to the progressive compaction effects associated with increasing depth of burial. A seismic ray propagating through a layer of gradual velocity change is continuously refracted to follow a curved ray path. For example, in the special case where velocity increases linearly with depth, the seismic ray paths describe arcs of circles. The deepest point reached by a ray travelling on a curved path is known as its *turning point*.

In such cases of continuous velocity change with depth, the travel−time curve for refracted rays that return to the surface along curved ray paths is itself curved, and the geometrical form of the curve may be analysed to derive information on the distribution of velocity as a function of depth (see, e.g. Dobrin & Savit 1988).

Velocity increase with depth may be significant in thick surface layers of clay due to progressive compaction and dewatering, but may also be significant in buried layers. Refracted arrivals from such buried layers are not true head waves since the associated rays do not travel along the top surface of the layer but along a curved path in the layer with a turning point at some depth below the interface. Such refracted waves are referred to as *diving waves* (Červený & Ravindra 1971). Methods of interpreting refraction data in terms of diving waves are generally complex, but include ray tracing techniques. Indeed, some ray tracing programmes require velocity gradients to be introduced into all layers of an interpretation model in order to generate diving waves rather than true head waves.

5.8 METHODOLOGY OF REFRACTION PROFILING

Many of the basic principles of refraction surveying have been covered in the preceding sections but in this section several aspects of the design of refraction profile lines are brought together in relation to the particular objectives of a refraction survey.

5.8.1 Field survey arrangements

Although the same principles apply to all scales of refraction profiling, the logistical problems of implementing a profile line increase as the required line length increases. Further, the problems of surveying on land are quite different from those encountered at sea. A consequence of these logistical differences is a very wide variety of survey arrangements for the implementation of refraction profile lines and these differences are illustrated by three examples.

For a small-scale refraction survey of a construction site to locate the water table or rockhead (both of which surfaces are generally good refractors), recordings out to an offset distance of about 100 m normally suffice, geophones being connected via a multicore cable to a portable 12- or 24-channel seismic recorder. A simple weight-dropping device (even a sledge hammer impacted on to a steel base plate) provides sufficient energy to traverse the short recording range. The dominant frequency of such a source exceeds 100 Hz and the required accuracy of seismic travel times is about 0.5 ms. Such a survey can be easily accomplished by two operators.

The logistical difficulties associated with the cable connection between a detector spread and a recording unit normally limit conventional refraction surveys to maximum shot-detector offsets of about 1 km and, hence, to depths of investigation of a few hundred metres. For larger scale refraction surveys it is necessary to dispense with a cable connection. At sea, such surveys can be carried out by a single vessel in conjunction with free-floating radio-transmitting sonobuoys (Fig. 5.16). Having deployed the sonobuoys, the vessel proceeds along the profile line repeatedly firing explosive charges or an air gun array. Seismic signals travelling back to the surface through the water layer are detected by a hydrophone suspended beneath each sonobuoy, amplified and transmitted back to the survey vessel where they are tape recorded along with the shot instant. By this means, refraction lines up to a few tens of kilometres may be implemented.

For large-scale marine surveys, use may be made of ocean bottom seismographs (OBSs) that are deployed on the sea bead and contain a digital recorder together with a high-precision clock unit to provide an accurate time base for the seismic recordings. Such instruments may be deployed for periods of up to a few days at a time. For the purposes of recovery, the OBSs are "popped-up" to surface by remotely triggering a release mechanism. Sea bed recording systems provide a better signal-to-noise ratio than hydrophones suspended in the water column and, in deep water, recording on the sea bed allows much better definition of shallow structures. In this type of survey the dominant frequency is typically in the range 10−50 Hz and travel times need to be known to about 10 ms.

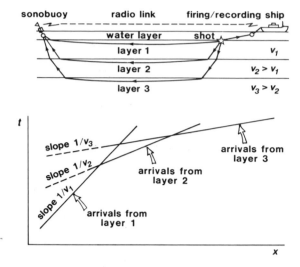

Fig. 5.16 Single-ship seismic refraction profiling.

A large-scale seismic refraction line on land to investigate deep crustal structure is typically 250–300 km long. Seismic events need to be recorded at a series of independently operated recording stations all receiving a radio-transmitted time code to provide a common time base for the recordings. Very large energy sources, such as depth charges (detonated at sea or in a lake) or large quarry blasts, are required in order that sufficient energy is transmitted over the length of the profile line. The dominant frequency of such sources is less than 10 Hz and the required accuracy of seismic travel times is about 50 ms. Such an experiment requires the active involvement of a large team of investigators.

Along long refraction lines, wide-angle reflection events are often detected together with the refracted phases and provide an additional source of information on subsurface structure; indeed, wide-angle reflection events are sometimes the most obvious arrivals and thus represent the primary interest (e.g. Brooks *et al.* 1984). Surveys specifically designed for the joint study of refracted and wide-angle reflection events are often referred to as *wide-angle surveys*.

5.8.2 **Recording scheme**

For complete mapping of refractors beneath a profile line it is important to arrange that head wave arrivals from all refractors of interest are obtained over the same portion of line. The importance of this can be seen by reference to Fig. 5.17 where it is

shown that a change in thickness of a surface low velocity layer would cause a change in the delay time associated with arrivals from a deeper refractor and may be erroneously interpreted as a change in refractor depth. The actual geometry of the shallow refractor could be mapped by means of shorter reversed profiles along the length of the main profile, to ensure that head waves from the shallow refractor were recorded at positions where the depth to the basal refractor was required. Knowledge of the disposition of the shallow refractor derived from the shorter profiles would then allow correction of travel times of arrivals from the deeper refractor.

The general design requirement is the formulation of an overall observational scheme as illustrated in Fig. 5.18. Such a scheme might include off-end shots into individual reversed profile lines, since off-end shots extend the length of refractor traversed by recorded head waves and provide insight into the structural causes of any observed complexities in the travel–time curves. Selection of detector spacing along the individual profile lines is determined by the required detail of the refractor geometry, the sampling interval of interpretation points on the refractor being approximately equal to the detector spacing. Thus, the horizontal resolution of the method is equivalent to the detector spacing.

5.8.3 **Weathering and elevation corrections**

The type of observational scheme illustrated in Fig. 5.18 is often implemented for the specific purpose of mapping the surface zone of weathering and associated low velocity across the length of a longer profile designed to investigate deeper structure. The velocity and thickness of the weathered layer are highly variable laterally and travel times of rays from underlying refractors need to be corrected for the variable delay introduced by the layer. This weathering correction is directly analogous to that applied in reflection seismology (see Section 4.6). The weathering correction is particularly important in shallow refraction surveying where the size of the correction is often a substantial percentage of the overall travel time of a refracted ray. In such cases, failure to apply an accurate weathering correction can lead to major error in interpreted depths to shallow refractors.

A weathering correction is applied by effectively replacing the weathered layer of velocity v_w with material of velocity v_1 equal to the velocity of the

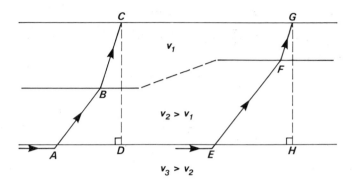

Fig. 5.17 Variation in the travel time of a head wave associated with variation in the thickness of a surface layer.

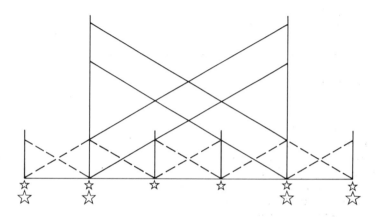

Fig. 5.18 A possible observational scheme to obtain shallow and deeper refraction coverage along a survey line. The inclined lines indicate the range of coverage from the individual shots shown.

underlying layer. For a ray critically refracted along the top of the layer immediately underlying the weathered layer, the weathering correction is simply the sum of the delay times at the shot and detector ends of the ray path. Application of this correction replaces the refracted ray path by a direct path from shot to detector in a layer of velocity v_1. For rays from a deeper refractor a different correction is required. Referring to Fig. 5.19, this correction effectively replaces ray path $ABCD$ by ray path AD. For a ray critically refracted in the nth layer the weathering correction t_w is given by

$$t_w = -(z_s + z_d)\{(v_n^2 - v_1^2)^{1/2}/v_1v_n - (v_n^2 - v_w^2)^{1/2}/v_wv_n\}$$

where z_s and z_d are the thicknesses of the weathered layer beneath the shot and detector respectively, and v_n is the velocity in the nth layer.

In addition to the weathering correction, a correction is also needed to remove the effect of differences in elevation of individual shots and detectors, and an elevation correction is therefore applied to reduce travel times to a common datum plane. The elevation

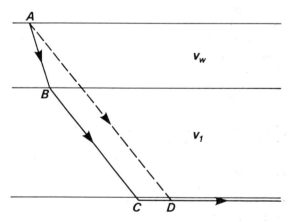

Fig. 5.19 The principle of the weathering correction in refraction seismology.

correction t_e for rays critically refracted in the nth layer is given by

$$t_e = -(h_s + h_d)\{(v_n^2 - v_1^2)^{1/2}/v_1v_n\}$$

where h_s and h_d are the heights above datum of the

shot point and detector location respectively.

In shallow water marine refraction surveying the water layer is conventionally treated as a weathered layer and a correction applied to replace the water layer by material of velocity equal to the velocity of the sea bed.

5.8.4 Display of refraction seismograms

In small-scale refraction surveys the individual seismograms are conventionally plotted out in their true time relationships by a multichannel oscillographic recorder similar to that employed to display seismic traces from land reflection spreads (see Fig. 4.10). From such displays, arrival times of refracted waves may be picked and, after suitable correction, utilized to plot the time−distance curves that form the basis of refraction interpretation.

Interpretation of large-scale refraction surveys is often as much concerned with later arriving phases, such as wide-angle reflections or S-wave arrivals, as with first arrivals and it is necessary to compile the individual seismograms into an overall record section on which the various seismic phases can be correlated from seismogram to seismogram. The optimal type of display is achieved using a *reduced time* scale in which any event at time t and offset distance x is plotted at the reduced time T where

$$T = t - x/v_R$$

and v_R is a scaling factor known as the *reduction velocity*. Thus, for example, a seismic arrival from deep in the Earth's crust with an overall travel time of 30 s to an offset distance of 150 km would, with a reduction velocity of $6\,\mathrm{km\,s}^{-1}$ have a reduced time of 5 s.

Plotting in reduced time has the effect of progressively moving seismic events forward as a function of offset and, therefore, rotating the associated time−distance curves towards the horizontal. For example, a time−distance curve with a reciprocal slope of $6\,\mathrm{km\,s}^{-1}$ on a $t - x$ graph would plot as a horizontal line on a $T - x$ graph using a reduction velocity of $6\,\mathrm{km\,s}^{-1}$. By appropriate choice of reduction velocity, seismic arrivals from a particular refractor of interest can be arranged to plot about a horizontal datum, so that relief on the refractor will show up directly as departures of the arrivals from a horizontal line. The use of reduced time also enables the display of complete seismograms with an expanded time scale appropriate for the analysis of later arriving phases. An example of a record section

from a crustal seismic experiment, plotted in reduced time, is illustrated in Fig. 5.20.

5.9 OTHER METHODS OF REFRACTION SURVEYING

Although the vast bulk of refraction surveying is carried out along profile lines, other spatial arrangements of shots and detectors may be utilized for particular purposes. Such arrangements include fan-shooting and irregularly distributed shots and recorders as used in the time term method.

Fan-shooting (Fig. 5.21) is a convenient method of accurately delineating a subsurface zone of anomalous velocity whose approximate position and size are already known. Detectors are distributed around a segment of arc approximately centred on one or more shot points, and travel times of refracted rays are measured to each detector. Through a homogeneous medium the travel times to detectors would be linearly related to range, but any ray path which encounters an anomalous velocity zone will be subject to a time lead or time lag depending upon the velocity of the zone relative to the velocity of the surrounding medium. Localized anomalous zones capable of detection and delineation by fan-shooting include salt domes, buried valleys and backfilled mine shafts.

An irregular, areal distribution of shots and detectors (Fig. 5.22(a)) represents a completely generalized approach to refraction surveying and facilitates mapping of the three-dimensional geometry of a subsurface refractor using the *time term method* of interpretation (Willmore & Bancroft 1960; Berry & West 1966). Rather than being an intrinsic aspect of the survey design, however, an areal distribution of shot points and recording sites may result simply from an opportunistic approach to refraction surveying in which freely available sources of seismic energy such as quarry blasts are utilized to derive subsurface information from seismic recordings.

The *time term method* uses the form of the travel−time equation containing delay times (equation (5.18)) and is subject to the same underlying assumptions as other interpretation methods using delay times. However, in the time term method a statistical approach is adopted to deal with a redundancy of data inherent in the method and to derive the best estimate of the interpretation parameters. Introducing an error term into the travel−time equation

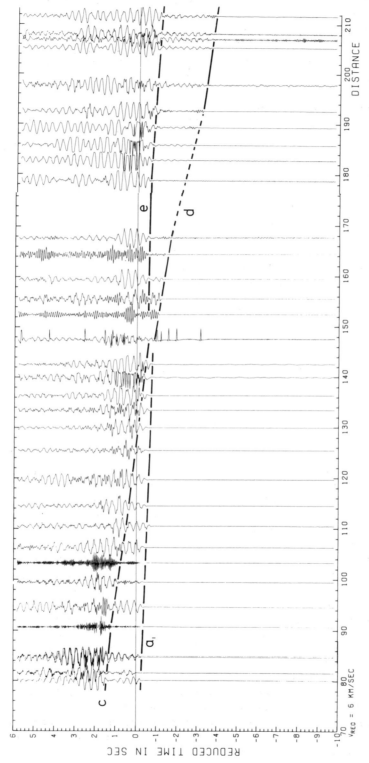

Fig. 5.20 Part of a time section from a large scale refraction profile, plotted in reduced time using a reduction velocity of $6\,\mathrm{km\,s^{-1}}$. The section was derived from the LISPB lithospheric seismic profile across Britain established in 1974. Phase a: head wave arrivals from a shallow crustal refractor with a velocity of about $6.3\,\mathrm{km\,s^{-1}}$; phases c and e: wide-angle reflections from lower crustal interfaces; phase d: head wave arrivals from the uppermost mantle (the P_n phase of earthquake seismology). (From Bamford *et al.* 1978.)

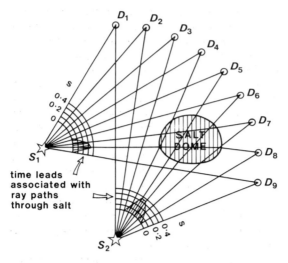

Fig. 5.21 Fan-shooting for the detection of localized zones of anomalous velocity.

$$t_{ij} = x_{ij}/v + a_i + a_j + \epsilon_{ij}$$

where t_{ij} is the travel time of head waves from the ith site to the jth site, x_{ij} is the offset distance between site i and site j, a_i and a_j are the delay times (time terms), v is the refractor velocity (assumed constant), and ϵ_{ij} is an error term associated with the measurement of t_{ij}.

If there are n sites there can be up to $n(n-1)$ observational linear equations of the above type, representing the situation of a shot and detector at each site and all sites sufficiently far apart for the observation of head waves from the underlying refractor. In practice there will be fewer observational equations than this because, normally, only a few of the sites are shot points and head wave arrivals are not recognized along every shot-detector path (Fig. 5.22(b)). There are $(n+1)$ unknowns, namely the individual delay times at the n sites and the refractor velocity v.

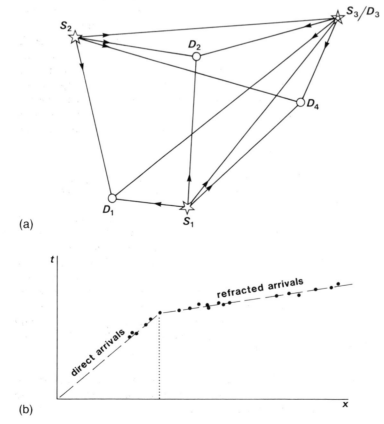

(a)

(b)

Fig. 5.22 (a) An example of the type of network of shots and detectors from which the travel times of refracted arrivals can be used in a time term analysis of the underlying refractor geometry. (b) The plot of travel time as a function of distance identifies the set of refracted arrivals that may be used in the analysis.

If the number m of observational equations equals the number of unknowns, the equations can be solved to derive the unknown quantities, although it is necessary either that at least one shot and detector position should coincide or that the delay time should be known at one site. In fact, with the time term approach to refraction surveying it is normally arranged for m to well exceed $(n + 1)$, and for several shot and detector positions to be interchanged. The resulting overdetermined set of equations is solved by deriving values for the individual delay times and refractor velocity that minimize the sum of squares of the errors ϵ_{ij}. Delay times can then be converted into local refractor depths using the same procedure as in the plus-minus method described earlier.

Although fan shooting involves surface shots and recorders, the method may be regarded as the historical precursor of an important group of modern exploration methods utilising shots and detectors located in boreholes. In these methods, known as *seismic tomography* or *seismic scanning*, subsurface zones are systematically investigated by transmitting very large numbers of seismic rays through them. An example is cross-hole seismics (see, e.g., Wong *et al.* 1987), in which shots generated at several depths down a borehole are recorded by detector arrays in an adjacent borehole to study velocity variations in the intervening section of ground. Suitable analysis of the set of travel times of rays transmitted along the dense pattern of intersecting ray paths enables detailed mapping of the velocity structure within the irradiated section. Use of high frequency sources permits high-resolution imaging of the velocity structure. Inversion of transmission times into velocity maps is achieved using tomographic imaging techniques similar in principle to those used in diagnostic medicine. The velocity information derived from seismic tomography may be used to predict spacial variations in, e.g., lithology, pore fluids, or rock fracturing, and the method is therefore of potential value in a wide range of exploration and engineering applications.

5.10 TWO-SHIP SEISMIC SURVEYING: COMBINED REFRACTION AND REFLECTION SURVEYING

Specialized methods of marine surveying involving the use of two survey vessels and multichannel recording include *expanding spread profiles* and *constant offset profiles* (Stoffa & Buhl 1979). These methods have been developed for the detailed study of the deep structure of the crust and upper mantle under continental margins and oceanic areas.

Expanding spread profiling (ESP) is designed to obtain detailed information relating to a localized region of the crust. The shot-firing vessel and recording vessel travel outwards at the same speed from a central position, obtaining reflected and refracted arrivals from subsurface interfaces out to large offsets. Thus, in addition to near-normal incidence reflections such as would be recorded in a conventional common depth point (CDP) reflection survey, wide-angle reflections and refracted arrivals are also recorded from the same section of crust. The combined reflection/refraction data allow derivation of a highly-detailed velocity-depth structure for the localized region.

Expanding spread profiles have also been carried out on land to investigate the crustal structure of continental areas (see, e.g., Wright *et al.* 1990).

In constant offset profiling (COP), the shot-firing and recording vessels travel along a profile line at a fixed, wide separation. Thus, wide-angle reflections and refractions are continuously recorded along the line. This survey technique facilitates the mapping of lateral changes in crustal structure over wide areas and allows continuous mapping of the types of refracting interface that do not give rise to good near-normal incidence reflections and which therefore cannot be mapped adequately using conventional reflection profiling. Such interfaces include zones of steep velocity gradient, in contrast to the first-order velocity discontinuities that constitute the best reflectors.

5.11 APPLICATIONS OF SEISMIC REFRACTION SURVEYING

Exploration using refraction methods covers a very wide range of applications. On the local scale, refraction surveys are widely used in foundation studies on construction sites to derive estimates of depth to rockhead beneath a cover of superficial material. Use of the plus-minus method or the generalized reciprocal method (Section 5.4) allows irregular rockhead geometries to be mapped in detail and thus reduces the need for test drilling with its associated high costs. Refraction surveys can also provide estimates of the elastic constants of local rock types, which have important engineering applications: use of special sources and geophones allows the separate recording of shear wave arrivals, and the combination of *P*- and *S*-wave velocity information enables calculation of Poisson's ratio (Section 3.3.1). If an estimate of density is available, the bulk modulus and shear modulus can also be calculated from *P*- and *S*-wave velocities. Such estimates of the elastic constants, based on the propagation of seismic waves, are referred to as dynamic, in contrast to the static estimates derived from load-testing of rock samples in the laboratory. Dynamic estimates tend to yield slightly higher values than loading tests.

The large difference in velocity between dry and wet sediments renders the water table a very effective refractor. Hence, refraction surveys find wide application in exploration programmes for underground water supplies in sedimentary sequences, often employed in conjunction with electrical resistivity methods (see Chapter 8).

The interpretation of seismic refraction profile data is most conveniently carried out using commercial software packages on personal computers. A wide range of good software is available for the plotting, automatic event picking and interpretation of such data.

The refraction method produces generalized models of subsurface structure with good velocity information, but it is unable to provide the amount of structural detail or the direct imaging of specific structures that are the hallmark of reflection seismology. The occasional need for better velocity information than can be derived from velocity analysis of reflection data alone (see Chapter 4), together with the relative ease of refraction surveying offshore,

gives the refraction method an important subsidiary role to reflection surveying in the exploration for hydrocarbons in some offshore areas.

Refraction and wide-angle surveys have been used extensively for regional investigation of the internal constitution and thickness of the Earth's crust. The information derived from such studies is complementary to the direct seismic imaging of crustal structure derived from large-scale reflection surveys of the type discussed in Section 4.12. Interpretation of large-scale refraction and wide-angle surveys is normally carried out by forward modelling of the travel times and amplitudes of recorded refracted and/or reflected phases using ray tracing techniques.

Large-scale surveys, using explosives as seismic sources, have been carried out to study crustal structure in most continental areas. An example is the LISPB experiment which was carried out in Britain in 1974 and produced the crustal section for northern Britain reproduced in Fig. 5.23.

Such experiments show that the continental crust is typically 30–40 km thick and that it is often internally layered. It is characterized by major regional variations in thickness and constitution which are often directly related to changes of surface geology. Thus, different orogenic provinces are often characterized by quite different crustal sections. Upper crustal velocities are usually in the range $5.8–6.3 \, km \, s^{-1}$ which, by analogy with velocity measurements of rock samples in the laboratory (see Section 3.4), may be interpreted as representing mainly granitic or granodioritic material. Lower

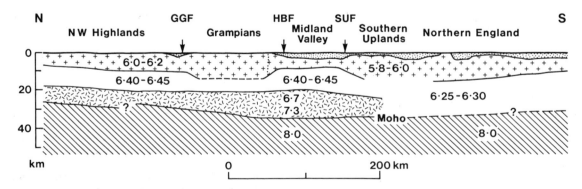

Fig. 5.23 Crustal cross-section across northern Britain based on interpretation of a large-scale seismic refraction experiment. Numbers refer to velocities in $km \, s^{-1}$. (After Bamford *et al.* 1978.)

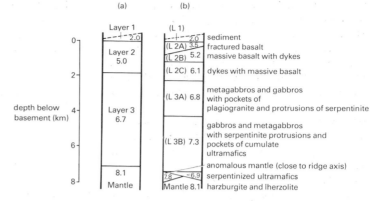

Fig. 5.24 Velocity structure of typical oceanic lithosphere in terms of layered structures proposed in 1965(a) and 1978(b), and its geological interpretation. (From Kearey & Vine 1990).

crustal velocities are normally in the range $6.5-7.0\,\mathrm{km\,s^{-1}}$ and may represent any of a variety of igneous and metamorphic rock types, including gabbro, gabbroic anorthosite and basic granulite. The latter rock type is regarded as the most probable major constituent of the lower crust on the basis of experimental studies of seismic velocities (Christensen & Fountain 1975).

Marine surveys, usually single-ship experiments, have shown the ocean basins to have a crust only $6-8\,\mathrm{km}$ thick, composed of three main layers with differing seismic velocities. This thickness and layering is maintained over vast areas beneath all the major oceans. The results of deep-sea drilling, together with the recognition of ophiolite complexes exposed on land as analogues of oceanic lithosphere, have enabled the nature of the individual seismic layers to be identified (Fig. 5.24).

5.12 **PROBLEMS**

1 A single-ended refraction profile designed to determine the depth to an underlying horizontal refractor reveals a top layer velocity of $3.0\,\mathrm{km\,s^{-1}}$ and a refractor velocity of $5.0\,\mathrm{km\,s^{-1}}$. The crossover distance is found to be $500\,\mathrm{m}$. What is the refractor depth?

2 What is the delay time for head wave arrivals from layer 3 in the following case?:

Layer	Depth (m)	Vel. ($\mathrm{km\,s^{-1}}$)
1	100	1.5
2	50	2.5
3	–	4.0

3 In order that both the horizontal-layer models given below should produce the same time–distance curves for head wave arrivals, what must be the thickness of the middle layer in Model 2?

	Vel. ($\mathrm{km\,s^{-1}}$)	Depth (km)
Model 1		
Layer 1	3.0	1.0
Layer 2	5.0	–
Model 2		
Layer 1	3.0	0.5
Layer 2	1.5	?
Layer 3	5.0	–

4 A single-ended refraction survey (Section 5.3) established to locate an underlying planar dipping refractor yields a top layer velocity of $2.2\,\mathrm{km\,s^{-1}}$ and a downdip apparent refractor velocity of $4.0\,\mathrm{km\,s^{-1}}$. When the shot point and geophones are moved forward by $150\,\mathrm{m}$, in the direction of refractor dip, head wave arrival times to any offset distance are increased by $5\,\mathrm{ms}$. Calculate the dip and true velocity of the refractor. If the intercept time of the refracted ray travel–time curve at the original shot-point is $20\,\mathrm{ms}$, what is the vertical depth to the refractor at that location?

5 A split-spread refraction profile (Section 5.3) with a central shot point is established to locate an underlying planar dipping refractor. The resultant time–distance curves yield a top layer velocity of $2.0\,\mathrm{km\,s^{-1}}$ and updip and downdip apparent velocities of $4.5\,\mathrm{km\,s^{-1}}$ and $3.5\,\mathrm{km\,s^{-1}}$, respectively. The common intercept time is $85\,\mathrm{ms}$. Calculate the true velocity and dip of the refractor and its vertical depth beneath the shot point.

6 The following data set was obtained from a reversed seismic refraction line $275\,\mathrm{m}$ long. The survey was carried out in a level area of alluvial cover to determine depths to the underlying bedrock surface.

Offset (m)	Travel time (ms)
Forward direction:	
12.5	6.0
25	12.5
37.5	19.0
50	25.0
75	37.0
100	42.5
125	48.5
150	53.0
175	57.0
200	61.5
225	66.0
250	71.0
275	76.5
Reverse direction:	
12.5	6.0
25	12.5
37.5	17.0
50	19.5
75	25.0
100	30.5
125	37.5
150	45.5
175	52.0
200	59.0
225	65.5
250	71.0
275	76.5

Carry out a plus-minus interpretation of the data and comment briefly on the resultant bedrock profile.

7 What subsurface structure is responsible for the travel–time curves shown in Fig. 5.25?

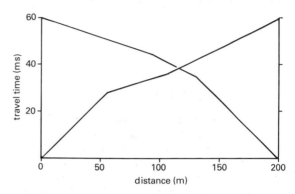

Fig. 5.25 Time–distance curves obtained in the forward and reverse directions along a refraction profile across an unknown subsurface structure.

FURTHER READING

Červený, V. & Ravindra, R. (1971) *Theory of Seismic Head Waves*. University of Toronto Press.

Dobrin, M.B. & Savit, C.H. (1988) *Introduction to Geophysical Prospecting*. (4th edn), McGraw-Hill, New York.

Giese, P., Prodehl, C. & Stein, A. (eds.) (1976) *Explosion Seismology in Central Europe*. Springer-Verlag, Berlin.

Palmer, D. (1980) *The Generalised Reciprocal Method of Seismic Refraction Interpretation*. Society of Exploration Geophysicists, Tulsa.

Palmer, D. (1986) *Handbook of Geophysical Exploration: Section 1, Seismic Exploration. Vol 13: Refraction Seismics*. Enpro Science Publications, Amsterdam.

Sjögren, B. (1984) *Shallow Refraction Seismics*. Chapman & Hall, London.

Stoffa, P.L. & Buhl, P. (1979) Two-ship multichannel seismic experiments for deep crustal studies: expanded spread and constant offset profiles. *J. Geophys. Res.*, **84**, 7645–60.

Willmore, P.L. & Bancroft, A.M. (1960) The time-term approach to refraction seismology. *Geophys. J.R. Astr. Soc.*, **3**, 419–32.

6 / Gravity surveying

6.1 INTRODUCTION

In gravity surveying, subsurface geology is investigated on the basis of variations in the Earth's gravitational field generated by differences of density between subsurface rocks. An underlying concept is the idea of a causative body, which is a rock unit of different density from its surroundings. A causative body represents a subsurface zone of anomalous mass and causes a localized perturbation in the gravitational field known as a gravity anomaly. A very wide range of geological situations give rise to zones of anomalous mass that produce significant gravity anomalies. On a small scale, buried relief on a bedrock surface, such as a buried valley, can give rise to measurable anomalies. On a larger scale, small negative anomalies are associated with salt domes, as discussed in Chapter 1. On a larger scale still, major gravity anomalies are generated by granite plutons or sedimentary basins. Interpretation of gravity anomalies allows an assessment to be made of the probable depth and shape of the causative body.

The ability to carry out gravity surveys in marine areas extends the scope of the method so that the technique may be employed in most areas of the world.

6.2 BASIC THEORY

The basis of the gravity survey method is Newton's Law of Gravitation, which states that the force of attraction F between two masses m_1 and m_2, whose dimensions are small with respect to the distance r between them, is given by

$$F = \frac{Gm_1m_2}{r^2} \tag{6.1}$$

where G is the Gravitational Constant ($6.67 \times 10^{-11}\,\mathrm{m^3\,kg^{-1}\,s^{-2}}$).

Consider the gravitational attraction of a spherical, non-rotating, homogeneous Earth of mass M and radius R on a small mass m on its surface. It is relatively simple to show that the mass of a sphere acts as though it were concentrated at the centre of the sphere and by substitution in equation (6.1)

$$F = \frac{GM}{R^2}\,m = mg \tag{6.2}$$

Force is related to mass by an acceleration and the term $g = GM/R^2$ is known as the gravitational acceleration or, simply, *gravity*. The weight of the mass is given by mg.

On such an Earth, gravity would be constant. However, the Earth's ellipsoidal shape, rotation, irregular surface relief and internal mass distribution cause gravity to vary over its surface.

The gravitational field is most usefully defined in terms of the *gravitational potential U*:

$$U = \frac{GM}{r} \tag{6.3}$$

Whereas the gravitational acceleration g is a vector quantity, having both magnitude and direction (vertically downwards), the gravitational potential U is a scalar, having magnitude only. The first derivative of U in any direction gives the component of gravity in that direction. Consequently a potential field approach provides computational flexibility. Equipotential surfaces can be defined on which U is constant. The sea-level surface, or *geoid*, is the most easily recognized equipotential surface, which is everywhere horizontal and orthogonal to the direction of gravity.

6.3 UNITS OF GRAVITY

The mean value of gravity at the Earth's surface is about $9.80\,\mathrm{m\,s^{-2}}$. Variations in gravity caused by density variations in the subsurface are of the order of $100\,\mathrm{\mu m\,s^{-2}}$. This unit of the micrometre per second per second is referred to as the *gravity unit* (gu). In gravity surveys on land an accuracy of ± 0.1 gu is readily attainable, corresponding to about one hundred millionth of the normal gravitational field. At sea the accuracy obtainable is considerably less, about ± 10 gu. The c.g.s. unit of gravity is the *milligal* ($1\,\mathrm{mgal} = 10^{-3}\,\mathrm{Gal} = 10^{-3}\,\mathrm{cm\ s^{-2}}$), equivalent to 10 gu.

6.4 MEASUREMENT OF GRAVITY

Since gravity is an acceleration, its measurement should simply involve determinations of length and time. However, such apparently simple measurements are not easily achievable at the precision and accuracy required in gravity surveying.

The measurement of an absolute value of gravity is extremely difficult and requires complex apparatus and a lengthy period of observation. Such measurement is classically made using large pendulums or falling body techniques (see, for example, Nettleton 1976, Whitcomb 1987).

The measurement of relative values of gravity, i.e. the differences of gravity between locations, is simpler and is the standard procedure in gravity surveying. Absolute gravity values at survey stations may be obtained by reference to the International Gravity Standardization Network (IGSN) of 1971 (Morelli *et al.* 1971), a network of stations at which the absolute values of gravity have been determined by reference to sites of absolute gravity measurements (see Section 6.7). By using a relative reading instrument to determine the difference in gravity between an IGSN station and a field location the absolute value of gravity at that location can be determined.

Previous generations of relative reading instruments were based on small pendulums or the oscillation of torsion fibres and, although portable, took considerable time to read. Modern instruments capable of rapid gravity measurements are known as *gravity meters* or *gravimeters*.

Gravimeters are basically spring balances carrying a constant mass. Variations in the weight of the mass caused by variations in gravity cause the length of the spring to vary and give a measure of the change in gravity. In Fig. 6.1 a spring of initial length s has been stretched by an amount δs as a result of an increase in gravity δg increasing the weight of the suspended mass m. The extension of the spring is proportional to the extending force (Hooke's Law), thus

$$m\delta g = k\delta s$$

and

$$\delta s = \frac{m}{k}\delta g \qquad (6.4)$$

where k is the elastic spring constant.

δs must be measured to a precision of $1:10^8$ in instruments suitable for gravity surveying on land.

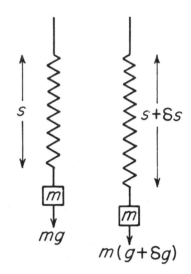

Fig. 6.1 Principle of stable gravimeter operation.

Although a large mass and a weak spring would increase the ratio m/k and, hence, the sensitivity of the instrument, in practice this would make the system liable to collapse. Consequently some form of optical, mechanical or electronic amplification of the extension is in practice required.

The necessity for the spring to serve a dual function, namely to support the mass and to act as the measuring device, severely restricted the sensitivity of early gravimeters, known as stable or static gravimeters. This problem is overcome in modern meters (unstable or astatic) which employ an additional force that acts in the same sense as the extension (or contraction) of the spring and consequently amplifies the movement directly.

An example of an unstable instrument is the LaCoste and Romberg gravimeter. The meter consists of a hinged beam, carrying a mass, supported by a spring attached immediately above the hinge (Fig. 6.2). The magnitude of the moment exerted by the spring on the beam is dependent upon the extension of the spring and the sine of the angle θ. If gravity increases, the beam is depressed and the spring further extended. Although the restoring force of the spring is increased, the angle θ is decreased to θ′. By suitable design of the spring and beam geometry the magnitude of the increase of restoring moment with increasing gravity can be made as small as desired. With ordinary springs the working range of such an instrument would be very small. However, by making use of a 'zero-length'

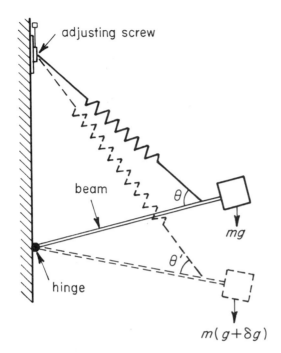

Fig. 6.2 Principle of the LaCoste and Romberg gravimeter.

spring which is pretensioned during manufacture so that the restoring force is proportional to the physical length of the spring rather than its extension, instruments can be fashioned with a very sensitive response over a wide range. The instrument is read by restoring the beam to the horizontal by altering the vertical location of the spring attachment with a micrometer screw. Thermal effects are removed by a battery-powered thermostatting system. The range of the instrument is 50 000 gu.

The other unstable instrument in common use is the Worden-type gravimeter. The necessary instability is provided by a similar mechanical arrangement, but in this case the beam is supported by two springs. The first of these springs acts as the measuring device, while the second alters the level of the 2000 gu reading range of the instrument. In certain specialized forms of this instrument the second spring is also calibrated, so that the overall reading range is similar to that of the LaCoste and Romberg gravimeter. Thermal effects are normally minimized by the use of quartz components and a bimetallic beam which compensates automatically for temperature changes. Consequently no thermostatting is required and it is simply necessary to house the instrument in

an evacuated flask. The restricted range of normal forms of the instrument, however, makes it unsuitable for intercontinental gravity ties or surveys in areas where gravity variation is extreme.

Gravimeters for general surveying use are capable of registering changes in gravity with an accuracy of 0.1 gu. Also available for more specialized surveys (Section 6.12) are gravimeters capable of detecting gravity changes as small as one microgal (10^{-8} m s^{-2}).

A shortcoming of gravimeters is the phenomenon of *drift*. This refers to a gradual change in reading with time, observable when the instrument is left at a fixed location. Drift results from the imperfect elasticity of the springs, which undergo anelastic creep with time. Drift can also result from temperature variations which, unless counteracted in some way, cause expansion or contraction of the measuring system and thus give rise to variations in measurements that are unrelated to changes in gravity. Drift is monitored by repeated meter readings at a fixed location throughout the day.

Gravity may be measured at discrete locations at sea using a remote-controlled land gravimeter, housed in a waterproof container, which is lowered over the side of the ship and, by remote operation, levelled and read on the sea bed. Measurements of comparable quality to readings on land may be obtained in this way, and the method has been used with success in relatively shallow waters. The disadvantage of the method is that the meter has to be lowered to the sea bed for each reading so that the rate of surveying is very slow. Moreover, in strong tidal currents, the survey ship needs to be anchored to keep it on station while the gravimeter is on the sea bed.

Gravity measurements may be made continuously at sea using a gravimeter modified for use on ships. Such instruments are known as shipborne, or shipboard, meters. The accuracy of measurements with a shipborne meter is considerably reduced compared to measurements on land because of the severe vertical and horizontal accelerations imposed on the shipborne meter by sea waves and the ship's motion. These external accelerations can cause variations in measured gravity of up to 10^6 gu and represent high amplitude noise from which a signal of much smaller gravity variations must be extracted. The effects of horizontal accelerations produced by waves, yawing of the ship and changes in its speed and heading can be largely eliminated by mounting the meter on a gyrostabilized, horizontal platform, so that the meter only responds to vertical accelerations. Deviations of the platform from the horizontal produce *off-levelling errors* which are normally less than 10 gu. External vertical accelerations resulting from wave motions cannot be distinguished from gravity but their effect can be diminished by heavily damping the suspension system and by averaging the reading over an interval considerably longer than the maximum period of the wave motions (about 8 s). As the ship oscillates vertically above and below the plane of the mean sea surface, the wave accelerations are equally negative and positive and are effectively removed by averaging over a few minutes. The operation is essentially low-pass filtering in which accelerations with periods of less than one to five minutes are rejected.

With shipborne meters employing a beam-supported sensor, such as the LaCoste and Romberg instrument, a further complication arises due to the influence of horizontal accelerations. The beam of the meter oscillates under the influence of the varying vertical accelerations caused by the ship's motions. When the beam is tilted out of the horizontal it will be further displaced by the turning force associated with any horizontal acceleration. For certain phase relationships between the vertical and horizontal components of motion of the ship, the horizontal accelerations may cause beam displacements that do not average out with time. Consider an example where the position of a meter in space describes a

circular motion under the influence of sea waves (Fig. 6.3). At time t_1, as shown in Fig. 6.3, the ship is moving down, displacing the beam upwards, and the horizontal component of motion is to the right, inducing an anticlockwise torque that decreases the upward displacement of the beam. At a slightly later time t_3 the ship is moving up, displacing the beam down, and the horizontal motion is to the left, again inducing an anticlockwise torque which, now, increases the downward displacement of the beam. In such a case, the overall effect of the horizontal accelerations is to produce a systematic error in the beam position. This effect is known as *cross-coupling*, and its magnitude is dependent on the damping characteristics of the meter and the amplitude and phase relationships of the horizontal and vertical motions. It leads to an error known as the *cross-coupling error* in the measured gravity value. In general, the cross-coupling error is small or negligible in good weather conditions but can become very large in high seas. Cross-coupling errors are corrected directly from the outputs of two horizontal accelerometers mounted on the stabilized platform.

The inability to compensate fully for extraneous accelerations reduces the accuracy of these shipborne measurements to 10 gu at best, the actual amount depending on prevailing sea conditions. Instrumental drift monitoring is also less precise as base ties are, of necessity, usually many days apart.

Cross-coupling is one of the major sources of

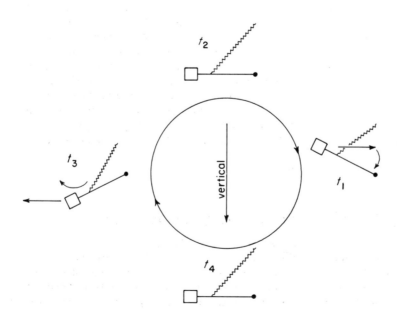

Fig. 6.3 Cross-coupling in a shipborne gravimeter.

error in measurements of gravity at sea made with instruments utilizing a beam-supported mass, and arises because of the directional nature of the system. No cross-coupling would occur if the sensor were symmetric about a vertical axis, and since the late 1960s new marine meters utilizing this feature have been developed.

The *vibrating string accelerometer* (Bowin *et al.* 1972) is based on the principle that the resonant frequency of a short, vertical string from which a mass is suspended is proportional to the square root of gravity. Changes in this frequency provide a measure of changes in gravity. Gravimeters based on this mechanism have never found much favour because of relatively low reported accuracies and erratic drift.

The most successful axially symmetric instrument to date is the *Bell gravimeter* (Bell & Watts 1986). The sensing element of the meter is the accelerometer shown in Fig. 6.4 which is mounted on a stable platform. The accelerometer, which is about 34 mm high and 23 mm in diameter, consists of a mass, wrapped in a coil, which is constrained to move only vertically between two permanent magnets. A DC current passed through the coil causes the mass to act as a magnet. In the null position, the weight of the mass is balanced by the forces exerted by the permanent magnets. When the mass moves vertically in response to a change in gravity or wave accelerations, the motion is detected by a servo loop which regulates the current in the coil, changing its magnetic moment so that it is driven back to the null position. The varying current is then a measure of changes in the vertical accelerations experienced by the sensor. As with beam-type meters, a weighted average filter is applied to the output in order to separate gravity changes from wave-generated accelerations.

Drift rates of the Bell gravimeter are low and uniform, and it has been demonstrated that the instrument is accurate to just a few gravity units, and is capable of discriminating anomalies with wavelengths of 1–2 km. This accuracy and resolution is considerably greater than that of earlier instruments, and it is anticipated that much smaller gravity anomalies will be detected than was previously possible. The factor preventing more widespread deployment of the meter is its large cost.

The measurement of gravity from aircraft is not at present satisfactory for other than reconnaissance surveys because of the large error in applying corrections. Eötvös corrections (Section 6.8.5) may be

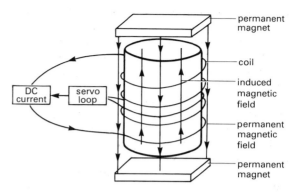

Fig. 6.4 Principle of the accelerometer unit of the Bell marine gravimeter. (After Bell & Watts 1986.)

as great as 16 000 gu at a speed of 200 knots, a 1% error in velocity or heading producing maximum errors of 180 gu and 250 gu, respectively. Vertical accelerations associated with the aircraft's motion with periods longer than the instrumental averaging time cannot readily be corrected. Such uncertainties can be overcome to a certain extent by the use of autopilots and automatic height stabilizers but the present precision of such systems is only of the order of 30 gu (Brozena & Peters 1988).

The calibration constants of gravimeters may vary with time and should be checked periodically. The most common procedure is to take readings at two or more locations where absolute or relative values of gravity are known. In calibrating Worden-type meters, these readings would be taken for several settings of the coarse adjusting screw so that the calibration constant is checked over as much of the full range of the instrument as possible. Such a procedure cannot be adopted for the LaCoste and Romberg gravimeter, where each different dial range has its own calibration constant. In this case checking can be accomplished by taking readings at different inclinations of the gravimeter on a tilt table, a task usually entrusted to the instrument's manufacturer.

6.5 GRAVITY ANOMALIES

Gravimeters effectively respond only to the vertical component of the gravitational attraction of an anomalous mass. Consider the gravitational effect of an anomalous mass δg, with horizontal and vertical components δg_x and δg_z, respectively, on the local gravity field g and its representation on a vector diagram (Fig. 6.5).

Solving the rectangle of forces

$$g + \delta g = ((g + \delta g_z)^2 + \delta g_x^2)^{1/2}$$
$$= (g^2 + 2g\delta g_z + \delta g_z^2 + \delta g_x^2)^{1/2}$$

Terms in δ^2 are insignificantly small and can thus be ignored. Binomial expansion of the equation then gives

$$g + \delta g \approx g + \delta g_z$$

so that

$$\delta g \approx \delta g_z \tag{6.5}$$

Consequently, measured perturbations in gravity effectively correspond to the vertical component of the attraction of the causative body. The local deflection of the vertical θ is given by

$$\theta = \tan^{-1} \frac{\delta g_x}{g}$$

and since $\delta g_x \ll g$, θ is usually insignificant. Very large mass anomalies such as mountain ranges can, however, produce measurable local vertical deflections.

6.6 GRAVITY ANOMALIES OF SIMPLE SHAPED BODIES

Consider the gravitational attraction of a point mass m at a distance r from the mass (Fig. 6.6).

The gravitational attraction Δg_r in the direction of the mass is given by

$$\Delta g_r = \frac{Gm}{r^2} \text{ from Newton's Law.}$$

Since only the vertical component of the attraction Δg_z is measured, the gravity anomaly Δg caused by the mass is

$$\Delta g = \frac{Gm}{r^2} \cos \theta$$

or

$$\Delta g = \frac{Gmz}{r^3} \tag{6.6}$$

Since a sphere acts as though its mass were concentrated at its centre, equation (6.6) also corresponds to the gravity anomaly of a sphere whose centre lies at a depth z.

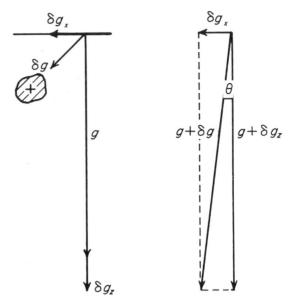

Fig. 6.5 Relationship between the gravitational field and the components of the gravity anomaly of a small mass.

Equation (6.6) can be used to build up the gravity anomaly of many simple geometric shapes by constructing them from a suite of small elements which correspond to point masses, and then summing (integrating) the attractions of these elements to derive the anomaly of the whole body.

Integration of equation (6.6) in a horizontal direction provides the equation for a line mass (Fig. 6.7) extending to infinity in this direction

$$\Delta g = \frac{2Gmz}{r^2} \tag{6.7}$$

Equation (6.7) also represents the anomaly of a horizontal cylinder, whose mass acts as though it is concentrated along its axis.

Integration in the second horizontal direction provides the gravity anomaly of an infinite horizontal sheet, and a further integration in the vertical direction between fixed limits provides the anomaly of an infinite horizontal slab

$$\Delta g = 2\pi G \rho t \tag{6.8}$$

where ρ is the density of the slab and t its thickness. Note that this attraction is independent of both the location of the observation point and the depth of the slab.

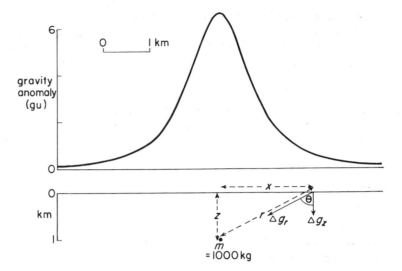

Fig. 6.6 The gravity anomaly of a point mass or sphere.

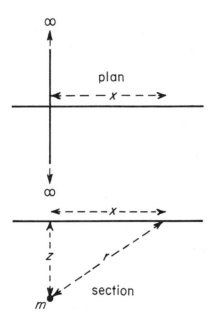

Fig. 6.7 Coordinates describing an infinite horizontal line mass.

A similar series of integrations, this time between fixed limits, can be used to determine the anomaly of a right rectangular prism.

In general, the gravity anomaly of a body of *any* shape can be determined by summing the attractions of all the mass elements which make up the body. Consider a small prismatic element of such a body of

density ρ, located at x', y', z', with sides of length $\delta x'$, $\delta y'$, $\delta z'$ (Fig. 6.8). The mass δm of this element, is given by

$$\delta m = \rho \delta x' \, \delta y' \, \delta z'$$

Consequently its attraction δg at a point outside the body (x, y, z), a distance r from the element, is derived from equation (6.6)

$$\delta g = G\rho \frac{(z' - z)}{r^3} \delta x' \, \delta y' \, \delta z'$$

The anomaly of the whole body Δg is then found by summing all such elements which make up the body

$$\Delta g = \Sigma \, \Sigma \, \Sigma \, G\rho \frac{(z' - z)}{r^3} \delta x' \, \delta y' \, \delta z' \qquad (6.9)$$

If $\delta x'$, $\delta y'$ and $\delta z'$ are allowed to approach zero, then

$$\Delta g = \int \int \int G\rho \frac{(z' - z)}{r^3} \, dx' \, dy' \, dz' \qquad (6.10)$$

where

$$r = ((x' - x)^2 + (y' - y)^2 + (z' - z)^2)^{1/2}$$

As shown before, the attraction of bodies of regular geometry can be determined by integrating equation (6.10) analytically. The anomalies of irregularly shaped bodies are calculated by numerical integration using equations of the form of equation (6.9).

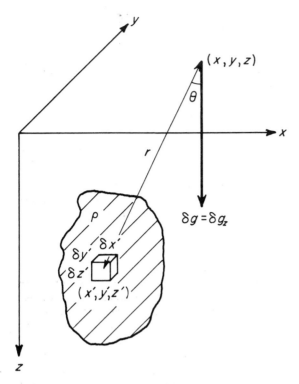

Fig. 6.8 The gravity anomaly of an element of a mass of irregular shape.

6.7 GRAVITY SURVEYING

The station spacing used in a gravity survey may vary from a few metres in the case of detailed mineral or geotechnical surveys to several kilometres in regional reconnaissance surveys. The station density should be greatest where the gravity field is changing most rapidly, as accurate measurement of gravity gradients is critical to subsequent interpretation. If absolute gravity values are required in order to interface the results with other gravity surveys, at least one easily accessible base station must be available where the absolute value of gravity is known. If the location of the nearest IGSN station is inconvenient, a gravimeter can be used to establish a local base by measuring the difference in gravity between the IGSN station and the local base. Because of instrumental drift this cannot be accomplished directly and a procedure known as *looping* is adopted. A series of alternate readings at recorded times is made at the two stations and drift curves constructed for each (Fig. 6.9). The differences in ordinate measurements (Δg_{1-4}) for the two

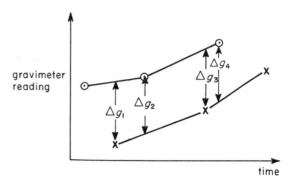

Fig. 6.9 The principle of looping. Crosses and circles represent alternate gravimeter readings taken at two base stations. The vertical separations between the drift curves for the two stations (Δg_{1-4}) provide an estimate of the gravity difference between them.

stations then may be averaged to give a measure of the drift-corrected gravity difference.

During a gravity survey the gravimeter is read at a base station at a frequency dependent on the drift characteristics of the instrument. At each survey station, location, time, elevation/water depth and gravimeter reading are recorded.

In order to obtain a reduced gravity value accurate to ± 1 gu, the reduction procedure described in the following section indicates that the gravimeter must be read to a precision of ± 0.1 gu, the latitude of the station must be known to ± 10 m and the elevation of the station must be known to ± 10 mm. The latitude of the station must consequently be determined from maps at a scale of $1:10\,000$ or smaller, or by the use of electronic position-fixing systems. Uncertainties in the elevations of gravity stations probably account for the greatest errors in reduced gravity values on land; at sea, water depths are easily determined with a precision depth recorder to an accuracy consistent with the gravity measurements. In well-surveyed land areas, the density of accurately-determined elevations at bench marks is normally sufficiently high that gravity stations can be sited at bench marks or connected to them by levelling surveys. Reconnaissance gravity surveys of less well-mapped areas require some form of independent elevation determination. Many such areas have been surveyed using anaeroid altimeters. The accuracy of heights determined by such instruments is dependent upon the prevailing climatic conditions and is of the order of $1-5$ m, leading to a relatively large uncertainty in the elevation corrections applied to the measured gravity values. The optimal equipment at present for surveys of this type is an inertial navigational system, which can provide elevations to ±0.5 m together with accurate locations. Such equipment is now available in a compact form suitable, for example, for mounting in helicopters, but its large cost inhibits its widespread usage. In the future it is probable that the full implementation of the Global Positioning System (GPS) (Davis *et al.* 1989) satellites will provide accurate elevations from a small, inexpensive receiver.

6.8 GRAVITY REDUCTION

Before the results of a gravity survey can be interpreted it is necessary to correct for all variations in the Earth's gravitational field which do not result from the differences of density in the underlying rocks. This process is known as *gravity reduction* or *reduction to the geoid*, as sea level is usually the most convenient datum level.

6.8.1 Drift correction

Correction for instrumental drift is based on repeated readings at a base station at recorded times throughout the day. The meter reading is plotted against time (Fig. 6.10) and drift is assumed to be linear between consecutive base readings. The drift correction at time t is d, which is subtracted from the observed value.

After drift correction the difference in gravity between an observation point and the base is found by multiplication of the difference in meter reading by the calibration factor of the gravimeter. Knowing this difference in gravity, the absolute gravity at the observation point g_{obs} can be computed from the known value of gravity at the base. Alternatively, readings can be related to an arbitrary datum, but this practice is not desirable as the results from different surveys cannot then be tied together.

6.8.2 Latitude correction

Gravity varies with latitude because of the non-spherical shape of the Earth and because the angular velocity of a point on the Earth's surface decreases from a maximum at the equator to zero at the poles (Fig. 6.11(a)). The centripetal acceleration generated by this rotation has a negative radial component that consequently causes gravity to decrease from pole to equator. The true shape of the

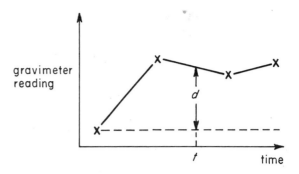

Fig. 6.10 A gravimeter drift curve constructed from repeated readings at a fixed location. The drift correction to be subtracted for a reading taken at time t is d.

Earth is an oblate spheroid or polar flattened ellipsoid (Fig. 6.11(b)) whose difference in equatorial and polar radii is some 21 km. Consequently, points near the equator are farther from the centre of mass of the Earth than those near the poles, causing gravity to increase from the equator to the poles. The amplitude of this effect is reduced by the differing subsurface mass distributions resulting from the equatorial bulge, the mass underlying equatorial regions being greater than that underlying polar regions.

The net effect of these various factors is that gravity at the poles exceeds gravity at the equator by some 51 860 gu, with the north–south gravity gradient at latitude ϕ being $8.12 \sin 2\phi$ gu km^{-1}.

Clairaut's formula relates gravity to latitude on the reference spheroid according to an equation of the form

$$g_\phi = g_0 (1 + k_1 \sin^2\phi - k_2 \sin^2 2\phi) \tag{6.11}$$

where g_ϕ is the predicted value of gravity at latitude ϕ, g_0 is the value of gravity at the equator and k_1, k_2 are constants dependent on the shape and speed of rotation of the Earth. Equation (6.11) is, in fact, an approximation of an infinite series. The values of g_0, k_1 and k_2 in current use define the Gravity Formula 1967 ($g_0 = 9\,780\,318$ gu, $k_1 = 0.0053024$; $k_2 = 0.0000059$; IAG 1971). Prior to 1967 less accurate constants were employed in the International Gravity Formula (1930). Results deduced using the earlier formula must be modified before incorporation into survey data reduced using the Gravity Formula 1967 by using the relationship g_ϕ (1967) − g_ϕ (1930) = $(136 \sin^2\phi - 172)$ gu.

An alternative, more accurate, representation of

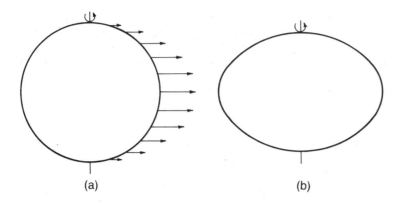

Fig. 6.11 (a) The variation in angular velocity with latitude around the Earth represented by vectors whose lengths are proportional to angular velocity. (b) An exaggerated representation of the shape of the Earth. The true shape of this oblate ellipsoid of revolution results in a difference in equatorial and polar radii of some 21 km.

the Gravity Formula 1967 (Mittermayer 1969), in which the constants are adjusted so as to minimize errors resulting from the truncation of the series is

$$g_\phi = 9\,780\,318.5\,(1 + 0.005278895\,\sin^2\phi + 0.000023462\,\sin^4\phi)\,\text{gu}$$

This form, however, is less suitable if the survey results are to incorporate pre-1967 data made compatible with the Gravity Formula 1967 using the above relationship.

The value g_ϕ gives the predicted value of gravity at sea level at any point on the Earth's surface and is subtracted from observed gravity to correct for latitude variation.

6.8.3 Elevation corrections

Correction for the differing elevations of gravity stations is made in three parts. The *free-air correction* (FAC) corrects for the decrease in gravity with height in free air resulting from increased distance from the centre of the Earth, according to Newton's Law. To reduce to datum an observation taken at height h (Fig. 6.12(a))

$$\text{FAC} = 3.086\,h\;\text{gu}\;(h\text{ in metres})$$

The FAC is positive for an observation point above datum to correct for the decrease in gravity with elevation.

The free-air correction accounts solely for variation in the distance of the observation point from the centre of the Earth; no account is taken of the gravitational effect of the rock present between the observation point and datum. The *Bouguer correction* (BC) removes this effect by approximating the rock layer beneath the observation point to an infinite horizontal slab with a thickness equal to

the elevation of the observation above datum (Fig. 6.12(b)). If ρ is the density of the rock, from equation (6.8)

$$\text{BC} = 2\pi G\rho h = 0.4191\,\rho h\;\text{gu}$$
$$(h\text{ in metres, }\rho\text{ in Mg m}^{-3})$$

On land the Bouguer correction must be subtracted, as the gravitational attraction of the rock between observation point and datum must be removed from the observed gravity value. The Bouguer correction of sea surface observations is positive to account for the lack of rock between surface and sea bed. The correction is equivalent to the replacement of the water layer by material of a specified rock density ρ_r. In this case

$$\text{BC} = 2\pi G\,(\rho_r - \rho_w)\,z$$

where z is the water depth and ρ_w the density of water.

The free-air and Bouguer corrections are often applied together as the *combined elevation correction*.

The Bouguer correction makes the assumption that the topography around the gravity station is flat. This is rarely the case and a further correction, the *terrain correction* (TC), must be made to account for topographic relief in the vicinity of the gravity station. This correction is always positive as may be appreciated from consideration of Fig. 6.12(c). The regions designated A form part of the Bouguer correction slab although they do not consist of rock. Consequently the Bouguer correction has overcorrected for these areas and their effect must be restored by a positive terrain correction. Region B consists of rock material that has been excluded from the Bouguer correction. It exerts an upward attraction at the observation point causing gravity to

Fig. 6.12 (a) The free-air correction for an observation at a height h above datum. (b) The Bouguer correction. The shaded region corresponds to a slab of rock of thickness h extending to infinity in both horizontal directions. (c) The terrain correction.

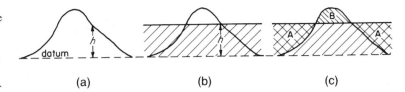

(a)   (b)   (c)

decrease. Its attraction must thus be corrected by a positive terrain correction.

Classically, terrain corrections are applied using a circular graticule divided by radial and concentric lines into a large number of compartments known, after its inventor, as a Hammer chart (Fig. 6.13). The outermost zone extends to almost 22 km, beyond which topographic effects are usually negligible. The graticule is laid on a topographic map with its centre on the gravity station and the average topographic elevation of each compartment is determined. The elevation of the gravity station is subtracted from these values, and the gravitational effect of each compartment is determined by reference to tables constructed using the formula for the gravitational effect of a sector of a vertical cylinder at its axis. The terrain correction is then computed by summing the gravitational contribution of all compartments. Table 6.1 shows the method of computation. Such operations are time-consuming as the topography of over 130 compartments has to be averaged for each station, but terrain correction is the one operation in gravity reduction that cannot be fully automated. Labour can be reduced by averaging

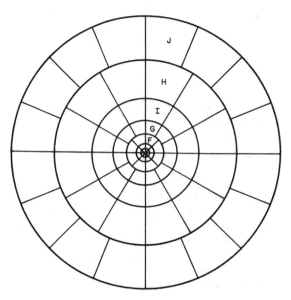

Fig. 6.13 A typical graticule used in the calculation of terrain corrections. A series of such graticules with zones varying in radius from 2 m to 21.9 km are used with topographic maps of varying scale.

Table 6.1 Terrain corrections.

Zone	r_1	r_2	n	Zone	r_1	r_2	n
B	2.0	16.6	4	H	1529.4	2614.4	12
C	16.6	53.3	6	I	2614.4	4468.8	12
D	53.3	170.1	6	J	4468.8	6652.2	16
E	170.1	390.1	8	K	6652.2	9902.5	16
F	390.1	894.8	8	L	9902.5	14740.9	16
G	894.8	1529.4	12	M	14740.9	21943.3	16

$$T = 0.4191 \, \frac{\rho}{n} \, (r_2 - r_1 + \sqrt{r_1^2 + z^2} - \sqrt{r_2^2 + z^2})$$

where T = terrain correction of compartment (gu); ρ = Bouguer correction density (Mg m^{-3}); n = number of compartments in zone; r_1 = inner radius of zone (m); r_2 = outer radius of zone (m); and z = modulus of elevation difference between observation point and mean elevation of compartment (m).

topography within a rectangular grid. Only a single digitization is required as the topographic effects may be calculated at any point within the grid by summing the effects of the right rectangular prisms defined by the grid squares and their elevation difference with the gravity station. This operation can effectively correct for the topography of areas distant from the gravity station and can be readily computerized. Correction for inner zones, however, must still be performed manually as any reasonable digitization scheme for a complete survey area and its environs must employ a sampling interval that is too large to provide an accurate representation of the terrain close to the station.

Terrain effects are low in areas of subdued topography, rarely exceeding 10 gu in flat-lying areas. In areas of rugged topography terrain effects are considerably greater, being at a maximum in steep-sided valleys, at the base or top of cliffs and at the summits of mountains.

Where terrain effects are considerably less than the desired accuracy of a survey, the terrain correction may be ignored. However, the usual necessity for this correction accounts for the bulk of time spent on gravity reduction and is thus a major contributor to the cost of a gravity survey.

6.8.4 **Tidal correction**

Gravity measured at a fixed location varies with time because of periodic variation in the gravitational effects of the Sun and Moon associated with their orbital motions, and correction must be made for this variation in a high precision survey. In spite of its much smaller mass, the gravitational attraction of the Moon is larger than that of the Sun because of its proximity. Also, these gravitational effects cause the shape of the solid Earth to vary in much the same way that the celestial attractions cause tides in the sea. These *solid Earth tides* are considerably smaller than oceanic tides and lag farther behind the lunar motion. They cause the elevation of an observation point to be altered by a few centimetres and thus vary its distance from the centre of mass of the Earth. The periodic gravity variations caused by the combined effects of Sun and Moon are known as *tidal variations*. They have a maximum amplitude of some 3 gu and a minimum period of about 12 hours.

If a gravimeter with a relatively high drift rate is used, base ties are made at an interval much smaller than the minimum Earth tide period and the tidal variations are automatically removed during the

drift correction. If a meter with a low drift rate is employed, base ties are normally made only at the start and end of the day so that the tidal variation has undergone a full cycle. In such a case, a separate tidal correction may need to be made. The tidal effects are predictable and published every year in the geophysical press.

6.8.5 **Eötvös correction**

The Eötvös correction (EC) is applied to gravity measurements taken on a moving vehicle such as a ship or an aircraft. Depending on the direction of travel, vehicular motion will generate a centripetal acceleration which either reinforces or opposes gravity. The correction required is

$$EC = 75.03V \sin \alpha \cos \phi + 0.04154 V^2 \, gu$$

where V is the speed of the vehicle in knots, α the heading and ϕ the latitude of the observation. In mid-latitudes the Eötvös correction is about $+75$ gu for each knot of E to W motion so that speed and heading must be accurately known.

SHould be 7.5 mGal at Equator.

6.8.6 **Free-air and Bouguer anomalies**

The *free-air anomaly* (FAA) and *Bouguer anomaly* (BA) may now be defined

$$FAA = g_{obs} - g_\phi + FAC \, (\pm EC) \qquad (6.12)$$

$$BA = g_{obs} - g_\phi + FAC \pm BC + TC \, (\pm EC) \qquad (6.13)$$

The Bouguer anomaly forms the basis for the interpretation of gravity data on land. In marine surveys Bouguer anomalies are conventionally computed for inshore and shallow water areas as the Bouguer correction removes the local gravitational effects associated with local changes in water depth. Moreover, the computation of the Bouguer anomaly in such areas allows direct comparison of gravity anomalies offshore and onshore and permits the combination of land and marine data into gravity contour maps. These may be used, for example, in tracing geological features across coastlines. The Bouguer anomaly is not appropriate for deeper water surveys, however, as in such areas the application of a Bouguer correction is an artificial device that leads to very large positive Bouguer anomaly values without significantly enhancing local gravity features of geological origin. Consequently the free-air anomaly is frequently used for interpretation in such areas. Moreover, the FAA provides a broad assessment of the degree of isostatic compensation of an area (e.g. Bott 1982).

Gravity anomalies are conventionally displayed on profiles or as isogal maps. Interpretation of the latter may be facilitated by utilizing digital image processing techniques similar to those used in the display of remotely sensed data. In particular, colour and shaded relief images may reveal structural features that may not be readily discernable on unprocessed maps (Plate 5a). This type of processing is equally appropriate to magnetic anomalies (Plate 5b; see for example Lee *et al.* 1990).

6.9 ROCK DENSITIES

Gravity anomalies result from the difference in density, or *density contrast*, between a body of rock and its surroundings. For a body of density ρ_1 embedded in material of density ρ_2, the density contrast $\Delta\rho$ is given by

$$\Delta\rho = \rho_1 - \rho_2$$

The sign of the density contrast determines the sign of the gravity anomaly.

Rock densities are among the least variable of all geophysical parameters. Most common rock types have densities in the range between 1.60 and $3.20\,\mathrm{Mg\,m^{-3}}$. The density of a rock is dependent on both its composition and porosity.

Variation in porosity is the main cause of density variation in sedimentary rocks. Thus, in sedimentary rock sequences, density tends to increase with depth, due to compaction, and with age, due to progressive cementation.

Most igneous and metamorphic rocks have negligible porosity, and composition is the main cause of density variation. Density generally increases as acidity decreases; thus there is a progression of density increase from acid through basic to ultrabasic igneous rock types. Density ranges for common rock types and ores are presented in Table 6.2.

A knowledge of rock density is necessary both for application of the Bouguer correction and for the interpretation of gravity data.

Density is commonly determined by direct measurements on rock samples. A sample is weighed in air and in water. The difference in weights provides the volume of the sample and so the dry density can be obtained. If the rock is porous the saturation density may be calculated by following the above procedure after saturating the rock with water. The density value employed in interpretation

Table 6.2. Approximate density ranges ($\mathrm{Mg\,m^{-3}}$) of some common rock types and ores.

Alluvium (wet)	1.96–2.00
Clay	1.63–2.60
Shale	2.06–2.66
Sandstone	
Cretaceous	2.05–2.35
Triassic	2.25–2.30
Carboniferous	2.35–2.55
Limestone	2.60–2.80
Chalk	1.94–2.23
Dolomite	2.28–2.90
Halite	2.10–2.40
Granite	2.52–2.75
Granodiorite	2.67–2.79
Anorthosite	2.61–2.75
Basalt	2.70–3.20
Gabbro	2.85–3.12
Gneiss	2.61–2.99
Quartzite	2.60–2.70
Amphibolite	2.79–3.14
Chromite	4.30–4.60
Pyrrhotite	4.50–4.80
Magnetite	4.90–5.20
Pyrite	4.90–5.20
Cassiterite	6.80–7.10
Galena	7.40–7.60

NB. The lower end of the density range quoted in many texts is often unreasonably extended by measurements made on samples affected by physical or chemical weathering.

then depends upon the location of the rock above or below the water table.

It should be stressed that the density of any particular rock type can be quite variable. Consequently it is usually necessary to measure several tens of samples of each particular rock type in order to obtain a reliable mean density and variance.

As well as these direct methods of density determination, there are several indirect (or *in situ*) methods. These usually provide a mean density of a particular rock unit which may be internally quite variable. *In situ* methods do, however, yield valuable information where sampling is hampered by lack of exposure or made impossible because the rocks concerned occur only at depth.

The measurement of gravity at different depths beneath the surface using a special borehole gravimeter (Section 11.11) or, more commonly, a standard gravimeter in a mineshaft, provides a measure of the mean density of the material between the observation levels. In Fig. 6.14 gravity has been measured at the surface and at a point underground at a depth h immediately below. If g_1 and g_2 are the values of gravity obtained at the two levels, then, applying free-air and Bouguer corrections one obtains

$$g_1 - g_2 = 3.086h - 4\pi G\rho h \qquad (6.14)$$

The Bouguer correction is double that employed on the surface as the slab of rock between the observation levels exerts both a downward attraction at the surface location and an upward attraction at the underground location. The density ρ of the medium separating the two observations can then be found from the difference in gravity. Density may also be measured in boreholes using a density (gamma–gamma) logger as discussed in Section 11.7.2.

Nettleton's method of density determination involves taking gravity observations over a small isolated topographic prominence. Field data are reduced using a series of different densities for the Bouguer and terrain corrections (Fig. 6.15). The density value that yields a Bouguer anomaly with the least correlation (positive or negative) with the topography is taken to represent the density of the

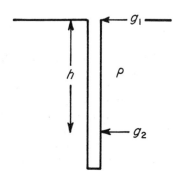

Fig. 6.14 Density determination by subsurface gravity measurements. The measured gravity difference $g_1 - g_2$ over a height difference h can be used to determine the mean density ρ of the rock separating the measurements.

prominence. The method is useful in that no borehole or mineshaft is required, and a mean density of the material forming the prominence is provided. A disadvantage of the method is that isolated relief features may be formed of anomalous materials which are not representative of the area in general.

Density information is also provided from the P-wave velocities of rocks obtained in seismic surveys.

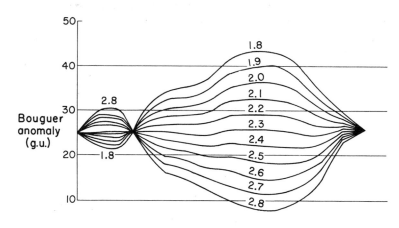

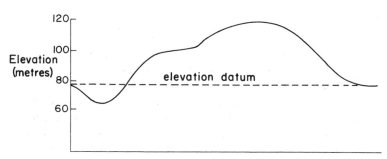

Fig. 6.15 Nettleton's method of density determination over an isolated topographic feature. Gravity reductions have been performed using densities ranging from 1.8 to 2.8 Mg m^{-3} for both Bouguer and terrain corrections. The profile corresponding to a value of 2.3 Mg m^{-3} shows least correlation with topography so this density is taken to represent the density of the feature. (After Dobrin & Savit 1988.)

Fig. 6.16 shows graphs of the logarithm of *P*-wave velocity against density for various rock types (Gardner *et al.* 1974), and the best-fitting linear relationship. Other workers (e.g. Birch 1960, 1961, Christensen & Fountain 1975) have derived similar relationships. The empirical velocity–density curve of Nafe & Drake (1963) indicates that densities estimated from seismic velocities are probably no more accurate than about $\pm 0.10\,\text{Mg m}^{-3}$. This however, is the only method available for the estimation of densities of deep lithologies that cannot be sampled directly.

6.10 INTERPRETATION OF GRAVITY ANOMALIES

6.10.1 The inverse problem

The interpretation of potential field anomalies (gravity, magnetic and electrical) is inherently ambiguous. The ambiguity arises because any given anomaly could be caused by an infinite number of possible sources. For example, concentric spheres of constant mass but differing density and radius would all produce the same anomaly, since their mass acts as though located at the centre of the sphere. This ambiguity represents the *inverse problem* of potential field interpretation, which states that although the anomaly of a given body may be calculated uniquely, there are an infinite number of bodies that could give rise to any specified anomaly. An important task in interpretation is to decrease this ambiguity by using all available external constraints on the nature and form of the anomalous body. Such constraints include geological information derived from surface outcrops, boreholes and mines, and from other, complementary, geophysical techniques (see, for example, Lines *et al.* 1988).

6.10.2 Regional fields and residual anomalies

Bouguer anomaly fields are often characterized by a broad, gently varying, regional anomaly on which may be superimposed higher wavenumber local anomalies (Fig. 6.17). Usually in gravity surveying it is the local anomalies that are of prime interest and the first step in interpretation is the removal of the *regional field* to isolate the *residual anomalies*. This may be performed graphically by sketching in a linear or curvilinear field by eye. Such a method is biased by the interpreter, but this is not necessarily

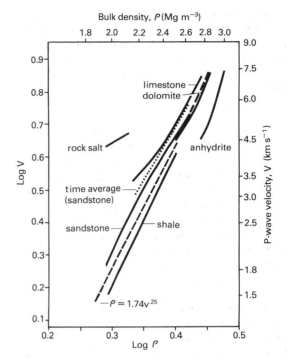

Fig. 6.16 Graphs of the logarithm of *P*-wave velocity against density for various rock types. Also shown is the best-fitting linear relationship between density and log velocity (after Gardner *et al.* 1974).

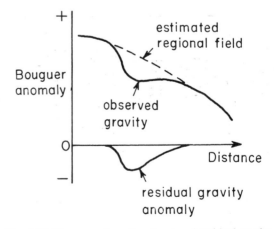

Fig. 6.17 The separation of regional and residual gravity anomalies from the observed Bouguer anomaly.

disadvantageous as geological knowledge can be incorporated into the selection of the regional field. Several analytical methods of regional field analysis are available and include trend surface analysis (fitting a polynomial to the observed data) and low-pass filtering (Section 6.12). Such procedures must be used critically as fictitious residual anomalies can sometimes arise when the regional field is subtracted from the observed data due to the mathematical procedures employed.

It is necessary before carrying out interpretation to differentiate between two-dimensional and three-dimensional anomalies. Two-dimensional anomalies are elongated in one horizontal direction so that the anomaly length in this direction is at least twice the anomaly width. Such anomalies may be interpreted in terms of structures which theoretically extend to infinity in the elongate direction by using profiles at right angles to the strike. Three-dimensional anomalies may have any shape and are considerably more difficult to interpret quantitatively.

Gravity interpretation proceeds via the methods of direct and indirect interpretation.

6.10.3 **Direct interpretation**

Direct interpretation provides, directly from the gravity anomalies, information on the anomalous body which is largely independent of the true shape of the body. Various methods are discussed below.

LIMITING DEPTH

Limiting depth refers to the maximum depth at which the top of a body could lie and still produce an observed gravity anomaly. Gravity anomalies decay with the inverse square of the distance from their source so that anomalies caused by deep structures are of lower amplitude and greater extent than those caused by shallow sources. This wavenumber–amplitude relationship to depth may be quantified to compute the maximum depth (or limiting depth) at which the top of the anomalous body could be situated.

(a) Half-width method. The half-width of an anomaly ($x_{1/2}$) is the horizontal distance from the anomaly maximum to the point at which the anomaly has reduced to half of its maximum value (Fig. 6.18(a)).

If the anomaly is three-dimensional, the initial assumption is made that it results from a point mass. Manipulation of the point mass formula (equation 6.6)), allows its depth to be determined in terms of the half-width

$$z = \frac{x_{1/2}}{(4^{1/3} - 1)^{1/2}}$$

Here, z represents the actual depth of the point mass or the centre of a sphere with the same mass. It is an overestimate of the depth to the top of the sphere, i.e. the limiting depth. Consequently, the limiting depth for any three-dimensional body is given by

$$z < \frac{x_{1/2}}{(4^{1/3} - 1)^{1/2}} \tag{6.15}$$

A similar approach is adopted for a two-dimensional anomaly, with the initial assumption that the anomaly results from a horizontal line mass (equation (6.7)). The depth to a line mass or to the centre of a horizontal cylinder with the same mass distribution is given by

$$z = x_{1/2}$$

For any two-dimensional body, the limiting depth is then given by

$$z < x_{1/2} \tag{6.16}$$

(b) Gradient–amplitude ratio method. This method requries the computation of the maximum anomaly amplitude (A_{max}) and the maximum horizontal gravity gradient (A'_{max}) (Fig. 6.18(b)). Again the initial assumption is made that a three-dimensional anomaly is caused by a point mass and a two-dimensional anomaly by a line mass. By differentiation of the relevant formulae, for any three-dimensional body

$$z < 0.86 \left| \frac{A_{max}}{A'_{max}} \right| \tag{6.17}$$

and for any two-dimensional body

$$z < 0.65 \left| \frac{A_{max}}{A'_{max}} \right| \tag{6.18}$$

(c) Second derivative methods. There are a number of limiting depth methods based on the computation of the maximum second horizontal derivative, or maximum rate of change of gradient, of a gravity anomaly (Smith 1959). Such methods provide rather more accurate limiting depth estimates than either the half-width or gradient–amplitude ratio methods if the observed anomaly is free from noise.

EXCESS MASS

The excess mass of a body can be uniquely deter-

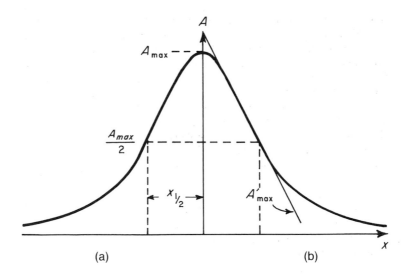

Fig. 6.18 Limiting depth calculations using (a) the half-width method and (b) the gradient–amplitude ratio.

(a)   (b)

mined from its gravity anomaly without making any assumptions about its shape, depth or density. Excess mass refers to the difference in mass between the body and the mass of country rock that would otherwise fill the space occupied by the body. The basis of this calculation is a formula derived from Gauss' theorem, and it involves a surface integration of the residual anomaly over the area in which it occurs. The survey area is divided into n grid squares of area Δa and the mean residual anomaly Δg found for each square. The excess mass M_e is then given by

$$M_e = \frac{1}{2\pi G} \sum_{i=1}^{n} \Delta g_i \Delta a_i \qquad (6.19)$$

Before using this procedure it is important that the regional field is removed so that the anomaly tails to zero. The method only works well for isolated anomalies whose extremities are well defined. Gravity anomalies decay slowly with distance from source and so these tails can cover a wide area and be important contributors to the summation.

To compute the actual mass M of the body, the densities of both anomalous body (ρ_1) and country rock (ρ_2) must be known

$$M = \frac{\rho_1 M_e}{(\rho_1 - \rho_2)} \qquad (6.20)$$

The method is of use in estimating the tonnage of ore bodies.

INFLECTION POINT

The locations of inflection points on gravity profiles, i.e. positions where the horizontal gravity gradient changes most rapidly, can provide useful information on the nature of the edge of an anomalous body. Over structures with outward dipping contacts, such as granite bodies (Fig. 6.19(a)), the inflection points (arrowed) lie near the base of the anomaly. Over structures with inward dipping contacts such as sedimentary basins (Fig. 6.19(b)), the inflection points lie near the uppermost edge of the anomaly.

APPROXIMATE THICKNESS

If the density contrast $\Delta\rho$ of an anomalous body is known, its thickness t may be crudely estimated from its maximum gravity anomaly Δg by making use of the slab formula (equation (6.8)).

$$t \approx \frac{\Delta g}{2\pi G \Delta \rho} \qquad (6.21)$$

This thickness will always be an underestimate for a body of restricted horizontal extent. The method is commonly used in estimating the throw of a fault from the difference in the gravity fields of the upthrown and downthrown sides.

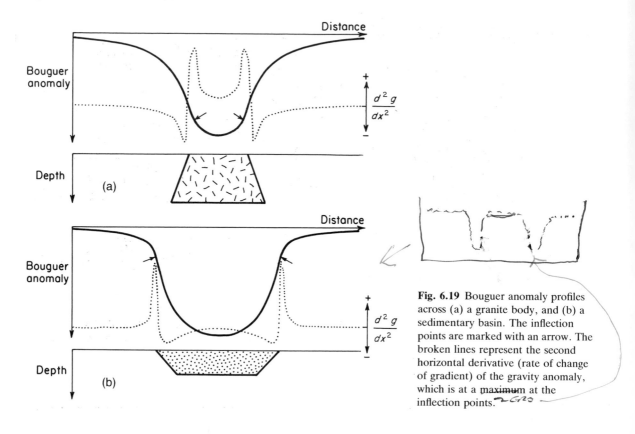

Fig. 6.19 Bouguer anomaly profiles across (a) a granite body, and (b) a sedimentary basin. The inflection points are marked with an arrow. The broken lines represent the second horizontal derivative (rate of change of gradient) of the gravity anomaly, which is at a maximum at the inflection points.

6.10.4 Indirect interpretation

In indirect interpretation, the causative body of a gravity anomaly is simulated by a model whose theoretical anomaly can be computed, and the shape of the model is altered until the computed anomaly closely matches the observed anomaly. Because of the inverse problem this model will not be a unique interpretation, but ambiguity can be decreased by using other constraints on the nature and form of the anomalous body.

A simple approach to indirect interpretation is the comparison of the observed anomaly with the anomaly computed for certain standard geometrical shapes whose size, position, form and density contrast are altered to improve the fit. Two-dimensional anomalies may be compared with anomalies computed for horizontal cylinders or half cylinders, and three-dimensional anomalies compared with those of spheres, vertical cylinders or right rectangular prisms. Combinations of such shapes may also be used to simulate an observed anomaly.

Fig. 6.20(a) shows a large, circular gravity anomaly situated near Darnley Bay, N.W.T., Canada. The anomaly is radially symmetrical and a profile across the anomaly (Fig. 6.20(b)) can be simulated by a model constructed from a suite of coaxial cylinders whose diameters decrease with depth so that the anomalous body has the overall form of an inverted cone. This study illustrates well the non-uniqueness of gravity interpretation. The nature of the causative body is unknown and so no information is available on its density. An alternative interpretation, again in the form of an inverted cone, but with an increased density contrast, is presented in Fig. 6.20(b). Both models provide adequate simulations of the observed anomaly, and cannot be distinguished using the information available.

The computation of anomalies over a model of irregular form is accomplished by dividing the model into a series of regularly-shaped compartments and calculating the combined effect of these compartments at each observation point. At one time this operation was performed by the use of graticules,

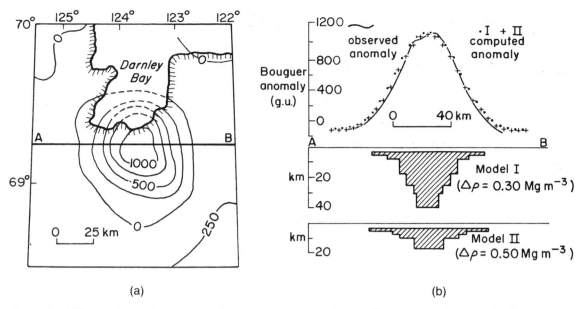

(a)   (b)

Fig. 6.20 (a) The circular gravity anomaly at Darnley Bay. N.W.T., Canada. Contour interval 250 g.u. (b) Two possible interpretations of the anomaly in terms of a model constructed from a suite of coaxial vertical cylinders. (After Stacey 1971.)

but nowadays the calculations are invariably performed by computers.

A two-dimensional gravity anomaly may be represented by a profile normal to the direction of elongation. This profile can be interpreted in terms of a model which maintains a constant cross-section to infinity in the horizontal direction perpendicular to the profile.

The basic unit for constructing the anomaly of a two-dimensional model is the semi-infinite slab with a sloping edge shown in Fig. 6.21, which extends to infinity into and out of the plane of the figure. The gravity anomaly of this slab Δg is given by

$$\Delta g = 2G\Delta\rho\,[\,-\{x_1\sin\theta + z_1\cos\theta\}\,\{\sin\theta\,\log_e(r_2/r_1)$$
$$+ \cos\theta\,(\phi_2 - \phi_1)\} + z_2\phi_2 - z_1\phi_1]$$
$$(6.22)$$

where $\Delta\rho$ is the density contrast of the slab, angles are expressed in radians and other parameters are defined as in Fig. 6.21 (Talwani *et al.* 1959). To calculate the anomaly of a two-dimensional body of irregular cross-section, the body is approximated by a polygon as shown in Fig. 6.22. The anomaly of the polygon is then found by proceeding around it summing the anomalies of the slabs bounded by edges where the depth increases and subtracting those where the depth decreases.

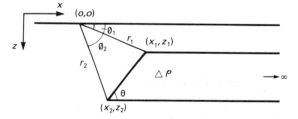

Fig. 6.21 Parameters used in defining the gravity anomaly of a semi-infinite slab with a sloping edge.

Fig. 6.23 illustrates a two-dimensional interpretation, in terms of a model of irregular geometry represented by a polygonal outline, of the Bodmin Moor granite of southwest England. The shape of the uppermost part of the model is controlled by the surface outcrop of granite, while the density contrasts employed are based on density measurements on rock samples. The interpretation shows unambiguously that the contacts of the granite slope outwards. Ambiguity is evident, however, in the interpretation of the gravity gradient over the northern flank of the granite. The model presented in Fig. 6.23 interprets the cause of this gradient as a northerly increase in the density of the granite; a

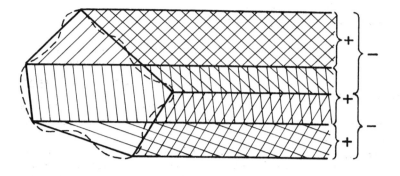

Fig. 6.22 The computation of gravity anomalies of two-dimensional bodies of irregular cross-section. The body is approximated by a polygon and the effects of semi-infinite slabs with sloping edges defined by the sides of the polygon are progressively added and subtracted until the anomaly of the polygon is obtained.

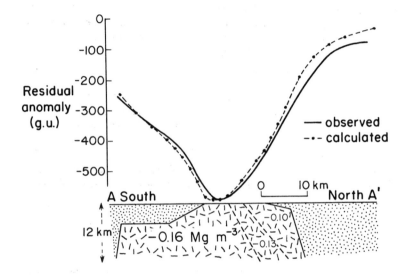

Fig. 6.23 A two-dimensional interpretation of the gravity anomaly of the Bodmin Moor granite, southwest England. See Fig. 6.27 for location. (After Bott & Scott 1964.)

possible alternative, however, would be a northerly thinning of a granite body of constant density contrast.

Two-dimensional methods can sometimes be extended to three-dimensional bodies by applying end-correction factors to account for the restricted extent of the causative body in the strike-direction (Cady 1980). The end-correction factors are, however, only approximations and full three-dimensional modelling is preferable.

The gravity anomaly of a three-dimensional body may be calculated by dividing the body into a series of horizontal slices and approximating each slice by a polygon (Talwani & Ewing 1960). Alternatively the body may be constructed out of a suite of right rectangular prisms.

However a model calculation is performed, indirect interpretation involves four steps:
1 Construction of a reasonable model.

2 Computation of its gravity anomaly.
3 Comparison of computed with observed anomaly.
4 Alteration of model to improve correspondence of observed and calculated anomalies and return to step **2**.

The process is thus iterative and the goodness of fit between observed and calculated anomalies is gradually improved. Step **4** can be performed manually for bodies of relatively simple geometry so that an interpretation is readily accomplished using interactive routines on a personal computer (Götze & Lahmeyer 1988). Bodies of complex geometry in two- or three-dimensions are not so simply dealt with and in such cases it is advantageous to employ techniques which perform the iteration automatically.

The most flexible of such methods is *non-linear optimization* (Al-Chalabi 1972). All variables (body points, density contrasts, regional field) may be allowed to vary within defined limits. The method then attempts to minimize some function F which defines the goodness of fit, e.g.

$$F = \sum_{i=1}^{n} (\Delta g_{\text{obs}_i} - \Delta g_{\text{calc}_i})^2$$

where Δg_{obs} and Δg_{calc} are a series of n observed and calculated values.

The minimization proceeds by altering the values of the variables within their stated limits to produce a successively smaller value for F for each iteration. The technique is elegant and successful but expensive in computer time.

Other such automatic techniques involve the simulation of the observed profile by a thin layer of variable density. This *equivalent layer* is then progressively expanded so that the whole body is of a uniform, specified density contrast. The body then has the form of a series of vertical prisms in either two or three dimensions which extend either above, below or symmetrically around the original equivalent layer. Such methods are less flexible than the non-linear optimization technique in that usually only a single density contrast may be specified and the model produced must either have a specified base or top or be symmetrical about a central horizontal plane.

6.11 ELEMENTARY POTENTIAL THEORY AND POTENTIAL FIELD MANIPULATION

Gravitational and magnetic fields are both potential fields. In general the potential at any point is defined as the work necessary to move a unit mass or pole from an infinite distance to that point through the ambient field. Potential fields obey Laplace's equation which states that the sum of the rates of change of the field gradient in three orthogonal directions is zero. In a normal Cartesian coordinate system with horizontal axes x, y and a vertical axis z, Laplace's equation is stated

$$\frac{\partial^2 A}{\partial x^2} + \frac{\partial^2 A}{\partial y^2} + \frac{\partial^2 A}{\partial z^2} = 0 \tag{6.23}$$

where A refers to a gravitational or magnetic field and is a function of (x, y, z).

In the case of a two-dimensional field there is no variation along one of the horizontal directions so that A is a function of x and z only and equation (6.23) simplifies to

$$\frac{\partial^2 A}{\partial x^2} + \frac{\partial^2 A}{\partial z^2} = 0 \tag{6.24}$$

Solution of this partial differential equation is easily performed by separation of variables

$$A_k(x, z) = (a \cos kx + b \sin kx)e^{kz} \tag{6.25}$$

where a and b are constants, the positive variable k is the spatial frequency or wavenumber, A_k is the potential field amplitude corresponding to that wavenumber and z is the level of observation. Equation (6.25) shows that a potential field can be represented in terms of sine and cosine waves whose amplitude is controlled exponentially by the level of observation.

Consider the simplest possible case where the two-dimensional anomaly measured at the surface $A(x, 0)$ is a sine wave

$$A(x, 0) = A_0 \sin kx \tag{6.26}$$

where A_0 is a constant and k the wavenumber of the sine wave. Equation (6.25) enables the general form of the equation to be stated for any value of z

$$A(x, z) = (A_0 \sin kx)e^{kz} \tag{6.27}$$

The field at a height h above the surface can then be determined by substitution in equation (6.27)

$$A(x, -h) = (A_0 \sin kx)e^{-kh} \tag{6.28}$$

and the field at depth d below the surface

$$A(x, d) = (A_0 \sin kx)e^{kd} \tag{6.29}$$

The sign of h and d is important as the z-axis is normally defined as positive downwards.

Equation (6.27) is an over-simplification in that a potential field is never a function of a single sine wave. Invariably such a field is composed of a range of wavenumbers. However, the technique is still valid as long as the field can be expressed in terms of all its component wavenumbers, a task easily performed by use of the Fourier transform (Section 2.3). If, then, instead of the terms $(a \cos kx + b \sin kx)$ in equation (6.24) or $(A_0 \sin kx)$ in equation (6.27), the full Fourier spectrum, derived by Fourier transformation of the field into the wavenumber domain, is substituted, the results of equations (6.28) and (6.29) remain valid.

These latter equations show that the field measured at the surface can be used to predict the field at any level above or below the plane of observation. This is the basis of the upward and downward field continuation methods in which the potential field above or below the original plane of measurement is calculated in order to accentuate the effects of deep or shallow structures respectively.

Upward continuation methods are employed in gravity interpretation to determine the form of regional gravity variation over a survey area, since the regional field is assumed to originate from relatively deep-seated structures. Fig. 6.24(a) is a Bouguer anomaly map of the Saguenay area in Quebec, Canada, and Fig. 6.24(b) represents the field continued upward to an elevation of 16 km. Comparison of the two figures clearly illustrates how the high wavenumber components of the observed field have been effectively removed by the continuation process. The upward continued field must result from relatively deep structures and consequently represents a valid regional field for the area. Upward continuation is also useful in the interpretation of magnetic anomaly fields (see Chapter 7) over areas containing many near-surface magnetic sources such as dykes and other intrusions. Upward continuation attenuates the high wavenumber anomalies associated with such features and enhances, relatively, the anomalies of the deeper-seated sources.

Downward continuation of potential fields is of more restricted application. The technique may be used in the resolution of the separate anomalies caused by adjacent structures whose effects overlap at the level of observation. On downward continuation, high wavenumber components are relatively enhanced and the anomalies show extreme fluctuations if the field is continued to a depth greater than that of its causative structure. The level at which these fluctuations commence provides an estimate of the limiting depth of the anomalous body. The effectiveness of this method is diminished if the potential field is contaminated with noise, as the noise is accentuated on downward continuation.

The selective enhancement of the low or high wavenumber components of potential fields may be achieved in a different but analogous manner by the application of *wavenumber filters*. Gravitational and magnetic fields may be processed and analysed in a similar fashion to seismic data, replacing frequency by wavenumber. Such processing is more complex than the equivalent seismic filtering as potential field data are generally arranged in two horizontal dimensions, i.e. contour maps, rather than a single dimension. However, it is possible to devise two-dimensional filters for the selective removal of high or low wavenumber components from the observed anomalies. The consequence of the application of such techniques is similar to upward or downward

continuation in that shallow structures are mainly responsible for the high wavenumber components of anomalies and deep structures for the low wavenumbers. However, it is not possible fully to isolate local or regional anomalies by wavenumber filtering because the wavenumber spectra of deep and shallow sources overlap.

Other manipulations of potential fields may be accomplished by the use of more complex filter operators (e.g. Gunn 1975). Vertical or horizontal derivatives of any order may be computed from the observed field. Such computations are not widely employed, but second horizontal derivative maps are occasionally used for interpretation as they accentuate anomalies associated with shallow bodies.

6.12 APPLICATIONS OF GRAVITY SURVEYING

Gravity studies are used extensively in the investigation of large and medium scale geological structures (Paterson & Reeves 1985). Early marine surveys, performed from submarines, indicated the existence of large positive and negative gravity anomalies associated with island arcs and oceanic trenches, respectively; subsequent shipborne work has demonstrated their lateral continuity and has shown that most of the major features of the Earth's surface can be delineated by gravity surveying. Gravity anomalies have also shown that most of these major relief features are in isostatic equilibrium, suggesting that the lithosphere is not capable of sustaining significant loads and yields isostatically to any change in surface loading. Fig. 6.25 shows the near-zero free-air anomalies over an ocean ridge which suggest that it is in isostatic equilibrium. The gravity interpretation, which is constrained by seismic refraction results, indicates that this compensation takes the form of a zone of mass deficiency in the underlying mantle. Its low seismic velocity and the high heat flow at the surface suggest that this is a region of partial melting and, perhaps, hydration. Gravity surveying can also be used in the study of ancient suture zones, which are interpreted as the sites of former plate boundaries within the continental lithosphere. These zones are often characterized by major linear gravity anomalies resulting from the different crustal sections juxtaposed across the sutures (Fig. 6.26).

On the medium scale, gravity anomalies can reveal the subsurface form of igneous intrusions such as granite batholiths and anorthosite massifs. For

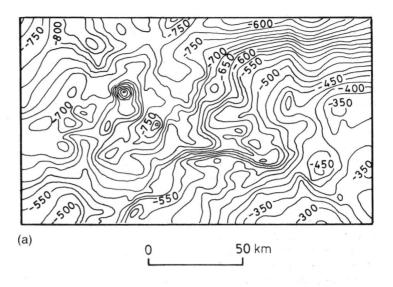

(a)

0   50 km

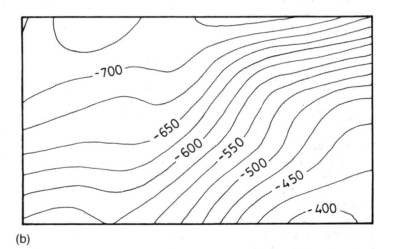

Fig. 6.24 (a) Observed Bouguer anomalies (gu) over the Saguenay area, Quebec, Canada. (b) The gravity field continued upward to an elevation of 16 km. (After Duncan & Garland 1977.)

(b)

example, gravity surveys in southwest England (Bott *et al.* 1958) have revealed a belt of large amplitude, negative Bouguer anomalies overlying a region of outcropping Variscan granites (Fig. 6.27). Modelling of the gravity anomalies (Fig. 6.23) has lead to the postulation of a continuous batholith chain some 10–15 km thick underlying southwest England (see, e.g. Brooks *et al.* 1983). Studies such as these have provided important constraints on the mechanism of emplacement, composition and origin of igneous bodies. Similarly, gravity surveying has been extensively used in the location of sedimentary basins, and their interpreted structures have provided important information on mechanisms of basin formation.

The gravity method was once extensively used by the petroleum industry for the location of possible hydrocarbon traps, but the subsequent vast improvement in efficiency and technology of seismic surveying has lead to the demise of gravity surveying as a primary exploration tool.

In commercial applications, gravity surveying is not used in reconnaissance exploration. This is because the method is relatively slow to execute, and therefore expensive, due to the necessity of accurately determined elevations and the length of the reduction procedure. Gravity methods do find application, however, as a follow-up technique used on a target defined by another, more cost-effective method. An important application of this type in

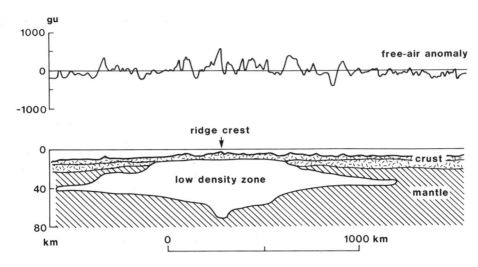

Fig. 6.25 Free-air anomaly profile across the mid-Atlantic ridge. (After Talwani *et al.* 1965.)

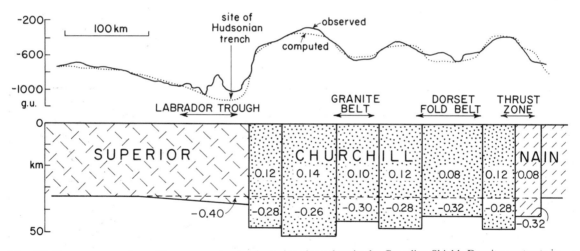

Fig. 6.26 Bouguer anomaly profile across a structural province boundary in the Canadian Shield. Density contrasts in $Mg\,m^{-3}$. (After Thomas & Kearey 1980.)

mineral exploration is the determination of ore tonnage by the excess mass method described in Section 6.10.3.

Gravity surveying may be used in hydrogeological investigations to determine the geometry of potential aquifers. Fig. 6.28 shows a Bouguer anomaly map of an area near Taltal, Chile (Van Overmeeren 1975). The region is extremely arid, with groundwater supply and storage controlled by deep geological features. The gravity minima revealed by the contours probably represent two buried valleys in

the alluvium overlying the granodioritic bedrock. Fig. 6.29 shows an interpretation of a profile over the minima. The bedrock topography was controlled by the results from a seismic refraction line which had been interpreted using the plus-minus method (Section 5.4). The seismic control allowed a mean density of the highly variable valley-fill deposits to be determined. On the basis of the geophysical results, two boreholes (Fig. 6.28) were sunk in the deepest parts of the valley fill and located groundwater ponded in the bedrock depressions.

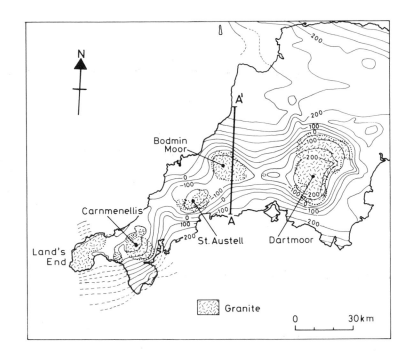

Fig. 6.27 Bouguer anomaly map of southwest England, showing a linear belt of large negative anomalies associated with the zone of Variscan granite outcrops. Contour interval 50 gu. (After Bott & Scott 1964.)

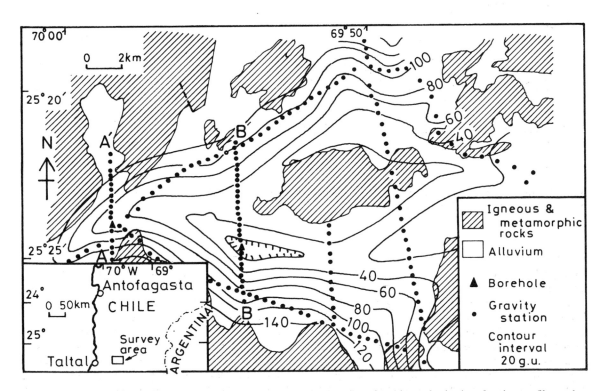

Fig. 6.28 Geological map of an area near Taltal, Chile, showing location of gravity and seismic refraction profiles and contoured Bouguer anomalies. (After Van Overmeeren 1975.)

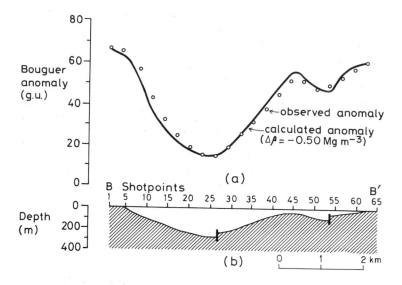

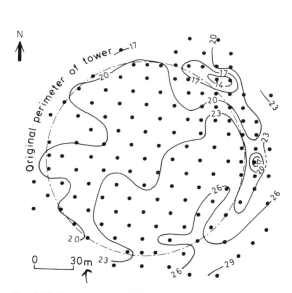

Fig. 6.29 Profile B–B′, Taltal area, Chile (see Fig. 6.28 for location). (a) Observed Bouguer anomaly and calculated anomaly for a model with a density contrast ($\Delta\rho$) of $-0.50\,\mathrm{Mg\,m^{-3}}$. (b) Gravity interpretation. (After Van Overmeeren 1975.)

In engineering and geotechnical applications, gravity surveying is sometimes used in the location of subsurface voids. Void detection has been made possible by the development of microgravimetric techniques which can detect gravity changes as small as a microgal. Arzi (1975) described a microgravity survey of the proposed site of a cooling tower serving a nuclear power plant, where it was suspected that solution cavities might be present in the dolomitic bedrock. Measurements were made on a 15 m grid at points whose elevations had been determined to ±3 mm, with base readings at 40 minute intervals. The soil thickness had been determined so that its effects could be computed and 'stripped' from the observations to remove gravity variations caused by undulating bedrock topography. The resulting Bouguer anomaly map is shown in Fig. 6.30. In the NE part of the site there are two minima near the proposed perimeter of the cooling tower, and subsequent drilling confirmed that they originated from buried cavities. Remedial work entailed the injection of grouting material into the cavities. A check on the effectiveness of the grouting was provided by a repeat gravity survey which, by an excess mass calculation (Section 6.10.3), showed that the change in the gravity field before and after grouting was caused by the replacement of voids by grouting material.

Microgravity surveys also find application in archaeological investigations, where they may be used in the detection of buried buildings, tombs and other artifacts. The technique has also been used

Fig. 6.30 Bouguer anomalies, uncorrected for topographic effects, over the cooling tower area. Contour interval 0.3 gu. (After Arzi 1975.)

to study the temporal movement of groundwater through a region.

An important recent development in gravity surveying is the design of portable instruments capable of measuring absolute gravity with high precision. Although the cost of such instruments is high it is possible that they will be used in the future to investigate large-scale mass movements in the Earth's interior and small cyclic gravity variations associated with neotectonic phenomena such as earthquakes and postglacial uplift.

Gravitational studies, both of the type described in this chapter and satellite observations, are important in geodesy, the study of the shape of the Earth. Gravity surveying also has military significance, since the trajectory of a missile is affected by gravity variation along its flight path.

6.13 PROBLEMS

1 Compare and contrast the LaCoste-Romberg and Worden-type gravimeters. State also the advantages and disadvantages of the two types of instrument.

2 What are the magnitudes of the terrain correction at gravity stations (a) at the top, (b) at the base, and (c) half-way up a vertical cliff 100 m high?

3 The table below shows data collected along a north—south gravity profile. Distances are measured from the south end of the profile, whose latitude is 51°12′24″N. The calibration constant of the Worden gravimeter used on the survey is 3.792 gu per dial unit. Before, during and after the survey, readings (marked *BS*) were taken at a base station where the value of gravity is 9 811 442.2 gu. This was done in order to monitor instrumental drift and to allow the absolute value of gravity to be determined at each observation point.

Station	Time	Dist. (m)	Elev. (m)	Reading
BS	0805			2934.2
1	0835	0	84.26	2946.3
2	0844	20	86.85	2941.0
3	0855	40	89.43	2935.7
4	0903	60	93.08	2930.4
1	0918			2946.5
BŠ	0940			2934.7
1	1009			2946.3
5	1024	80	100.37	2926.6
6	1033	100	100.91	2927.9
7	1044	120	103.22	2920.0
8	1053	140	107.35	2915.1
1	1111			2946.5
BS	1145			2935.2
1	1214			2946.2
9	1232	160	110.10	2911.5
10	1242	180	114.89	2907.2
11	1300	200	118.96	2904.0
1	1315			2946.3
BS	1350			2935.5

(a) Perform a gravity reduction of the survey data and comment on the accuracy of each step. Use a density of 2.70 Mg m^{-3} for the Bouguer correction.

(b) Draw a series of sections illustrating the variation in topography, observed gravity, free-air anomaly and Bouguer anomaly along the profile. Comment on the sections.

(c) What further information would be required before a full interpretation could be made of the Bouguer anomaly?

4 Two survey vessels with shipborne gravity meters are steaming at 6 knots in opposite directions along an east—west course. If the difference in gravity read by the two meters is 635 gu as the ships pass, what is the latitude?

5 The gravity anomaly Δg of an infinite horizontal slab of thickness t and density contrast $\Delta \rho$ is given by

$$\Delta g = 2\pi G \Delta \rho t$$

where the gravitational constant G is 6.67×10^{-11} m^3 kg^{-1} s^{-2}.

(a) Scale this equation to provide Δg in gu when $\Delta \rho$ is expressed in Mg m^{-3} and t in m.

(b) This equation is used to provide a preliminary estimate of the gravity anomaly of a body of specified thickness. Using this equation, calculate the gravity anomaly of (a) a granite 12 km thick of density 2.67 Mg m^{-3}; and (b) a sandstone body 4 km thick of density 2.30 Mg m^{-3}, where the density of the surrounding metamorphic rocks is 2.80 Mg m^{-3}. Are the anomalies so calculated liable to be over- or under-estimates?

6 Show that the half-width of the gravity anomaly caused by a horizontal cylinder is equal to the depth of the axis of the cylinder.

7 Figure 6.31 is a Bouguer anomaly map, contoured at an interval of 50 gu, of a drift covered area.

(a) On the map, sketch in what you consider to be the regional field and then remove it from the observed field to isolate residual anomalies, which can be represented on the map as contours drawn in a different colour.

(b) Construct gravity profiles along line $A-A'$ illustrating the observed, regional and residual anomalies.

(c) Perform a direct interpretation of the residual anomaly, obtaining as much information as possible on the depth, thickness and shape of the source.

(d) The bedrock constitutes part of a Precambrian shield. Speculate on the nature of the anomalous body, giving reasons for your ideas.

8 Contour the gravity data on the map shown in Fig. 6.32 using an interval of 10 gu. Draw a representative profile.

(a) Use limiting-depth calculations based on the half-width and gradient-amplitude methods to determine the depth to the centre of mass of the anomalous body. Comment on any difference between the depth estimates provided by the two methods.

(b) Determine the mass deficiency present using the formula for the gravity anomaly of a point mass. If the anomaly is caused by a salt dome of density 2.22 Mg m^{-3}

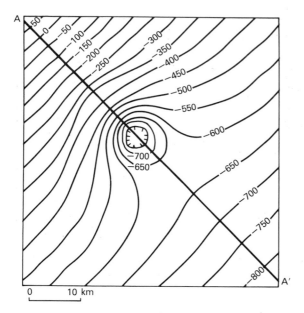

Fig. 6.31 Bouguer anomaly map pertaining to Question 7. Contour interval 50 gu.

within sediments of density $2.60\,\mathrm{Mg\,m^{-3}}$, calculate the volume and mass of salt present and the depth to the top of the salt dome. Compute the actual gravity anomaly of the salt and comment on any differences with the observed anomaly.

(c) What is the lowest possible density contrast of the anomalous body?

(d) Determine the mass deficiency present using a method based on Gauss's Theorem. Comment on the accuracy of the value obtained and compare it with the answer to (b). Calculate the actual mass present assuming the same densities as in (a).

9 The map in Fig. 6.33 shows Bouguer anomalies over a gabbro intrusion in a schist terrain. In the eastern part of the map, horizontally bedded Mesozoic sediments unconformably overlie the schists. A seismic refraction line has been established over the sediments in the location shown. Time–distance data and typical velocities and densities are given below.

Interpret the geophysical results using the following scheme:

(a) Use the refraction data to determine the thickness and possible nature of the Mesozoic rocks beneath the seismic line.

(b) Use this interpretation to calculate the gravity anomaly of the Mesozoic rocks at this location. Correct the observed gravity anomaly for the effect of the Mesozoic rocks.

(c) Determine the maximum gravity anomaly of the gabbro. Assuming the gabbro to have the form of a vertical cylinder, determine the depth to its base.

The gravity anomaly Δg of a vertical cylinder of density contrast $\Delta\rho$, radius r, length L, depth to top z_U and depth to base z_L is given by

$$\Delta g = 2\pi G\Delta\rho(L - (z_L^2 + r^2)^{1/2} + (z_U^2 + r^2)^{1/2})$$

where G is the gravitational constant.

State any assumptions and possible causes of error in your interpretation.

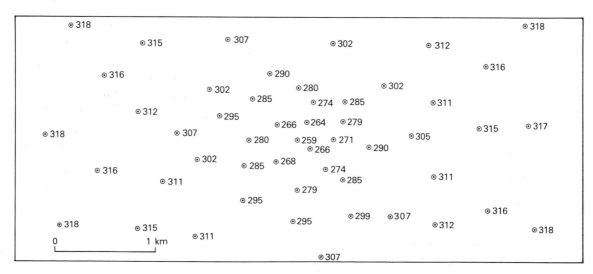

Fig. 6.32 Bouguer anomaly observations pertaining to Question 8. Values in gu.

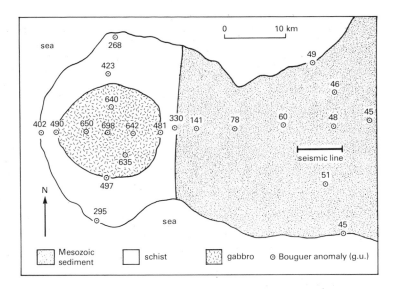

Fig. 6.33 Map of geophysical observations pertaining to Question 9. Bouguer anomaly values in gu.

Typical densities and seismic velocities

	ρ (Mg m^{-3})	Veloc. (km s^{-1})
Jur./Cret.	2.15	1.20−1.80
Trias	2.35	2.40−3.00
Schist	2.75	3.60−4.90
Gabbro	2.95	

Jur. = Jurassic; Cret. = Cretaceous.

Seismic data

Dist. (m)	Time (s)
530	0.349
600	0.391
670	0.441
1130	0.739
1200	0.787
1270	0.831
1800	1.160
1870	1.177
1940	1.192
2730	1.377
2800	1.393
2870	1.409
3530	1.563
3600	1.582
3670	1.599

FURTHER READING

Baranov, W. (1975) *Potential Fields and Their Transformations in Applied Geophysics.* Gebrüder Borntraeger, Berlin.

Bott, M.H.P. (1973) Inverse methods in the interpretation of magnetic and gravity anomalies. *In*: Alder, B., Fernbach, S. & Bolt, B.A. (eds.), *Methods in Computational Physics*, **13**, 133−62.

Dehlinger, P. (1978) *Marine Gravity.* Elsevier, Amsterdam.

LaCoste, L.J.B. (1967) Measurement of gravity at sea and in the air. *Rev. Geophys.*, **5**, 477−526.

LaCoste, L.J.B., Ford, J., Bowles, R. & Archer, K. (1982) Gravity measurements in an airplane using state-of-the-art navigation and altimetry. *Geophysics*, **47**, 832−7.

Milsom, J. (1989) *Field Geophysics.* Open University Press, Milton Keynes.

Nettleton, L.L. (1971) *Elementary Gravity and Magnetics for Geologists and Seismologists.* Monograph Series No. 1. Society of Exploration Geophysicists, Tulsa.

Nettleton, L.L. (1976) *Gravity and Magnetics in Oil Exploration.* McGraw-Hill, New York.

Ramsey, A.S. (1964) *An Introduction to the Theory of Newtonian Attraction.* Cambridge University Press, Cambridge.

Torge, W. (1989) *Gravimetry.* Walter de Gruyter, Berlin.

Tsuboi, C. (1983) *Gravity.* Allen & Unwin, London.

10 Over a typical ocean spreading centre, the free-air gravity anomaly is approximately zero and the Bouguer anomaly is large and negative. Why?

7 / Magnetic surveying

7.1 INTRODUCTION

The aim of a magnetic survey is to investigate subsurface geology on the basis of anomalies in the Earth's magnetic field resulting from the magnetic properties of the underlying rocks. Although most rock-forming minerals are effectively non-magnetic, certain rock types contain sufficient magnetic minerals to produce significant magnetic anomalies. Similarly, man-made ferrous objects also generate magnetic anomalies. Magnetic surveying thus has a broad range of applications, from small-scale engineering or archaeological surveys to detect buried metallic objects, to large-scale surveys carried out to investigate regional geological structure.

Magnetic surveys can be performed on land, at sea and in the air. Consequently the technique is widely employed, and the speed of operation of airborne surveys makes the method very attractive in the search for types of ore deposit that contain magnetic minerals.

7.2 BASIC CONCEPTS

Within the vicinity of a bar magnet a magnetic flux is developed which flows from one end of the magnet to the other (Fig. 7.1). This flux can be mapped from the directions assumed by a small compass needle suspended within it. The points within the magnet where the flux converges are known as the *poles* of the magnet. A freely-suspended bar magnet similarly aligns in the flux of the Earth's magnetic field. The pole of the magnet which tends to point in the direction of the Earth's north pole is called the north-seeking or positive pole, and this is balanced by a south-seeking or negative pole of identical strength at the opposite end of the magnet.

The force F between two magnetic poles of strengths m_1 and m_2 separated by a distance r is given by

$$F = \frac{\mu_0 m_1 m_2}{4\pi\mu_R r^2} \qquad (7.1)$$

where μ_0 and μ_R are constants corresponding to the *magnetic permeability of vacuum* and the *relative magnetic permeability* of the medium separating the poles (see later). The force is attractive if the poles are of different sign and repulsive if they are of like sign.

The *magnetic field B* due to a pole of strength m at a distance r from the pole is defined as the force exerted on a unit positive pole at that point

$$B = \frac{\mu_0 m}{4\pi\mu_R r^2} \qquad (7.2)$$

Magnetic fields can be defined in terms of *magnetic potentials* in a similar manner to gravitational fields. For a single pole of strength m, the magnetic potential V at a distance r from the pole is given by

$$V = \frac{\mu_0 m}{4\pi\mu_R r} \qquad (7.3)$$

The magnetic field component in any direction is then given by the partial derivative of the potential in that direction.

In the SI system of units, magnetic parameters are defined in terms of the flow of electrical current (see, for example, Reilly 1972). If a current is passed through a coil consisting of several turns of wire, a *magnetic flux* flows through and around the coil annulus which arises from a *magnetizing force H*. The magnitude of H is proportional to the number of turns in the coil and the strength of the current, and inversely proportional to the length of the wire, so that H is expressed in $A\,m^{-1}$. The density of the magnetic flux, measured over an area perpendicular to the direction of flow, is known as the *magnetic induction* or *magnetic field B* of the coil. B is proportional to H and the constant of proportionality μ is known as the *magnetic permeability*. Lenz's law of induction relates the rate of change of magnetic flux in a circuit to the voltage developed within it, so that B is expressed in $volt\,s\,m^{-2}$ (Weber (Wb) m^{-2}). The unit of the $Wb\,m^{-2}$ is designated the *tesla* (T). Permeability is consequently expressed in $Wb\,A^{-1}\,m^{-1}$ or Henry (H) m^{-1}. The c.g.s. unit of magnetic field strength is the *Gauss* (G), numerically equivalent to 10^{-4}T.

The tesla is too large a unit in which to express

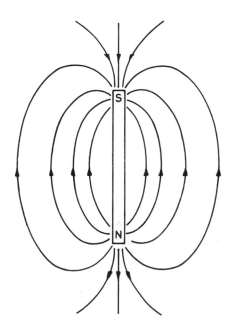

Fig. 7.1 The magnetic flux surrounding a bar magnet.

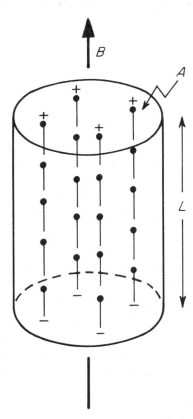

Fig. 7.2 Schematic representation of an element of material in which elementary dipoles align in the direction of an external field B to produce an overall induced magnetization.

the small magnetic anomalies caused by rocks, and a subunit, the *nanotesla* (nT), is employed ($1\,\text{nT} = 10^{-9}\,\text{T}$). The c.g.s. system employs the numerically equivalent *gamma* (γ), equal to $10^{-5}\,\text{G}$.

Common magnets exhibit a pair of poles and are therefore referred to as dipoles. The *magnetic moment M* of a dipole with poles of strength m a distance l apart is given by

$$M = ml \tag{7.4}$$

The magnetic moment of a current-carrying coil is proportional to the number of turns in the coil, its cross-sectional area and the magnitude of the current so that magnetic moment is expressed in $A\,m^2$.

When a material is placed in a magnetic field it may acquire a magnetization in the direction of the field which is lost when the material is removed from the field. This phenomenon is referred to as *induced magnetization* or *magnetic polarization*, and results from the alignment of elementary dipoles (see below) within the material in the direction of the field. As a result of this alignment the material has magnetic poles distributed over its surface which correspond to the ends of the dipoles (Fig. 7.2). The intensity of induced magnetization J_i of a material is defined as the dipole moment per unit volume of material:

$$J_i = \frac{M}{LA} \tag{7.5}$$

where M is the magnetic moment of a sample of length L and cross-sectional area A. J_i is consequently expressed in $A\,m^{-1}$. In the c.g.s. system intensity of magnetization is expressed in $\text{emu}\,\text{cm}^{-3}$ (emu = electromagnetic unit), where $1\,\text{emu}\,\text{cm}^{-3} = 1000\,A\,m^{-1}$.

The induced intensity of magnetization is proportional to the strength of the magnetizing force H of the inducing field:

$$J_i = kH \tag{7.6}$$

where k is the *magnetic susceptibility* of the material. Since J_i and H are both measured in $A\,m^{-1}$, susceptibility is dimensionless in the SI system. In the c.g.s. system susceptibility is similarly dimensionless, but a consequence of rationalizing the SI system is that SI susceptibility values are a factor 4π greater than corresponding c.g.s. values.

In a vacuum the magnetic field strength B and magnetizing force H are related by $B = \mu_0 H$ where μ_0 is the permeability of vacuum ($4\pi \times 10^{-7}$ H m^{-1}). Air and water have very similar permeabilities to μ_0 and so this relationship can be taken to represent the Earth's magnetic field when it is undisturbed by magnetic materials. When a magnetic material is placed in this field, the resulting magnetization gives rise to an additional magnetic field in the region occupied by the material, whose strength is given by $\mu_0 J_i$. Within the body the total magnetic field, or magnetic induction, B is given by

$$B = \mu_0 H + \mu_0 J_i$$

Substituting equation (7.6)

$$B = \mu_0 H + \mu_0 k H = (1 + k)\mu_0 H = \mu_R \mu_0 H$$

where μ_R is a dimensionless constant known as the *relative magnetic permeability*. The magnetic permeability μ is thus equal to the product of the relative permeability and the permeability of vacuum, and has the same dimensions as μ_0. μ_R for air and water is thus close to unity.

All substances are magnetic at an atomic scale. Each atom acts as a dipole due to both the spin of its electrons and the orbital path of the electrons around the nucleus. Quantum theory allows two electrons to exist in the same state (or electron shell) provided that their spins are in opposite directions. Two such electrons are called paired electrons and their spin magnetic moments cancel. In *diamagnetic* materials all electron shells are full and no unpaired electrons exist. When placed in a magnetic field the orbital paths of the electrons rotate so as to produce a magnetic field in opposition to the applied field. Consequently the susceptibility of diamagnetic substances is weak and negative. In *paramagnetic* substances the electron shells are incomplete so that a magnetic field results from the spin of their unpaired electrons. When placed in an external magnetic field the dipoles corresponding to the unpaired electron spins rotate to produce a field in the same sense as the applied field so that the susceptibility is positive. This is still, however, a relatively weak effect.

In small grains of certain paramagnetic substances whose atoms contain several unpaired electrons, the dipoles associated with the spins of the unpaired electrons are magnetically coupled between adjacent atoms. The grain is then said to constitute a single *magnetic domain*. Depending on the degree of overlap of the electron orbits, this coupling may be either parallel or antiparallel. In *ferromagnetic* materials the dipoles are parallel (Fig. 7.3), giving rise to a very strong spontaneous magnetization which can exist even in the absence of an external magnetic field, and a very high susceptibility. Ferromagnetic substances include iron, cobalt and nickel, and rarely occur naturally in the Earth's crust. In *antiferromagnetic* materials such as haematite, the dipole coupling is antiparallel with equal numbers of dipoles in each direction. The magnetic fields of the dipoles are self-cancelling so that there is no external magnetic effect. However, defects in the crystal lattice structure of an antiferromagnetic material may give rise to a small net magnetization, called *parasitic antiferromagnetism*. In *ferrimagnetic* materials such as magnetite, the dipole coupling is similarly antiparallel, but the numbers of dipoles in each direction are unequal. Consequently ferrimagnetic materials can exhibit a strong spontaneous magnetization and a high susceptibility. Virtually all the minerals responsible for the magnetic properties of common rock types (Section 7.3) fall into this category.

The strength of the magnetization of ferromagnetic and ferrimagnetic substances decreases with temperature and disappears at the *Curie temperature*. Above this temperature interatomic

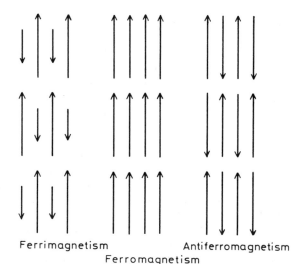

Ferrimagnetism   Antiferromagnetism
Ferromagnetism

Fig. 7.3 Schematic representation of the strength and orientation of elementary dipoles within ferrimagnetic, ferromagnetic and antiferromagnetic domains.

distances are increased to separations which preclude electron coupling, and the material behaves as an ordinary paramagnetic substance.

In larger grains, the total magnetic energy is decreased if the magnetization of each grain subdivides into individual volume elements (magnetic domains) with diameters of the order of a micrometre, within which there is parallel coupling of dipoles. In the absence of any external magnetic field the domains become oriented in such a way as to reduce the magnetic forces between adjacent domains. The boundary between two domains, the *Bloch wall*, is a narrow zone in which the dipoles cant over from one domain direction to the other.

When a multidomain grain is placed in a weak external magnetic field, the Bloch wall unrolls and causes a growth of those domains magnetized in the direction of the field at the expense of domains magnetized in other directions. This induced magnetization is lost when the applied field is removed as the domain walls rotate back to their original configuration. When stronger fields are applied domain walls unroll irreversibly across small imperfections in the grain so that those domains magnetized in the direction of the field are permanently enlarged. The inherited magnetization remaining after removal of the applied field is known as *remanent*, or *permanent*, *magnetization* J_r. The application of even stronger magnetic fields causes all possible domain wall movements to occur and the material is then said to be *magnetically saturated*.

Primary remanent magnetization may be acquired either as an igneous rock solidifies and cools through the Curie temperature of its magnetic minerals (thermoremanent magnetization, TRM) or as the magnetic particles of a sediment align within the Earth's field during sedimentation (detrital remanent magnetization, DRM). Secondary remanent magnetizations may be impressed later in the history of a rock as magnetic minerals recrystallize or grow during diagenesis or metamorphism (chemical remanent magnetization, CRM). Remanent magnetization may develop slowly in a rock standing in an ambient magnetic field as the domain magnetizations relax into the direction of the field (viscous remanent magnetization, VRM).

Any rock containing magnetic minerals may possess both induced and remanent magnetizations J_i and J_r. The relative intensities of induced and remanent magnetizations are commonly expressed in terms of the *Königsberger ratio*, $J_r:J_i$. These may be in different directions and may differ significantly in magnitude. The magnetic effects of such a rock arise from the resultant J of the two magnetization vectors (Fig. 7.4). The magnitude of J controls the amplitude of the magnetic anomaly and the orientation of J influences its shape.

7.3 ROCK MAGNETISM

Most common rock-forming minerals exhibit a very low magnetic susceptibility and rocks owe their magnetic character to the generally small proportion of magnetic minerals that they contain. There are only two geochemical groups which provide such minerals. The iron−titanium−oxygen group possesses a solid solution series of magnetic minerals from magnetite (Fe_3O_4) to ulvöspinel (Fe_2TiO_4). The other common iron oxide, haematite (Fe_2O_3) is antiferromagnetic and thus does not give rise to magnetic anomalies (see Section 7.12) unless a parasitic antiferromagnetism is developed. The iron−sulphur group provides the magnetic mineral pyrrhotite (FeS_{1+x}, $0 < x < 0.15$) whose magnetic susceptibility is dependent upon the actual composition.

By far the most common magnetic mineral is magnetite, which has a Curie temperature of 578 °C. Although the size, shape and dispersion of the magnetite grains within a rock affect its magnetic character, it is reasonable to classify the magnetic behaviour of rocks according to their overall magnetite content. A histogram illustrating the susceptibilities of common rock types is presented in Fig. 7.5.

Basic igneous rocks are usually highly magnetic due to their relatively high magnetite content. The proportion of magnetite in igneous rocks tends to

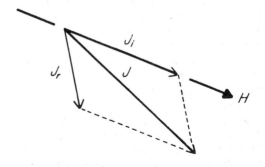

Fig. 7.4 Vector diagram illustrating the relationship between induced (J_i), remanent (J_r) and total (J) magnetization components.

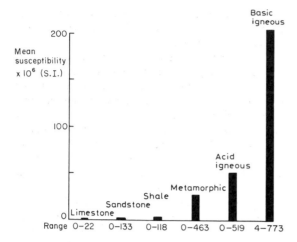

Fig. 7.5 Histogram showing mean values and ranges in susceptibility of common rock types. (After Dobrin 1976.)

decrease with increasing acidity so that acid igneous rocks, although variable in their magnetic behaviour, are usually less magnetic than basic rocks. Metamorphic rocks are also variable in their magnetic character. If the partial pressure of oxygen is relatively low, magnetite becomes resorbed and the iron and oxygen are incorporated into other mineral phases as the grade of metamorphism increases. Relatively high oxygen partial pressure can, however, result in the formation of magnetite as an accessory mineral in metamorphic reactions.

In general the magnetite content and, hence, the susceptibility of rocks is extremely variable and there can be considerable overlap between different lithologies. It is not usually possible to identify with certainty the causative lithology of any anomaly from magnetic information alone. However, sedimentary rocks are effectively non-magnetic unless they contain a significant amount of magnetite in the heavy mineral fraction. Where magnetic anomalies are observed over sediment-covered areas the anomalies are generally caused by an underlying igneous or metamorphic basement, or by intrusions into the sediments.

Common causes of magnetic anomalies include dykes, faulted, folded or truncated sills and lava flows, massive basic intrusions, metamorphic basement rocks and magnetite ore bodies. Magnetic anomalies range in amplitude from a few tens of nT over deep metamorphic basement to several hundred nT over basic intrusions and may reach an amplitude of several thousand nT over magnetite ores.

7.4 THE GEOMAGNETIC FIELD

Magnetic anomalies caused by rocks are localized effects superimposed on the normal magnetic field of the Earth (geomagnetic field). Consequently, knowledge of the behaviour of the geomagnetic field is necessary in both the reduction of magnetic data to a suitable datum and in the interpretation of the resulting anomalies. The geomagnetic field is geometrically more complex than the gravity field of the Earth and exhibits irregular variation in both orientation and magnitude with latitude, longitude and time.

At any point on the Earth's surface a freely suspended magnetic needle will assume a position in space in the direction of the ambient geomagnetic field. This will generally be at an angle to both the vertical and geographic north. In order to describe the magnetic field vector, use is made of descriptors known as the geomagnetic elements (Fig. 7.6). The *total field vector B* has a vertical component Z, and a horizontal component H in the direction of magnetic north. The dip of B is the *inclination I* of the field and the horizontal angle between geographic and magnetic north is the *declination D*. B varies in strength from about 25 000 nT in equatorial regions to about 70 000 nT at the poles.

In the northern hemisphere the magnetic field generally dips downward towards the north and becomes vertical at the north magnetic pole (Fig. 7.7). In the southern hemisphere the dip is generally upwards towards the north. The line of zero inclination approximates the geographic equator, and is known as the magnetic equator.

About 90% of the Earth's field can be represented by the field of a theoretical magnetic dipole at the centre of the Earth inclined at about 11.5° to the axis of rotation. The magnetic moment of this fictitious *geocentric dipole* can be calculated from the observed field. If this dipole field is subtracted from the observed magnetic field, the residual field can then be approximated by the effects of a second, smaller, dipole. The process can be continued by fitting dipoles of ever decreasing moment until the observed geomagnetic field is simulated to any required degree of accuracy. The effects of each fictitious dipole contribute to a function known as a harmonic and the technique of successive approximations of the observed field is known as spherical harmonic analysis—the equivalent of Fourier analysis in spherical polar coordinates. The method has been used to compute the formula of the International Geo-

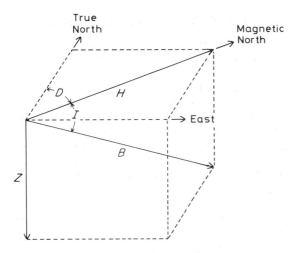

Fig. 7.6 The geomagnetic elements.

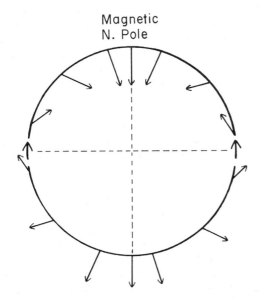

Fig. 7.7 The variation of the inclination of the total magnetic field with latitude based on a simple dipole approximation of the geomagnetic field. (After Sharma 1976.)

for latitude correction (see Section 6.8.2) as a large number of harmonics is employed (Barraclough & Malin 1971, Peddie 1983).

The geomagnetic field cannot in fact result from permanent magnetism in the Earth's deep interior. The required dipolar magnetic moments are far greater than is considered realistic and the prevailing high temperatures are far in excess of the Curie temperature of any known magnetic material. The cause of the geomagnetic field is attributed to a dynamo action produced by the circulation of charged particles in coupled convective cells within the outer, fluid, part of the Earth's core. The exchange of dominance between such cells is believed to produce the periodic changes in polarity of the geomagnetic field revealed by palaeomagnetic studies. The circulation patterns within the core are not fixed and change slowly with time. This is reflected in a slow, progressive, temporal change in all the geomagnetic elements and is known as *secular variation*. Such variation is predictable and a well-known example is the gradual rotation of the north magnetic pole around the geographic pole.

Magnetic effects of external origin cause the geomagnetic field to vary on a daily basis to produce *diurnal variations*. Under normal conditions (*Q* or quiet days) the diurnal variation is smooth and regular and has an amplitude of about 20–80 nT, being at a maximum in polar regions. Such variation results from the magnetic field induced by the flow of charged particles within the ionosphere towards the magnetic poles, as both the circulation patterns and diurnal variations vary in sympathy with the tidal effects of the Sun and Moon.

Some days (*D* or disturbed days) are distinguished by far less regular diurnal variations and involve large, short-term disturbances in the geomagnetic field, with amplitudes of up to 1000 nT, known as magnetic storms. Such days are usually associated with intense solar activity and result from the arrival in the ionosphere of charged solar particles. Magnetic surveying should be discontinued during such storms because of the impossibility of correcting the data collected for the rapid and high-amplitude changes in the magnetic field.

7.5 MAGNETIC ANOMALIES

All magnetic anomalies caused by rocks are superimposed on the geomagnetic field in the same way that gravity anomalies are superimposed on the Earth's gravitational field. The magnetic case is more

magnetic Reference Field (IGRF) which defines the theoretical undisturbed magnetic field at any point on the Earth's surface. In magnetic surveying, the IGRF is used to remove from the magnetic data those magnetic variations attributable to this theoretical field. The formula is considerably more complex than the equivalent Gravity Formula used complex, however, as the geomagnetic field varies not only in amplitude, but also in direction, whereas the gravitational field is everywhere, by definition, vertical.

Describing the normal geomagnetic field by a vector diagram (Fig. 7.8(a)), the geomagnetic elements are related

$$B^2 = H^2 + Z^2 \tag{7.7}$$

A magnetic anomaly is now superimposed on the Earth's field causing a change ΔB in the strength of the total field vector B. Let the anomaly produce a vertical component ΔZ and a horizontal component ΔH at an angle α to H (Fig. 7.8(b)). Only that part of ΔH in the direction of H, namely $\Delta H'$, will contribute to the anomaly

$$\Delta H' = \Delta H \cos \alpha \tag{7.8}$$

Using a similar vector diagram to include the magnetic anomaly (Fig. 7.8(c))

$$(B + \Delta B)^2 = (H + \Delta H')^2 + (Z + \Delta Z)^2$$

.If this equation is expanded, the equality of equation (7.7) substituted and the insignificant terms in Δ^2 ignored, the equation reduces to

$$\Delta B = \Delta Z(Z/B) + \Delta H'(H/B)$$

Substituting equation (7.8) and angular descriptions of geomagnetic element ratios

$$\Delta B = \Delta Z \sin I + \Delta H \cos I \cos \alpha \tag{7.9}$$

where I is the inclination of the geomagnetic field.

This approach can be used to calculate the magnetic anomaly caused by a small isolated magnetic pole of strength m, defined as the effect of this pole on a unit positive pole at the observation point. The pole is situated at depth z, a horizontal distance x and radial distance r from the observation point

(Fig. 7.9). The force of repulsion ΔB_r on the unit positive pole in the direction r is given by substitution in equation (7.1), with $\mu_R = 1$

$$\Delta B_r = \frac{Cm}{r^2}$$

where $C = \mu_0/4\pi$.

If it is assumed that the profile lies in the direction of magnetic north so that the horizontal component of the anomaly lies in this direction, the horizontal (ΔH) and vertical (ΔZ) components of this force can be computed by resolving in the relevant directions

$$\Delta H = \frac{Cm}{r^2} \cos \theta = \frac{Cmx}{r^3} \tag{7.10}$$

$$\Delta Z = \frac{-Cm}{r^2} \sin \theta = \frac{-Cmz}{r^3} \tag{7.11}$$

The vertical field anomaly is negative as, by convention, the z-axis is positive downwards. Plots of the form of these anomalies are shown in Fig. 7.9. The horizontal field anomaly is a positive/negative couplet and the vertical field anomaly is centred over the pole.

The total field anomaly ΔB is then obtained by substituting the expressions of equations (7.10) and (7.11) in equation (7.9), where $\alpha = 0$. If the profile were not in the direction of magnetic north, the angle α would represent the angle between magnetic north and the profile direction.

7.6 MAGNETIC SURVEYING INSTRUMENTS

Since the early 1900s a variety of surveying instruments has been designed that is capable of

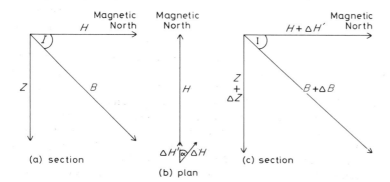

(a) section

(b) plan

(c) section

Fig. 7.8 Vector representation of the geomagnetic field with and without a superimposed magnetic anomaly.

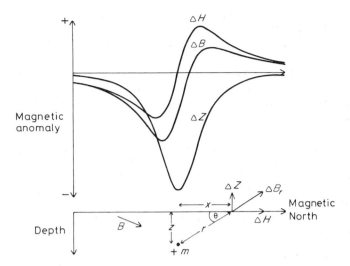

Fig. 7.9 The horizontal (ΔH), vertical (ΔZ) and total field (ΔB) anomalies due to an isolated positive pole.

measuring the geomagnetic elements Z, H and B. Most modern survey instruments, however, are designed to measure B only. The precision normally required is $\pm 1\,nT$ which is approximately one part in 5×10^5 of the background field, a considerably lower requirement of precision than is necessary for gravity measurements (see Chapter 6).

In early magnetic surveys the geomagnetic elements were measured using *magnetic variometers*. There were several types, including the torsion head magnetometer and the Schmidt vertical balance, but all consisted essentially of bar magnets suspended in the Earth's field. Such devices required accurate levelling and a stable platform for measurement so that readings were time-consuming and limited to sites on land.

Since the 1940s, a new generation of instruments has been developed which provides virtually instantaneous readings and requires only coarse orientation so that magnetic measurements can be taken on land, at sea and in the air.

The first such device to be developed was the *fluxgate magnetometer*, which found early application during the second world war in the detection of submarines from the air. The instrument employs two identical ferromagnetic cores of such high permeability that the geomagnetic field can induce a magnetization that is a substantial proportion of their saturation value (see Section 7.2). Identical primary and secondary coils are wound in opposite directions around the cores (Fig. 7.10). An alternating current of $50-1000\,Hz$ is passed through the primary coils (Fig. 7.10(a)), generating an alternating magnetic field. In the absence of any external magnetic field, the cores are driven to saturation near the peak of each half-cycle of the current (Fig. 7.10(b)). The alternating magnetic field in the cores induces an alternating voltage in the secondary coils which is at a maximum when the field is changing most rapidly (Fig. 7.10(c)). Since the coils are wound in opposite directions, the voltage in the coils is equal and of opposite sign so that their combined output is zero. In the presence of an external magnetic field, such as the Earth's field, which has a component parallel to the axis of the cores, saturation occurs earlier for the core whose primary field is reinforced by the external field and later for the core opposed by the external field. The induced voltages are now out of phase as the cores reach saturation at different times (Fig. 7.10(d)). Consequently the combined output of the secondary coils is no longer zero but consists of a series of voltage pulses (Fig. 7.10(e)), the magnitude of which can be shown to be proportional to the amplitude of the external field component.

The instrument can be used to measure Z or H by aligning the cores in these directions, but the required accuracy of orientation is some eleven seconds of arc to achieve a reading accuracy of $\pm 1\,nT$. Such accuracy is difficult to obtain on the ground and impossible when the instrument is mobile. The total geomagnetic field can, however, be measured to $\pm 1\,nT$ with far less precise orientation as the field changes much more slowly as a function of orien-

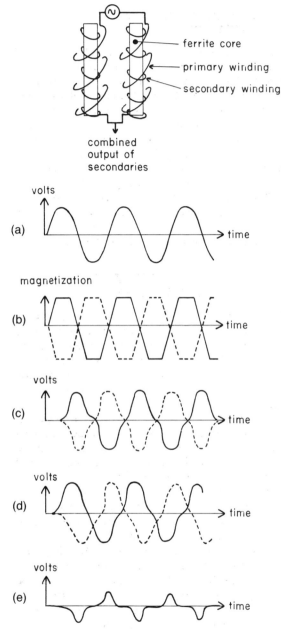

Fig. 7.10 Principle of the fluxgate magnetometer. Solid and broken lines in (b)–(d) refer to the responses of the two cores.

tation about the total field direction. Airborne versions of the instrument employ orienting mechanisms of various types to maintain the axis of the instrument in the direction of the geomagnetic field. This is accomplished by making use of the feedback signal generated by additional sensors whenever the instrument moves out of orientation to drive servo-motors which realign the cores into the desired direction.

The fluxgate magnetometer is a continuous-reading instrument and is relatively insensitive to magnetic field gradients along the length of the cores. The instrument may be temperature sensitive, requiring correction.

The most commonly used magnetometer for both survey work and observatory monitoring is currently the *nuclear precession* or *proton magnetometer*. The sensing device of the proton magnetometer is a container filled with a liquid rich in hydrogen atoms, such as kerosene or water, surrounded by a coil (Fig. 7.11(a)). The hydrogen nuclei (protons) act as small dipoles and normally align parallel to the ambient geomagnetic field B_e (Fig. 7.11(b)). A current is passed through the coil to generate a magnetic field B_p fifty to a hundred times larger than the geomagnetic field, and in a different direction, causing the protons to realign in this new direction (Fig. 7.11(c)). The current to the coil is then switched off so that the polarizing field is rapidly removed. The protons return to their original alignment with B_e by spiralling, or precessing, in phase around this direction (Fig. 7.11(d)) with a period of about 0.5 ms, taking some 1–3 s to achieve their original orientation. The frequency f of this precession is given by

$$f = \frac{\gamma_p B_e}{2\pi}$$

where γ_p is the gyromagnetic ratio of the proton, an accurately known constant. Consequently measurement of f, about 2 kHz, provides a very accurate measurement of the strength of the total geomagnetic field. f is determined by measurement of the alternating voltage of the same frequency induced to flow in the coil by the precessing protons.

Field instruments provide absolute readings of the total magnetic field accurate to $\pm 1\,\text{nT}$ although much greater precision can be attained if necessary. The sensor does not have to be accurately oriented, although it should ideally lie at an appreciable angle to the total field vector. Consequently, readings may be taken by sensors towed behind ships or aircraft without the necessity of orienting mechanisms. Aeromagnetic surveying with proton magnetometers may suffer from the slight disadvantage that readings are not continuous due to the finite cycle period. Small anomalies may be missed since an

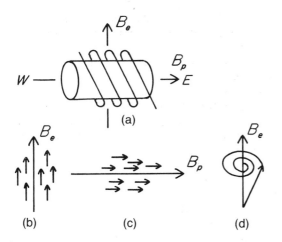

Fig. 7.11 Principle of the proton magnetometer.

aircraft travels a significant distance between the discrete measurements, which may be spaced at intervals of a few seconds. This problem has been largely obviated by modern instruments with recycling periods of the order of a second. The proton magnetometer is sensitive to acute magnetic gradients which may cause protons in different parts of the sensor to precess at different rates with a consequent adverse effect on signal strength.

The sensing elements of fluxgate or proton magnetometers can be used in pairs to measure either horizontal or vertical magnetic field gradients. *Magnetic gradiometers* are differential magnetometers in which the spacing between the sensors is fixed and small with respect to the distance of the causative body whose magnetic field gradient is to be measured. Magnetic gradients can be measured, albeit less conveniently, with a magnetometer by taking two successive measurements at close vertical or horizontal spacings. Magnetic gradiometers are employed in surveys of shallow magnetic features as the gradient anomalies tend to resolve complex anomalies into their individual components, which can be used in the determination of the location, shape and depth of the causative bodies. The method has the further advantages that regional and temporal variations in the geomagnetic field are automatically removed. Marine and airborne versions of the instrument are discussed by Wold & Cooper (1989) and Hood & Teskey (1989), respectively.

Since modern magnetic instruments require no precise levelling, a magnetic survey on land invariably proceeds more rapidly than a gravity survey.

7.7 GROUND MAGNETIC SURVEYS

Ground magnetic surveys are usually performed over relatively small areas on a previously defined target. Consequently, station spacing is commonly of the order of 10–100 m, although smaller spacings may be employed where magnetic gradients are high. Readings should not be taken in the vicinity of metallic objects such as railway lines, cars, roads, fencing, houses etc. which might perturb the local magnetic field. For similar reasons, operators of magnetometers should not carry metallic objects.

Base station readings are not necessary for monitoring instrumental drift as fluxgate and proton magnetometers do not drift, but may be used to monitor diurnal variations (see Section 7.9).

7.8 AEROMAGNETIC AND MARINE SURVEYS

The vast majority of magnetic surveys are carried out in the air, with the sensor towed in a housing known as a 'bird' to remove the instrument from the magnetic effects of the aircraft or fixed in a 'stinger' in the tail of the aircraft, in which case inboard coil installations compensate for the aircraft's magnetic field.

Aeromagnetic surveying is rapid and cost effective, typically costing some 40% less per line kilometre than a ground survey. Vast areas can be surveyed rapidly without the cost of sending a field party into the survey area and data can be obtained from areas inaccessible to ground survey.

The most difficult problem in airborne surveys is position fixing. Where available, electronic positioning systems are employed. Without these it is necessary to use aerial photography. Terrain photographs are taken simultaneously with the magnetic readings so that the location can subsequently be determined by reference to topographic maps.

Marine magnetic surveying techniques are similar to those of airborne surveying. The sensor is towed in a 'fish' at least 2½ ship's lengths behind the vessel to remove its magnetic effects. Marine surveying is obviously slower than aeromagnetic surveying, but is frequently carried out in conjunction with several other geophysical methods, such as gravity surveying and continuous seismic profiling, which cannot be employed in the air.

7.9 REDUCTION OF MAGNETIC OBSERVATIONS

The reduction of magnetic data is necessary to remove all causes of magnetic variation from the observations other than those arising from the magnetic effects of the subsurface.

7.9.1 Diurnal variation correction

The effects of diurnal variation may be removed in several ways. On land a method similar to gravimeter drift monitoring may be employed in which the magnetometer is read at a fixed base station periodically throughout the day. The differences observed in base readings are then distributed among the readings at stations occupied during the day according to the time of observation. It should be remembered that base readings taken during a gravity survey are made to correct for both the drift of the gravimeter and tidal effects: magnetometers do not drift and base readings are taken solely to correct for temporal variation in the measured field. Such a procedure is inefficient as the instrument has to be returned periodically to a base location and is not practical in marine or airborne surveys. These problems may be overcome by use of a base magnetometer, a continuous-reading instrument which records magnetic variations at a fixed location within or close to the survey area. This method is preferable on land as the survey proceeds faster and the diurnal variations are fully charted. Where the survey is of regional extent the records of a magnetic observatory may be used. Such observatories continuously record changes in all the geomagnetic elements. However, diurnal variations differ quite markedly from place to place and so the observatory used should not be more than about 100 km from the survey area.

Diurnal variation during an aeromagnetic survey may alternatively be assessed by arranging numerous crossover points in the survey plan (Fig. 7.12). Analysis of the differences in readings at each crossover, representing the field change over a series of different time periods, allows the whole survey to be corrected for diurnal variation by a process of network adjustment, without the necessity of a base instrument.

Diurnal variations, however recorded, must be examined carefully. If large, high-frequency variations are apparent, resulting from a magnetic storm, the survey results should be discarded.

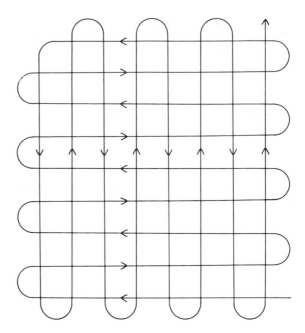

Fig. 7.12 A typical flight plan for an aeromagnetic survey.

7.9.2 Geomagnetic correction

The magnetic equivalent of the latitude correction in gravity surveying is the *geomagnetic correction* which removes the effect of a geomagnetic reference field from the survey data. The most rigorous method of geomagnetic correction is the use of the IGRF (Section 7.4), which expresses the undisturbed geomagnetic field in terms of a large number of harmonics and includes temporal terms to correct for secular variation. The complexity of the IGRF requires the calculation of corrections by computer. It must be realized, however, that the IGRF is imperfect as the harmonics employed are based on observations at relatively few, scattered, magnetic observatories. Consequently the IGRF in areas remote from observatories can be substantially in error.

Over the area of a magnetic survey the geomagnetic reference field may be approximated by a uniform gradient defined in terms of latitudinal and longitudinal gradient components. For example, the geomagnetic field over the British Isles is approximated by the following gradient components: $2.13\,nT\,km^{-1}\,N$; $0.26\,nT\,km^{-1}\,W$. For any survey area the relevant gradient values may be assessed from magnetic maps covering a much larger region.

The appropriate regional gradients may also be obtained by employing a single dipole approximation of the Earth's field and using the well-known equations for the magnetic field of a dipole to derive local field gradients:

$$Z = \frac{\mu_0}{4\pi} \frac{2M}{R^3} \cos\theta, \; H = \frac{\mu_0}{4\pi} \frac{M}{R^3} \sin\theta \qquad (7.12)$$

$$\frac{\partial Z}{\partial\theta} = -2H, \; \frac{\partial H}{\partial\theta} = \frac{Z}{2} \qquad (7.13)$$

where Z and H are the vertical and horizontal field components, θ the colatitude in radians, R the radius of the Earth, M the magnetic moment of the Earth and $\partial Z/\partial\theta$ and $\partial H/\partial\theta$ the rate of change of Z and H with colatitude.

An alternative method of removing the regional gradient over a relatively small survey area is by use of trend analysis. A trend line (for profile data) or trend surface (for areal data) is fitted to the observations using the least squares criterion, and subsequently subtracted from the observed data to leave the local anomalies as positive and negative residuals (Fig. 7.13).

7.9.3 Elevation and terrain corrections

The vertical gradient of the geomagnetic field is only some $0.03\,\text{nT}\,\text{m}^{-1}$ at the poles and $-0.015\,\text{nT}\,\text{m}^{-1}$ at the equator, so an *elevation correction* is not usually applied. The influence of topography can be significant in ground magnetic surveys but is not completely predictable as it depends upon the magnetic properties of the topographic features. Therefore, in magnetic surveying *terrain corrections* are rarely applied.

Having applied diurnal and geomagnetic corrections, all remaining magnetic field variations should be caused solely by spatial variations in the magnetic properties of the subsurface and are referred to as magnetic anomalies.

7.10 INTERPRETATION OF MAGNETIC ANOMALIES

7.10.1 Introduction

The interpretation of magnetic anomalies is similar in its procedures and limitations to gravity interpretation as both techniques utilize natural potential fields based on inverse square laws of attraction.

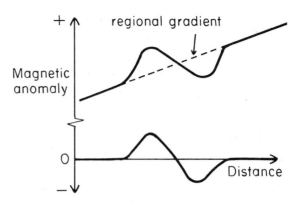

Fig. 7.13 The removal of a regional gradient from a magnetic field by trend analysis. The regional field is approximated by a linear trend.

There are several differences, however, which increase the complexity of magnetic interpretation.

Whereas the gravity anomaly of a causative body is entirely positive or negative, depending on whether the body is more or less dense than its surroundings, the magnetic anomaly of a finite body invariably contains positive and negative elements arising from the dipolar nature of magnetism (Fig. 7.14). Moreover, whereas density is a scalar, intensity of magnetization is a vector, and the direction of magnetization in a body closely controls the shape of its magnetic anomaly. Thus bodies of identical shape can give rise to very different magnetic anomalies. For the above reasons magnetic anomalies are often much less closely related to the shape of the causative body than are gravity anomalies.

The intensity of magnetization of a rock is largely dependent upon the amount, size, shape and distribution of its contained ferrimagnetic minerals and these represent only a small proportion of its constituents. By contrast, density is a bulk property. Intensity of magnetization can vary by a factor of 10^6 between different rock types, and is thus considerably more variable than density, where the range is commonly $1.50-3.50\,\text{Mg}\,\text{m}^{-3}$.

Magnetic anomalies are independent of the distance units employed. For example, the same magnitude anomaly is produced by, say, a $3\,\text{m}$ cube (on a metre scale) as a $3\,\text{km}$ cube (on a kilometre scale) with the same magnetic properties. The same is not true of gravity anomalies.

The problem of ambiguity in magnetic interpretation is the same as for gravity, i.e. the same

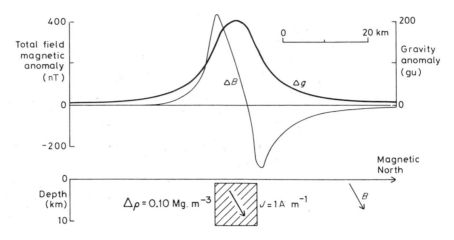

Fig. 7.14 Gravity (Δg) and magnetic (ΔB) anomalies over the same two-dimensional body.

inverse problem is encountered. Thus, just as with gravity, all external controls on the nature and form of the causative body must be employed to reduce the ambiguity. An example of this problem is illustrated in Fig. 7.15, which shows two possible interpretations of a magnetic profile across the Barbados Ridge in the eastern Caribbean. In both cases the regional variations are attributed to the variation in depth of a 1 km thick oceanic crustal layer 2. The high amplitude central anomaly, however, can be explained either by the presence of a detached sliver of oceanic crust (Fig. 7.15(a)) or a rise of metamorphosed sediments at depth (Fig. 7.15(b)).

Much qualitative information may be derived from a magnetic contour map. This applies especially to aeromagnetic maps which often provide major clues as to the geology and structure of a broad region from an assessment of the shapes and trends of anomalies. Sediment-covered areas with relatively deep basement are typically represented by smooth magnetic contours reflecting basement structures and magnetization contrasts. Igneous and metamorphic terrains generate far more complex magnetic anomalies, and the effects of deep geological features may be obscured by high wavenumber anomalies of near-surface origin. In most types of terrain an aeromagnetic map can be a useful aid to reconnaissance geological mapping. Such qualitative interpretations may be greatly facilitated by the use of digital image processing techniques (Section 6.8.6).

In carrying out quantitative interpretation of magnetic anomalies, both direct and indirect methods may be employed, but the former are much more limited than for gravity interpretation and no equivalent general equations exist for total field anomalies.

7.10.2 Direct interpretation

Limiting depth is the most important parameter derived by direct interpretation, and this may be deduced from magnetic anomalies by making use of their property of decaying rapidly with distance from source. Magnetic anomalies caused by shallow structures are more dominated by high wavenumber components than those resulting from deeper sources. This effect may be quantified by computing the power spectrum of the anomaly as it can be shown, for certain types of source body, that the log-power spectrum has a linear gradient whose magnitude is dependent upon the depth of the source (Spector & Grant 1970). Such techniques of spectral analysis provide rapid depth estimates from regularly-spaced digital field data: no geomagnetic or diurnal corrections are necessary as these remove only low wavenumber components and do not affect the depth estimates which are controlled by the high wavenumber components of the observed field. Fig. 7.16 shows a magnetic profile across the Aves Ridge in the eastern Caribbean. In this region the configuration of the sediment/basement interface is reasonably well known from both seismic reflection and refraction surveys. The magnetic anomalies clearly show their most rapid fluctuation over areas of relatively shallow basement, and this observation is

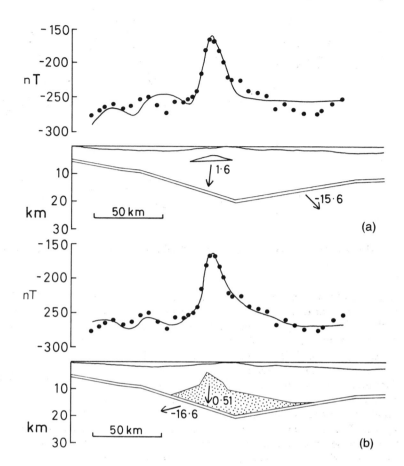

Fig. 7.15 An example of ambiguity in magnetic interpretation. The arrows correspond to the directions of magnetization vectors, whose magnitude is given in A m^{-1}. (After Westbrook 1975.)

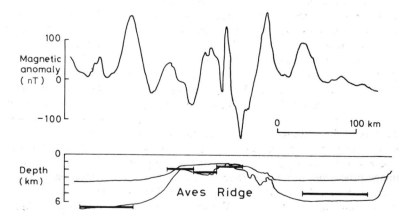

Fig. 7.16 Magnetic anomalies over the Aves Ridge, eastern Caribbean. Lower diagram illustrates bathymetry and basement/sediment interface. Horizontal bars indicate depth estimates of the magnetic basement derived by spectral analysis of the magnetic data.

quantified by the power spectral depth estimates (horizontal bars) which show excellent correlation with the known basement relief.

7.10.3 Indirect interpretation

Indirect interpretation of magnetic anomalies is similar to gravity interpretation in that an attempt is made to match the observed anomaly with that calculated for a model by iterative adjustments to the model. Simple magnetic anomalies may be simulated by a single dipole. Such an approximation to the magnetization of a real geological body is often valid for highly magnetic ore bodies whose direction of magnetization tends to align with their long dimension (Fig. 7.17). In such cases the anomaly is calculated by summing the effects of both poles at the observation points, employing equations (7.10), (7.11) and (7.9). More complicated magnetic bodies, however, require a different approach.

The magnetic anomaly of most regularly-shaped bodies can be calculated by building up the bodies from a series of dipoles parallel to the magnetization direction (Fig. 7.18). The poles of the magnets are negative on the surface of the body where the magnetization vector enters the body and positive where it leaves the body. Thus any uniformly-magnetized body can be represented by a set of magnetic poles distributed over its surface. Consider one of these elementary magnets of length l and

cross-sectional area δA in a body with intensity of magnetization J and magnetic moment M. From equation (7.5)

$$M = J\delta Al \qquad (7.14)$$

If the pole strength of the magnet is m, from equation (7.4) $m = M/l$, and substituting in equation (7.14)

$$m = J\delta A \qquad (7.15)$$

If $\delta A'$ is the area of the end of the magnet and θ the angle between the magnetization vector and a direction normal to the end face

$$\delta A = \delta A' \cos \theta$$

Substituting in equation (7.15)

$$m = J\delta A' \cos \theta$$

thus

the pole strength per unit area $= J \cos \theta \qquad (7.16)$

A consequence of the distribution of an equal number of positive and negative poles over the surface of a magnetic body is that an infinite horizontal layer produces no magnetic anomaly since the effects of the poles on the upper and lower surfaces are self-cancelling. Consequently, magnetic anomalies are not produced by continuous sills or lava flows. Where, however, the horizontal structure is truncated, the vertical edge will produce a magnetic anomaly (Fig. 7.19).

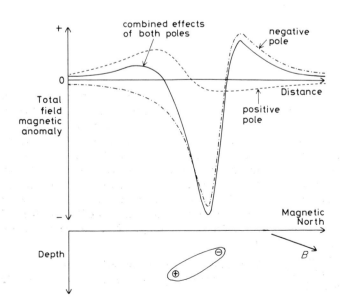

Fig. 7.17 The total field magnetic anomaly of an elongate body approximated by a dipole.

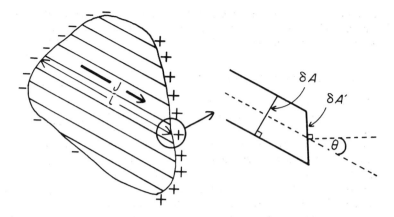

Fig. 7.18 The representation of the magnetic effects of an irregularly shaped body in terms of a number of elements parallel to the magnetization direction. Inset shows in detail the end of one such element.

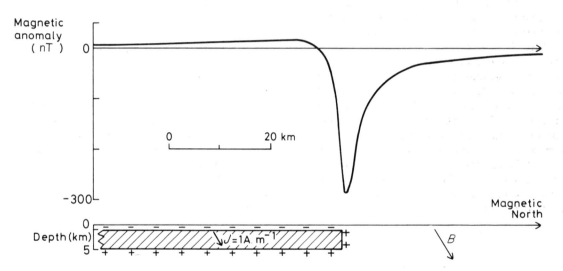

Fig. 7.19 The total field magnetic anomaly of a faulted sill.

The magnetic anomaly of a body of regular shape is calculated by determining the pole distribution over the surface of the body using equation (7.16). Each small element of the surface is then considered and its vertical and horizontal component anomalies are calculated at each observation point using equations (7.10) and (7.11). The effects of all such elements are summed (integrated) to produce the vertical and horizontal anomalies for the whole body and the total field anomaly is calculated using equation (7.9). The integration can be performed analytically for bodies of regular shape, while irregularly-shaped bodies may be split into regular shapes and the integration performed numerically.

In two-dimensional modelling, an approach similar to gravity interpretation can be adopted (Section 6.10.4) in which the cross-sectional form of the body is approximated by a polygonal outline. The anomaly of the polygon is then computed by adding or subtracting the anomalies of semi-infinite slabs with sloping edges corresponding to the sides of the polygon (Fig. 7.20). In the magnetic case, the horizontal ΔH, vertical ΔZ and total field ΔB anomalies (nT) of the slab shown in Fig. 7.20 are given by (Talwani *et al.* 1965)

$$\Delta Z = 200 \sin \theta \, [J_x \{ \sin \theta \log_e(r_2/r_1) + \phi \cos \theta \} + J_z \{ \cos \theta \log_e(r_2/r_1) - \phi \sin \theta \}] \qquad (7.17a)$$

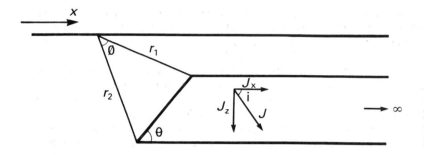

Fig. 7.20 Parameters used in defining the magnetic anomaly of a semi-infinite slab with a sloping edge.

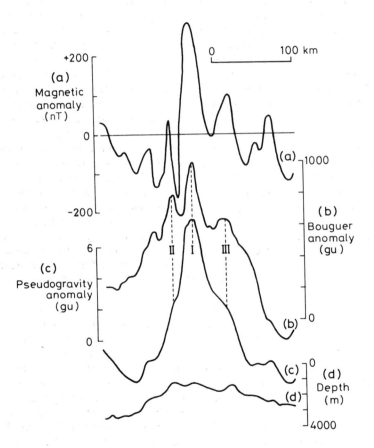

Fig. 7.21 (a) Observed magnetic anomalies over the Aves Ridge, eastern Caribbean. (b) Bouguer gravity anomalies with long wavelength regional field removed. (c) Pseudogravity anomalies computed for induced magnetization and a density:magnetization ratio of unity. (d) Bathymetry.

$$\Delta H = 200 \sin \theta [J_x \{\phi \sin \theta - \cos \theta \log_e(r_2/r_1)\} + J_z \{\phi \cos \theta + \sin \theta \log_e(r_2/r_1)\}] \sin \alpha \qquad (7.17b)$$

$$\Delta B = \Delta Z \sin I + \Delta H \cos I \qquad (7.17c)$$

where angles are expressed in radians, $J_x (= J \cos i)$ and $J_z (= J \sin i)$ are the horizontal and vertical components of the magnetization J, α is the horizontal angle between the direction of the profile and magnetic north, and I is the inclination of the geomagnetic field. Examples of this technique have been presented in Fig. 7.15. An important difference from gravity interpretation is the increased stringency with which the two-dimensional approximation should be applied. It can be shown that two-dimensional magnetic interpretation is much more sensitive to errors associated with variation along strike than is the case with gravity interpretation; the length–width ratio of a magnetic anomaly should be at least 10:1 for a two-dimensional approximation to be valid, in contrast to gravity interpretation where a 2:1 length–width ratio is sufficient to validate two-dimensional interpretation.

Three-dimensional modelling of magnetic anomalies is complex. Probably the most convenient methods are to approximate the causative body by a cluster of right rectangular prisms or by a series of horizontal slices of polygonal outline.

Because of the dipolar nature of magnetic anomalies, trial and error methods of indirect interpretation are difficult to perform manually since anomaly shape is not closely related to the geometry of the causative body. Consequently the automatic methods of interpretation described in Section 6.11.4 are widely employed.

The continuation and filtering operations used in gravity interpretation and described in Section 6.12 are equally applicable to magnetic fields. A further processing operation that may be applied to magnetic anomalies is known as *reduction to the pole*, and involves the conversion of the anomalies into their equivalent form at the north magnetic pole (Baranov & Naudy 1964). This process usually simplifies the magnetic anomalies as the ambient field is then vertical and bodies with magnetizations which are solely induced produce anomalies that are axisymmetric. The existence of remanent magnetization, however, commonly prevents reduction to the pole from producing the desired simplification in the resultant pattern of magnetic anomalies.

7.11 POTENTIAL FIELD TRANSFORMATIONS

The formulae for the gravitational potential caused by a point mass and the magnetic potential due to an isolated pole were presented in equations (6.3) and (7.3). A consequence of the similar laws of attraction governing gravitating and magnetic bodies is that these two equations have the variable of inverse distance $(1/r)$ in common. Elimination of this term between the two formulae provides a relationship between the gravitational and magnetic potentials known as *Poisson's equation*. In reality the relationship is more complex than implied by equations (6.3) and (7.3) as isolated magnetic poles do not exist. However, the validity of the relationship between the two potential fields remains. Since gravity or magnetic fields can be determined by differentiation of the relevant potential in the required direction, Poisson's equation provides a method of transforming magnetic fields into gravitational fields and *vice versa* for bodies in which the ratio of

intensity of magnetization to density remains constant. Such transformed fields are known as *pseudogravitational* and *pseudomagnetic* fields (Garland 1951).

One application of this technique is the transformation of magnetic anomalies into pseudogravity anomalies for the purposes of indirect interpretation, as the latter are significantly easier to interpret than their magnetic counterpart. The method is even more powerful when the pseudofield is compared with a corresponding measured field. For example the comparison of gravity anomalies with the pseudogravity anomalies derived from magnetic anomalies over the same area can show whether the same geological bodies are the cause of the two types of anomaly. Performing the transformation for different orientations of the magnetization vector provides an estimate of the true vector orientation since this will produce a pseudogravity field which most closely approximates the observed gravity field. The relative amplitudes of these two fields then provide a measure of the ratio of intensity of magnetization to density. These potential field transformations provide an elegant means of comparing gravity and magnetic anomalies over the same area and sometimes allow greater information to be derived about their causative bodies than would be possible if the techniques were treated in isolation. A computer program which performs pseudofield transformations is given in Gilbert & Galdeano (1985).

Figs 7.21(a) and (b) show magnetic and residual gravity anomaly profiles across the Aves Ridge, a submarine prominence in the eastern Caribbean which runs parallel to the island arc of the Lesser Antilles. The pseudogravity profile calculated from the magnetic profile assuming induced magnetization is presented in Fig. 7.21(c). It is readily apparent that the main pseudogravity peak correlates with peak I on the gravity profile and that peaks II and III correlate with much weaker features on the pseudofield profile. The data thus suggest that the density features responsible for the gravity maxima are also magnetic, with the causative body of the central peak having a significantly greater susceptibility than the flanking bodies.

7.12 APPLICATIONS OF MAGNETIC SURVEYING

Magnetic surveying is a rapid and cost-effective technique and represents one of the most widely-

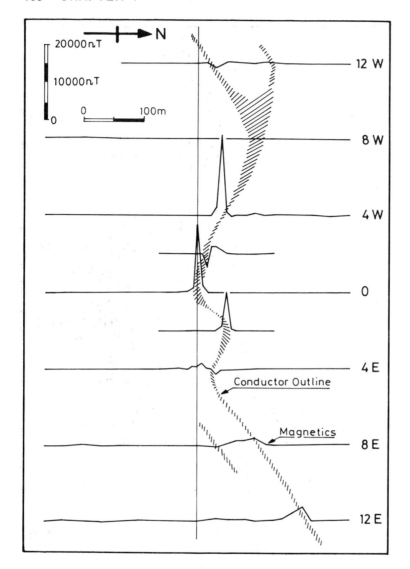

Fig. 7.22 Vertical field ground magnetic anomaly profiles over a massive sulphide ore body in Quebec, Canada. The shaded area represents the location of the ore body inferred from EM measurements. (After White 1966.)

used geophysical methods in terms of line-length surveyed (Paterson & Reeves 1985).

Magnetic surveys are used extensively in the search for metalliferous mineral deposits, a task accomplished rapidly and economically by airborne methods. Magnetic surveys are capable of locating massive sulphide deposits (Fig. 7.22), especially when used in conjunction with electromagnetic methods (Section 9.12). However, the principal target of magnetic surveying is iron ore. The ratio of magnetite to haematite must be high for the ore to produce significant anomalies, as haematite is commonly non-magnetic (Section 7.2). Fig. 7.23 shows total field magnetic anomalies from an airborne survey of the Northern Middleback Range, South Australia, in which it is seen that the haematitic ore bodies are not associated with the major anomalies. Fig. 7.24 shows the results from an aeromagnetic survey of part of the Eyre Peninsula of South Australia which reveal the presence of a large anomaly elongated east–west. Subsequent ground traverses were performed over this anomaly using both magnetic and gravity methods (Fig. 7.25) and it was found that the magnetic and gravity profiles

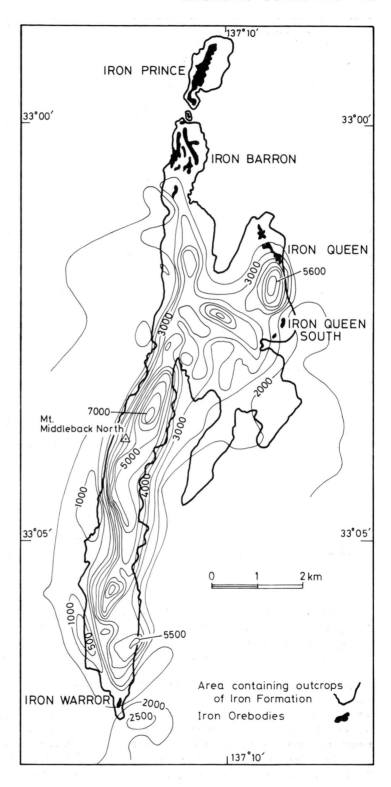

Fig. 7.23 Aeromagnetic anomalies over the Northern Middleback Range, South Australia. The iron ore bodies are of hematitic composition. Contour interval 500 nT. (After Webb 1966.)

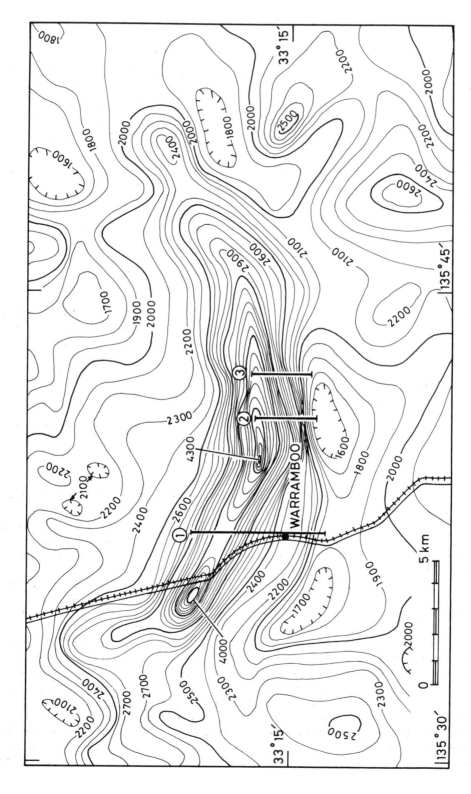

Fig. 7.24 High-level aeromagnetic anomalies over part of the Eyre Peninsula, South Australia. Contour interval 100 nT. (After Webb 1966.)

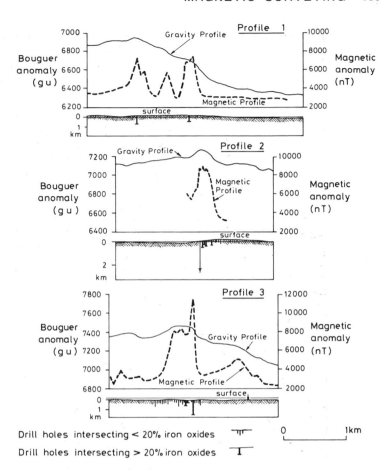

Fig. 7.25 Gravity and magnetic ground profiles over part of the Eyre Peninsula, South Australia, at the locations shown in Fig. 7.24 (After Webb 1966.)

Drill holes intersecting < 20% iron oxides

Drill holes intersecting > 20% iron oxides

0   1km

exhibit coincident highs. Subsequent drilling on these highs revealed the presence of a magnetite-bearing ore body at shallow depth with an iron content of about 30%.

In geotechnical and archaeological investigations, magnetic surveys may be used to delineate zones of faulting in bedrock and to locate buried, metallic, man-made features such as pipelines, old mine workings and buildings. Fig. 7.26 shows a total magnetic field contour map of the site of a proposed apartment block in Bristol, England. The area had been exploited for coal in the past and stability problems would arise from the presence of old shafts and buried workings (Clark 1986). Lined shafts of up to 2 m diameter were subsequently found beneath anomalies A and D, while other isolated anomalies such as B and C were known, or suspected, to be associated with buried metallic objects.

In academic studies, magnetic surveys can be used in regional investigations of large-scale crustal features, although the sources of major magnetic anomalies tend to be restricted to rocks of basic or ultrabasic composition. Moreover, magnetic surveying is of limited use in the study of the deeper geology of the continental crust because the Curie isotherm for common ferrimagnetic minerals lies at a depth of about 20 km and the sources of major anomalies are consequently restricted to the upper part of the continental crust.

Although the contribution of magnetic surveying to knowledge of continental geology has been modest, magnetic surveying in oceanic areas has had a profound influence on the development of plate tectonic theory (Kearey & Vine 1990) and on views of the formation of oceanic lithosphere. Early magnetic surveying at sea showed that the oceanic crust is characterized by a pattern of linear magnetic anomalies (Fig. 7.27) attributable to strips of oceanic

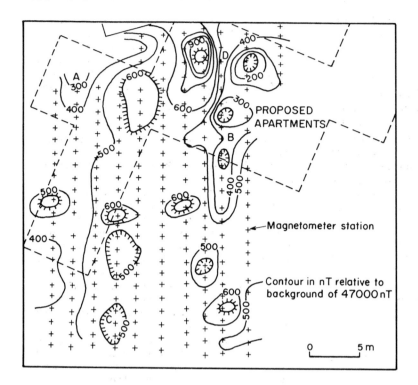

Fig. 7.26 Magnetic anomaly contour map of a site in Bristol, England. Contour interval 100 nT. (After Hooper & McDowell 1977.)

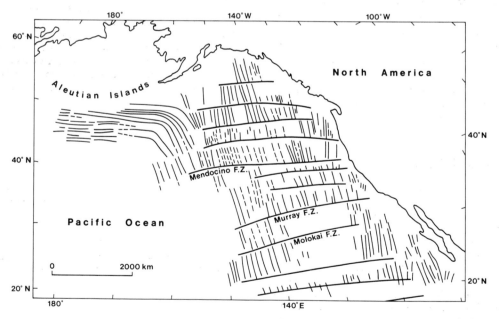

Fig. 7.27 Pattern of linear magnetic anomalies and major fracture zones in the northeast Pacific Ocean.

crust alternately magnetized in a normal and reverse direction (Mason & Raff 1961). The bilateral symmetry of these linear magnetic anomalies about oceanic ridges and rises (Vine & Matthews 1963) led directly to the theory of sea floor spreading and the establishment of a time scale for polarity transitions of the geomagnetic field (Heirtzler *et al.* 1968). Consequently, oceanic crust can be dated on the basis of the pattern of magnetic polarity transitions preserved within it.

Transform faults disrupt the pattern of linear magnetic anomalies (see Fig. 7.27) and their distribution can therefore be mapped magnetically. Since these faults lie along arcs of small circles to the prevailing pole of rotation at the time of transform fault movement, individual regimes of spreading during the evolution of an ocean basin can be identified by detailed magnetic surveying. Such studies have been carried out in all the major oceans and show the evolution of an ocean basin to be a complex process involving several discrete phases of spreading, each with a distinct pole of rotation.

Magnetic surveying is a very useful aid to geological mapping. Over extensive regions with a thick sedimentary cover, structural features may be revealed if magnetic horizons such as ferruginous sandstones and shales, tuffs and lava flows are present within the sedimentary sequence. In the absence of magnetic sediments, magnetic survey data can provide information on the nature and form of the crystalline basement. Both cases are applicable to petroleum exploration in the location of structural traps within sediments or features of basement topography which might influence the overlying sedimentary sequence. The magnetic method may also be used to assist a programme of reconnaissance geological mapping based on widely-spaced grid samples, since aeromagnetic anomalies can be employed to delineate geological boundaries between sampling points.

7.13 PROBLEMS

1 Discuss the advantages and disadvantages of aeromagnetic surveying.
2 How and why do the methods of reduction of gravity and magnetic data differ?
3 Compare and contrast the techniques of interpretation of gravity and magnetic anomalies.
4 Assuming that the magnetic moment of the Earth is 8×10^{22} A m^2, its radius 6370 km and that its magnetic field conforms to an axial dipole model, calculate the geomagnetic elements at 60 °N and 75 °S. Calculate also the total field magnetic gradients in nT km^{-1} N at these latitudes.
5 Using equations (7.17a,b,c), derive expressions for the horizontal, vertical and total field magnetic anomalies of a vertical dyke of infinite depth striking at an angle α to magnetic north.

Given that geomagnetic inclination I is related to latitude θ by $\tan I = 2 \tan \theta$, use these formulae to calculate the magnetic anomalies of east−west striking dykes of width 40 m, depth 20 m and intensity of magnetization 2 A m^{-1}, at a latitude of 45°, in the following cases:
(a) In the northern hemisphere with induced magnetization.
(b) In the northern hemisphere with reversed magnetization.
(c) In the southern hemisphere with normal magnetization.
(d) In the southern hemisphere with reversed magnetization.
How would the anomalies change if the width and depth were increased to 400 m and 200 m, respectively?

6 (a) Calculate the vertical, horizontal and total field magnetic anomaly profiles across a dipole which strikes in the direction of the magnetic meridian and dips to the south at 30° with the negative pole at the northern end 5 m beneath the surface. The length of the dipole is 50 m and the strength of each pole is 300 A m. The local geomagnetic field dips to the north at 70°.
(b) What is the effect on the profiles if the dipole strikes 25°E of the magnetic meridian?
(c) If the anomalies calculated in (a) actually originate from a cylinder whose magnetic moment is the same as the dipole and whose diameter is 10 m, calculate the intensity of magnetization of the cylinder.
(d) Fig. 7.28 shows a total field magnetic anomaly profile across buried volcanic rocks to the south of Bristol, England. Does the profile constructed in (a) represent a reasonable simulation of this anomaly? If so, calculate the dimensions and intensity of magnetization of a possible magnetic source. What other information would be needed to provide a more detailed interpretation of the anomaly?

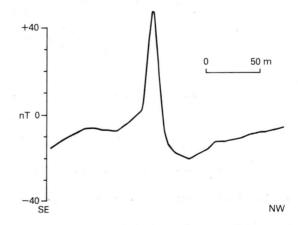

Fig. 7.28 Total field magnetic profile across buried volcanic rocks south of Bristol, England. (After Kearey & Allison 1980.)

FURTHER READING

Arnaud Gerkens, J.C.d'. (1989) *Foundations of Exploration Geophysics*. Elsevier, Amsterdam.

Baranov, W. (1975) *Potential Fields and Their Transformation in Applied Geophysics*. Gebrüder Borntraeger, Berlin.

Bott, M.H.P. (1973) Inverse methods in the interpretation of magnetic and gravity anomalies. *In*: Alder B., Fernbach, S. & Bolt, B.A. (eds.) *Methods in Computational Physics*, **13**, 133–62.

Garland, G.D. (1951) Combined analysis of gravity and magnetic anomalies *Geophysics*, **16**, 51–62.

Gunn, P.J. (1975) Linear transformations of gravity and magnetic fields. *Geophys. Prosp.*, **23**, 300–12.

Kanasewich, E.R. & Agarwal, R.G. (1970) Analysis of combined gravity and magnetic fields in wave number domain. *J. geophys. Res.*, **75**, 5702–12.

Nettleton, L.L. (1971) *Elementary Gravity and Magnetics for Geologists and Seismologists*. Monograph Series No. 1. Society of Exploration Geophysicists, Tulsa.

Nettleton, L.L. (1976) *Gravity and Magnetics in Oil Exploration*. McGraw-Hill, New York.

Sharma, P. (1976) *Geophysical Methods in Geology*. Elsevier, Amsterdam.

Stacey, F.D. & Banerjee, S.K. (1974) *The Physical Principles of Rock Magnetism*. Elsevier, Amsterdam.

Tarling, D.H. (1983) *Palaeomagnetism*. Chapman & Hall, London.

Vacquier, V., Steenland, N.C., Henderson, R.G. & Zeitz, I. (1951) Interpretation of aeromagnetic maps. *Geol. Soc. Am. Mem.*, **47**.

8 / Electrical surveying

8.1 INTRODUCTION

There are many methods of electrical surveying. Some make use of naturally-occurring fields within the Earth while others require the introduction of artificially-generated currents into the ground. The resistivity method is used in the study of horizontal and vertical discontinuities in the electrical properties of the ground, and also in the detection of three-dimensional bodies of anomalous electrical conductivity. It is routinely used in engineering and hydrogeological investigations to investigate the shallow subsurface geology. The induced polarization method makes use of the capacitative action of the subsurface to locate zones where conductive minerals are disseminated within their host rocks. The self-potential method makes use of natural currents flowing in the ground that are generated by electrochemical processes to locate shallow bodies of anomalous conductivity.

Electrical methods utilize direct currents or low frequency alternating currents to investigate the electrical properties of the subsurface, in contrast to the electromagnetic methods discussed in the next chapter that use alternating electromagnetic fields of higher frequency to this end.

8.2 RESISTIVITY METHOD

8.2.1 Introduction

In the resistivity method, artificially-generated electric currents are introduced into the ground and the resulting potential differences are measured at the surface. Deviations from the pattern of potential differences expected from homogeneous ground provide information on the form and electrical properties of subsurface inhomogeneities.

8.2.2 Resistivities of rocks and minerals

The *resistivity* of a material is defined as the resistance in ohms between the opposite faces of a unit cube of the material. For a conducting cylinder of resistance δR, length δL and cross-sectional area δA (Fig. 8.1) the resistivity ρ is given by

$$\rho = \frac{\delta R \delta A}{\delta L} \qquad (8.1)$$

The SI unit of resistivity is the ohm-metre (ohm m) and the reciprocal of resistivity is termed *conductivity* (units: siemens (S) per metre; $1\,\mathrm{S\,m^{-1}} = 1\,\mathrm{ohm^{-1}\,m^{-1}}$).

Resistivity is one of the most variable of physical properties. Certain minerals such as native metals and graphite conduct electricity via the passage of electrons. Most rock-forming minerals are, however, insulators, and electrical current is carried through a rock mainly by the passage of ions in pore waters. Thus most rocks conduct electricity by electrolytic rather than electronic processes. It follows that porosity is the major control of the resistivity of rocks, and that resistivity generally increases as porosity decreases. However, even crystalline rocks with negligible intergranular porosity are conductive along cracks and fissures. Fig. 8.2 shows the range of resistivities expected for common rock types. It is apparent that there is considerable overlap between different rock types and, consequently, identification of a rock type is not possible solely on the basis of resistivity data.

Strictly, equation (8.1) refers to electronic conduction but it may still be used to describe the

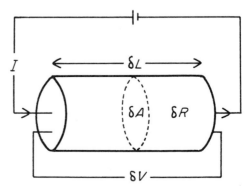

Fig. 8.1 The parameters used in defining resistivity.

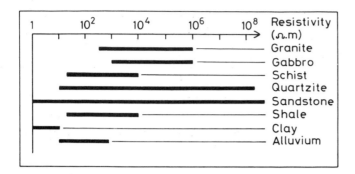

Fig. 8.2 The approximate range of resistivity values of common rock types.

effective resistivity of a rock, i.e. the resistivity of the rock and its pore water. The effective resistivity can also be expressed in terms of the resistivity and volume of the pore water present according to an empirical formula given by Archie (1942)

$$\rho = a\phi^{-b}f^{-c}\rho_w \tag{8.2}$$

where ϕ is the porosity, f the fraction of pores containing water of resistivity ρ_w and a, b and c are empirical constants. ρ_w can vary considerably according to the quantities and conductivities of dissolved materials.

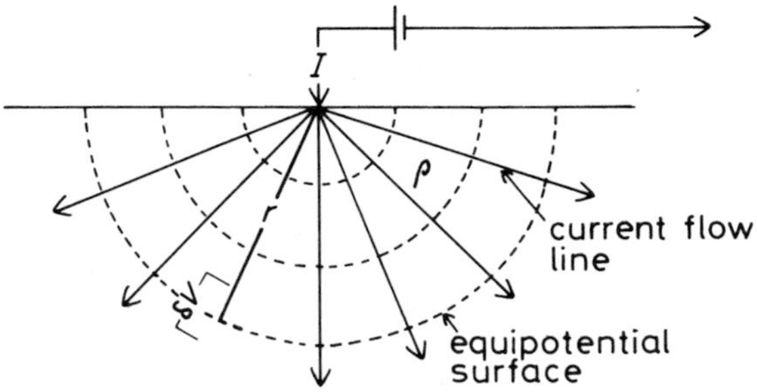

8.2.3 Current flow in the ground

Consider the element of homogeneous material shown in Fig. 8.1. A current I is passed through the cylinder causing a potential drop $-\delta V$ between the ends of the element.

Ohm's law relates the current, potential difference and resistance such that $-\delta V = \delta RI$, and from equation (8.1) $\delta R = \rho\delta L/\delta A$. Substituting

$$\frac{\delta V}{\delta L} = -\frac{\rho I}{\delta A} = -\rho i \tag{8.3}$$

$\delta V/\delta L$ represents the potential gradient through the element in volt m^{-1} and i the current density in $A\,m^{-2}$. In general the current density in any direction within a material is given by the negative partial derivative of the potential in that direction divided by the resistivity.

Now consider a single current electrode on the surface of a medium of uniform resistivity ρ (Fig. 8.3). The circuit is completed by a current sink at a large distance from the electrode. Current flows radially away from the electrode so that the current distribution is uniform over hemispherical shells centred on the source. At a distance r from the

Fig. 8.3 Current flow from a single surface electrode.

electrode the shell has a surface area of $2\pi r^2$, so the current density i is given by

$$i = \frac{I}{2\pi r^2} \tag{8.4}$$

From equation (8.3), the potential gradient associated with this current density is

$$\frac{\partial V}{\partial r} = -\rho i = -\frac{\rho I}{2\pi r^2} \tag{8.5}$$

The potential V_r at distance r is then obtained by integration

$$V_r = \int \partial V = -\int \frac{\rho I \partial r}{2\pi r^2} = \frac{\rho I}{2\pi r} \tag{8.6}$$

The constant of integration is zero since $V_r = 0$ when $r = \infty$.

Equation (8.6) allows the calculation of the potential at any point on or below the surface of a homogeneous half space. The hemispherical shells in Fig. 8.3 mark surfaces of constant voltage and are termed *equipotential surfaces*.

Now consider the case where the current sink is a finite distance from the source (Fig. 8.4). The potential V_C at an internal electrode C is the sum of the potential contributions V_A and V_B from the current source at A and the sink at B

$$V_C = V_A + V_B$$

From equation (8.6)

$$V_C = \frac{\rho I}{2\pi}\left(\frac{1}{r_A} - \frac{1}{r_B}\right) \qquad (8.7)$$

Similarly

$$V_D = \frac{\rho I}{2\pi}\left(\frac{1}{R_A} - \frac{1}{R_B}\right) \qquad (8.8)$$

Absolute potentials are difficult to monitor so the potential difference ΔV between electrodes C and D is measured

$$\Delta V = V_C - V_D = \frac{\rho I}{2\pi}\left\{\left(\frac{1}{r_A} - \frac{1}{r_B}\right) - \left(\frac{1}{R_A} - \frac{1}{R_B}\right)\right\}$$

Thus

$$\rho = \frac{2\pi \Delta V}{I\left\{\left(\dfrac{1}{r_A} - \dfrac{1}{r_B}\right) - \left(\dfrac{1}{R_A} - \dfrac{1}{R_B}\right)\right\}} \qquad (8.9)$$

Where the ground is uniform, the resistivity calculated from equation (8.9) should be constant and independent of both electrode spacing and surface location. When subsurface inhomogeneities exist, however, the resistivity will vary with the relative positions of the electrodes. Any computed value is then known as the *apparent resistivity* ρ_a and will be a function of the form of the inhomogeneity. Equation (8.9) is the basic equation for calculating the apparent resistivity for any electrode configuration.

In homogeneous ground the depth of current penetration increases as the separation of the current electrodes is increased, and Fig. 8.5 shows the proportion of current flowing beneath a given depth Z as the ratio of electrode separation L to depth increases. When $L = Z$ about 30% of the current flows below Z and when $L = 2Z$ about 50% of the current flows below Z. The current electrode separation must be chosen so that the ground is energized to the required depth, and should be at least equal to this depth. This places practical limits on the depths of penetration attainable by normal resistivity methods due to the difficulty in laying long lengths of cable and the generation of sufficient power.

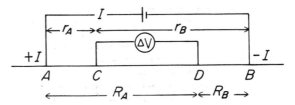

Fig. 8.4 The generalized form of the electrode configuration used in resistivity measurements.

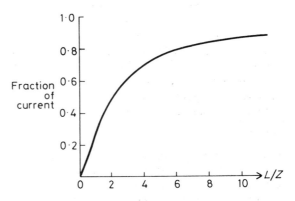

Fig. 8.5 The fraction of current penetrating below a depth Z for a current electrode separation L. (After Telford *et al.* 1976.)

Depths of penetration of about 1 km are the limit for normal equipment.

Two main types of procedure are employed in resistivity surveys.

Vertical electrical sounding (VES), also known as 'electrical drilling' or 'expanding probe', is used mainly in the study of horizontal or near-horizontal interfaces. The current and potential electrodes are maintained at the same relative spacing and the whole spread is progressively expanded about a fixed central point. Consequently, readings are taken as the current reaches progressively greater depths. The technique is extensively used in geotechnical surveys to determine overburden thickness and also in hydrogeology to define horizontal zones of porous strata.

Constant separation traversing (CST), also known as 'electrical profiling', is used to determine lateral variations of resistivity. The current and potential electrodes are maintained at a fixed separation and progressively moved along a profile. This method is employed in mineral prospecting to locate faults or shear zones and to detect localized bodies of anomalous conductivity. It is also used in geotechnical surveys to determine variations in bedrock depth and the presence of steep discontinuities. Results from a series of CST traverses with a fixed electrode spacing can be employed in the production of resistivity contour maps.

8.2.4 Electrode spreads

Many configurations of electrodes have been designed (Habberjam 1979) and although several are occasionally employed in specialized surveys, only two are in common use. The *Wenner configuration* is the simpler in that current and potential electrodes are maintained at an equal spacing *a* (Fig. 8.6). Substitution of this condition into equation (8.9) yields

$$\rho_a = 2\pi a \frac{\Delta V}{I} \qquad (8.10)$$

During VES the spacing *a* is gradually increased about a fixed central point and in CST the whole spread is moved along a profile with a fixed value of *a*. The efficiency of performing vertical electrical sounding can be greatly increased by making use of a multicore cable to which a number of electrodes are permanently attached at standard separations (Barker 1981). A sounding can then be rapidly accomplished by switching between different sets of four electrodes. Such a system has the additional advantage that, by measuring ground resistances at two electrode array positions, the effects of near-surface lateral resistivity variations can be substantially reduced.

In surveying with the Wenner configuration all four electrodes need to be moved between successive readings. This labour is partially overcome by the use of the *Schlumberger configuration* (Fig. 8.6) in which the inner, potential electrodes have a spacing 2*l* which is a small proportion of that of the outer, current electrodes (2*L*). In CST surveys with the Schlumberger configuration several lateral movements of the potential electrodes may be accommodated without the necessity of moving the current electrodes. In VES surveys the potential electrodes remain fixed and the current electrodes are expanded symmetrically about the centre of the spread. With very large values of *L* it may, however, be necessary to increase *l* also in order to maintain a measurable potential.

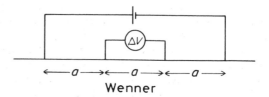

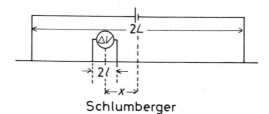

Fig. 8.6 The Wenner and Schlumberger electrode configurations.

For the Schlumberger configuration

$$\rho_a = \frac{\pi}{2l} \frac{(L^2 - x^2)^2}{(L^2 + x^2)} \frac{\Delta V}{I} \qquad (8.11)$$

where *x* is the separation of the mid-points of the potential and current electrodes. When used symmetrically, $x = 0$, so

$$\rho_a = \frac{\pi L^2}{2l} \frac{\Delta V}{I} \qquad (8.12)$$

8.2.5 Resistivity surveying equipment

Resistivity survey instruments are designed to measure the resistance of the ground, i.e. the ratio ($\Delta V/I$) in equations (8.10), (8.11) and (8.12), to a very high accuracy. They must be capable of reading to the very low levels of resistance commonly encountered in resistivity surveying. Apparent resistivity values are computed from the resistance measurements using the formula relevant to the electrode configuration in use.

Most modern resistivity meters employ low-frequency alternating current rather than direct current for two main reasons. Firstly, if direct current were employed there would eventually be a build up of anions around the negative electrode and cations around the positive electrode, i.e. electrolytic polarization would occur, and this would inhibit the arrival of further ions at the electrodes. Periodic reversal of the current prevents such an accumulation of ions and thus overcomes electrolytic polarization. Secondly, the use of alternating current overcomes the effects of telluric currents (see Chapter 9), which are natural electric currents in the ground that flow parallel to the Earth's surface and cause regional potential gradients. The use of alternating current nullifies their effects since at each current reversal the telluric currents alternately increase or decrease the measured potential difference by equal amounts. Summing the results over several cycles thus removes telluric effects (Fig. 8.7). The frequency of the alternating current used in resistivity surveying depends upon the required depth of penetration (see equation (9.2)). For penetration of the order of 10 m, a frequency of 100 Hz is suitable, and this is decreased to less than 10 Hz for depths of investigation of about 100 m. For very deep ground penetration direct currents must be used, and more complex measures adopted to overcome electrolytic polarization and telluric current effects.

Resistivity meters are designed to measure potential differences when no current is flowing. Such a null method is used to overcome the effects of contact resistance of the electrodes with the ground. The potential between the potential electrodes is balanced by the potential tapped from a variable resistance. No current then flows in the resistivity circuit so that contact resistance will not register, and the variable resistance reading represents the true resistance of the ground (equal to the ratio $\Delta V/I$ in the relevant equations).

Previous generations of resistivity meters required the nulling of a displayed voltage by manual manipulation of a resistor bank. Modern instruments are available with microprocessor-controlled electronic circuitry which accomplishes this operation internally and, moreover, performs checks on the circuitry before display of the result.

8.2.6 Interpretation of resistivity data

Electrical surveys are among the most difficult of all the geophysical methods to interpret quantitatively because of the complex theoretical basis of the technique. In resistivity interpretation, mathematical analysis is most highly developed for VES, less well for CST over two-dimensional structures and least well for CST over three-dimensional bodies. The resistivity method utilizes a potential field and consequently suffers from similar ambiguity problems to the gravitational and magnetic methods.

Since a potential field is involved, the apparent

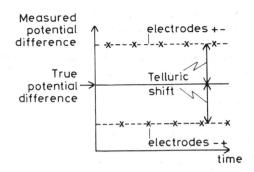

Fig. 8.7 The use of alternating current to remove the effects of telluric currents during a resistivity measurement. Summing the measured potential difference over several cycles provides the true potential difference.

resistivity signature of any structure should be computed by solution of Laplace's equation (Section 6.11) and insertion of the boundary conditions for the particular structure under consideration, or by integrating it directly. In practice such solutions are invariably complex. Consequently a simplified approach is initially adopted here in which electric fields are assumed to act in a manner similar to light. It should be remembered, however, that such an optical analogue is not strictly valid in all cases.

8.2.7 Vertical electrical sounding interpretation

Consider a Wenner electrode spread above a single horizontal interface between media with resistivities ρ_1 (upper) and ρ_2 (lower) with $\rho_1 > \rho_2$ (Fig. 8.8). On passing through the interface the current flow lines are deflected towards the interface in a fashion similar to refracted seismic waves (Chapter 3) since the less-resistive lower layer provides a more attractive path for the current. When the electrode separation is small, most of the current flows in the upper layer with the consequence that the apparent resistivity tends towards ρ_1. As the electrode separation is gradually increased more and more current flows within the lower layer and the apparent resistivity then approaches ρ_2. A similar situation obtains when $\rho_2 > \rho_1$, although in this case the apparent resistivity approaches ρ_2 more gradually as the more resistive lower layer is a less attractive path for the current.

Where three horizontal layers are present the apparent resistivity curves are more complex

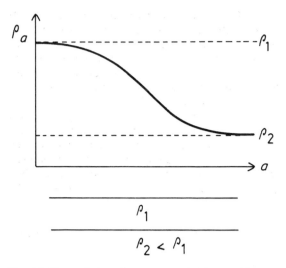

Fig. 8.8 The variation of apparent resistivity ρ_a with electrode separation a over a single horizontal interface between media with resistivities ρ_1 and ρ_2.

(Fig. 8.9). Although the apparent resistivity approaches ρ_1 and ρ_3 for small and large electrode spacings, the presence of the intermediate layer causes a deflection of the apparent resistivity curve at intermediate spacings. If the resistivity of the intermediate layer is greater or less than the resistivities of the upper and lower layers the apparent resistivity curve is either bell-shaped or basin-shaped (Fig. 8.9(a)). A middle layer with a resistivity intermediate between ρ_1 and ρ_3 produces apparent

resistivity curves characterized by a progressive increase or decrease in resistivity as a function of electrode spacing (Fig. 8.9(b)). The presence of four or more layers further increases the complexity of apparent resistivity curves.

Simple examination of the way in which apparent resistivity varies with electrode spacing may thus provide estimates of the resistivities of the upper and lowest layers and indicate the relative resistivities of any intermediate layers. In order to compute layer thicknesses it is necessary to be able to calculate the apparent resistivity of a layered structure. The first computation of this type was performed by Hummel in the 1930s using an optical analogue to calculate the apparent resistivity signature of a simple two-layered model.

Referring to Fig. 8.10, current I is introduced into the ground at point C_0 above a single interface at depth z between an upper medium 1 of resistivity ρ_1 and a lower medium 2 of resistivity ρ_2. The two parallel interfaces between medium 1 and 2 and between medium 1 and the air produce an infinite series of images of the source, located above and below the surface. Thus C_1 is the image of C_0 in the medium 1/2 interface at depth $2z$, C_1' is the image of C_1 in the medium 1/air interface at height $2z$, C_2 is the image of C_1' in the medium 1/2 interface at depth $4z$, etc. Each image in the medium 1/2 interface is reduced in intensity by a factor k, the reflection coefficient of the interface. (There is no reduction in intensity of images in the medium 1/air interface, as its reflection coefficient is unity). A consequence of the progressive reduction in intensity

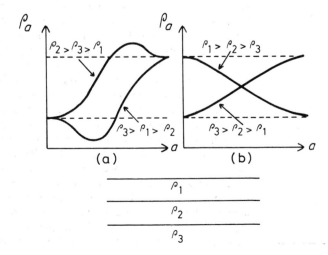

(a)   (b)

Fig. 8.9 The variation of apparent resistivity ρ_a with electrode separation a over three horizontal layers.

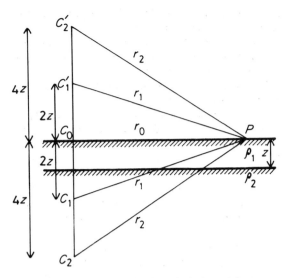

Fig. 8.10 Parameters used in the calculation of the potential due to a single surface electrode above a single horizontal interface using the method of images.

is that only a few images have to be considered in arriving at a reasonable estimate of the potential at point P.

Table 8.1 summarizes this argument.

The potential V_p at point P is the sum of the contributions of all sources. Employing equation (8.6)

$$V_p = \frac{I\rho_1}{2\pi r_0} + \frac{2kI\rho_1}{2\pi r_1} + \frac{2k^2 I\rho_1}{2\pi r_2} + \ldots + \frac{2k^i I\rho_1}{2\pi r_i} + \ldots$$

Thus

$$V_p = \frac{I\rho_1}{2\pi}\left(\frac{1}{r_0} + 2\sum_{n=1}^{\infty}\frac{k^n}{r_n}\right) \tag{8.13}$$

where

$$r_n = (r_0^2 + (2nz)^2)^{1/2}$$

Table 8.1. Distribution and intensity of electrical sources due to a single horizontal interface

Source	Intensity	Depth/height	Distance
C_0	I	0	r_0
C_1	kI	$2z$	r_1
C_1'	kI	$2z$	r_1
C_2	$k^2 I$	$4z$	r_2
C_2'	$k^2 I$	$4z$	r_2
.	.	.	.
.	.	.	.

The first term in the brackets of equation (8.13) refers to the normal potential attaining if the subsurface were homogeneous, and the second term to the disturbing potential caused by the interface. The series is convergent as the dimming factor, or reflection coefficient, k is less than unity ($k = (\rho_2 - \rho_1)/(\rho_2 + \rho_1)$, cf. Section 3.6.1).

Knowledge of the potential resulting at a single point from a single current electrode allows the computation of the potential difference ΔV between two electrodes, resulting from two current electrodes, by the addition and subtraction of their contribution to the potential at these points. For the Wenner system with spacing a

$$\Delta V = \frac{I\rho_1}{2\pi a}(1 + 4F) \tag{8.14}$$

where

$$F = \sum_{n=1}^{\infty} k^n \left[(1 + 4n^2 z^2/a^2)^{-1/2} - (4 + 4n^2 z^2/a^2)^{-1/2}\right] \tag{8.15}$$

Relating this to the apparent resistivity ρ_a measured by the Wenner system (equation (8.10))

$$\rho_a = \rho_1(1 + 4F) \tag{8.16}$$

Consequently the apparent resistivity can be computed for a range of electrode spacings.

Similar computations can be performed for multilayer structures, although the calculations are more easily executed using recurrence formulae and filtering techniques designed for this purpose (see later). Field data can then be compared with graphs (master curves) representing the calculated effects of layered models derived by such methods, a technique known as *curve matching*. Fig. 8.11 shows an interpretation using a set of master curves for vertical electrical sounding with a Wenner spread over two horizontal layers. The master curves are prepared in dimensionless form for a number of values of the reflection coefficient k by dividing the calculated apparent resistivity values ρ_a by the upper layer resistivity ρ_1 (the latter derived from the field curve at electrode spacings approaching zero), and by dividing the electrode spacings a by the upper layer thickness z_1. The curves are plotted on logarithmic paper, which has the effect of producing a more regular appearance as the fluctuations of resistivity then tend to be of similar wavelength over the entire length of the curves. The field curve to be interpreted is plotted on transparent logarithmic paper with the same modulus as the master curves. It is then shifted

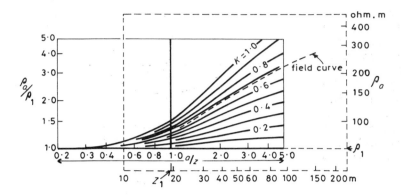

Fig. 8.11 The interpretation of a two-layer apparent resistivity graph by comparison with a set of master curves. The upper layer resistivity ρ_1 is $68\,\Omega\,m$ and its thickness z_1 is 19.5 m. (After Griffiths & King 1981.)

over the master curves, keeping the coordinate axes parallel, until a reasonable match is obtained with one of the master curves or with an interpolated curve. The point at which $\rho_a/\rho_1 = a/z = 1$ on the master sheet gives the true values of ρ_1 and z_1 on the relevant axes. ρ_2 is obtained from the k-value of the best-fitting curve.

Curve matching is simple for the two-layer case since only a single sheet of master curves is required. When three layers are present much larger sets of curves are required to represent the increased number of possible combinations of resistivities and layer thicknesses. Curve matching is simplified if the master curves are arranged according to curve type (Fig. 8.9), and sets of master curves for both Wenner and Schlumberger electrode configurations are available (Orellana & Mooney 1966, 1972). The number of master curves required for full interpretation of a four-layer field curve is prohibitively large although limited sets have been published.

The interpretation of resistivity curves over multi-layered structures may alternatively be performed by *partial curve matching* (Bhattacharya & Patra 1968). The method involves the matching of successive portions of the field curve by a set of two-layer curves. After each segment is fitted the interpreted resistivities and layer thickness are combined by use of auxilliary curves into a single layer with an equivalent thickness z_e and resistivity ρ_e. This equivalent layer then forms the upper layer in the interpretation of the next segment of the field curve with another two-layer curve (Fig. 8.12). Similar techniques are available in which successive use is made of three-layer master curves.

The curve matching methods described above are not now widely used because of the general availability of the more sophisticated interpretational

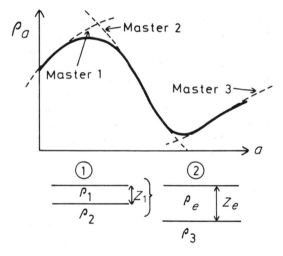

Fig. 8.12 The technique of partial curve matching. A two-layer curve is fitted to the early part of the graph and the resistivities ρ_1 and ρ_2 and thickness z_1 of the upper layer determined. ρ_1, ρ_2 and z_1 are combined into a single equivalent layer of resistivity ρ_e and thickness z_e which then forms the upper layer in the interpretation of the next segment of the graph with a second two-layer curve.

techniques described below. Curve matching methods might still be used, however, to obtain interpretations in the field in the absence of computing facilities, or to derive an approximate model that is to be used as a starting point for one of the more complex routines.

Equation (8.13) represents the potential at the surface resulting from a single point of current injection over two horizontal layers as predicted by the method of images. In general, however, the potential arising from any number of horizontal layers is derived by solution of Laplace's equation (Section 6.11). The equation in this case is normally represented in cylindrical coordinates as electrical fields have cylindrical symmetry with respect to the vertical line through the current source (Fig. 8.13). The solution and application of the relevant boundary conditions are complex (e.g. Koefoed 1979), but show that the potential V at the surface over a series of horizontal layers, the uppermost of which has a resistivity ρ_1, at a distance r from a current source of strength I is given by

$$V = \frac{\rho_1 I}{2\pi} \int_0^\infty K(\lambda) \, J_0(\lambda r) d\lambda \qquad (8.17)$$

λ is the variable of integration. $J_0(\lambda r)$ is a specialized function known as a Bessel function of order zero whose behaviour is known completely. $K(\lambda)$ is known as a kernel function and is controlled by the thicknesses and resistivities of the underlying layers. The kernel function can be built up relatively simply for any number of layers using *recurrence relationships* (Koefoed 1979) which progressively add the effects of successive layers in the sequence. A useful additional parameter is the resistivity transform $T(\lambda)$ defined by

$$T_i(\lambda) = \rho_i K_i(\lambda) \qquad (8.18)$$

where $T_i(\lambda)$ is the resistivity transform of the ith layer which has a resistivity ρ_i and a kernel function $K_i(\lambda)$. $T(\lambda)$ can similarly be constructed using recurrence relationships.

By methods analogous to those used to construct equation (8.16), a relationship between the apparent resistivity and resistivity transform can be derived. For example, this relationship for a Wenner spread with electrode spacing a is

$$\rho_a = 2a \int_0^\infty T(\lambda)[J_0(\lambda a) - J_0(2\lambda a)]d\lambda \qquad (8.19)$$

The resistivity transform function has the dimensions of resistivity and the variable λ has the dimensions of inverse length. It has been found that if $T(\lambda)$ is plotted as a function of λ^{-1} the relationship is similar to the variation of apparent resistivity with electrode spacing for the same sequence of horizontal layers. Indeed only a simple filtering operation is required to transform the $T(\lambda):\lambda^{-1}$ relationship (resistivity transform) into the $\rho_a:a$ relationship (apparent resistivity function). Such a filter is known as an indirect filter. The inverse operation, i.e. the

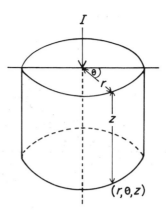

Fig. 8.13 Cylindrical polar coordinates.

determination of the resistivity transform from the apparent resistivity function, can be performed using a direct filter.

Apparent resistivity curves over multilayered models can be computed relatively easily by determining the resistivity transform from the layer parameters using a recurrence relationship and then filtering the transform to derive the apparent resistivity function. Such a technique is considerably more efficient than the method used in the derivation of equation (8.13).

This method leads to a form of interpretation similar to the indirect interpretation of gravity and magnetic anomalies, in which field data are compared with data calculated for a model whose parameters are varied in order to simulate the field observations. This comparison can be made either between observed and calculated apparent resistivity profiles or the equivalent resistivity transforms, the latter method requiring the derivation of the resistivity transform from the field resistivity data by direct filtering. Such techniques lend themselves well to automatic iterative processes of interpretation in which a computer performs the adjustments necessary to a layered model derived by an approximate interpretation method in order to improve the correspondence between observed and calculated functions.

In addition to this indirect modelling there are also a number of direct methods of interpreting resistivity data which derive the layer parameters directly from the field profiles (e.g. Zohdy 1989). Such methods usually involve the following steps:
1 Determination of the resistivity transform of the field data by direct filtering.

2 Determination of the parameters of the upper layer by fitting the early part of the resistivity transform curve with a synthetic two-layer curve.

3 Subtraction of the effects of the upper layer by reducing all observations to the base of the previously determined layer by the use of a *reduction equation* (the inverse of a recurrence relationship).

Steps **2** and **3** are then repeated so that the parameters of successively deeper layers are determined. Such methods suffer from the drawback that errors increase with depth so that any error made early in the interpretation becomes magnified. The direct interpetation methods consequently employ various techniques to suppress such error magnification.

The indirect and direct methods described above have now largely superseded curve-matching techniques and provide considerably more accurate interpretations.

Interpretation of VES data suffers from non-uniqueness arising from problems known as *equivalence* and *suppression*. The problem of equivalence (see, for example, van Overmeeren 1989) is illustrated by the fact that identical bell-shaped or basin-shaped resistivity curves (Fig. 8.9(a)) can be obtained for different layered models. Identical bell-shaped curves are obtained if the product of the thickness z and resistivity ρ, known as the transverse resistance, of the middle layer remains constant. For basin-shaped curves the equivalence function of the middle layer is z/ρ, known as the longitudinal conductance. The problem of suppression applies to resistivity curves in which apparent resistivity progressively increases or decreases as a function of electrode spacing (Fig. 8.9(b)). In such cases the addition of an extra intermediate layer causes a slight horizontal shift of the curve without altering its overall shape. In the interpretation of relatively noisy field data such an intermediate layer may not be detected.

It is the conventional practice in VES interpretation to make the assumption that layers are horizontal and isotropic. Deviations from these assumptions result in errors in the final interpretation.

The assumption of isotropy can be incorrect for individual layers. For example, in sediments such as clay or shale the resistivity perpendicular to the layering is usually greater than in the direction of the layering. Anisotropy cannot be detected in subsurface layers during vertical electrical sounding and normally results in too large a thickness being assigned to the layers. Other anisotropic effects are depth dependent, e.g. the reduction with depth of the degree of weathering, and the increase with depth of both compaction of sediments and salinity of pore fluids. The presence of a vertical contact, such as a fault, gives rise to lateral inhomogeneity which can greatly affect the interpretation of an electrical sounding in its vicinity.

If the layers are dipping, the basic theory discussed above is invalid. Using the optical analogue, the number of images produced by a dipping interface is finite, the images being arranged around a circle (Fig. 8.14). Because the intensity of the images progressively decreases, only the first few need to be considered in deriving a reasonable estimate of the resulting potential. Consequently the effect of dip can probably be ignored for inclinations up to about 20°, which provide a sufficient number of images.

Topography can influence electrical surveys as current flow lines tend to follow the ground surface. Equipotential surfaces are thus distorted and anomalous readings can result.

8.2.8 Constant separation traversing interpretation

Constant separation traverses are obtained by moving an electrode spread with fixed electrode separation along a traverse line, the array of electrodes being aligned either in the direction of the traverse (longitudinal traverse) or at right angles to it (transverse traverse). The former technique is more efficient as only a single electrode has to be moved from one end of the spread to the other, and the electrodes reconnected, between adjacent readings.

Figure 8.15 shows a transverse traverse across a single vertical contact between two media of resistivities ρ_1 and ρ_2. The apparent resistivity curve varies smoothly from ρ_1 to ρ_2 across the contact.

A longitudinal traverse over a similar structure shows the same variation from ρ_1 to ρ_2 at its extremities, but the intermediate parts of the curve exhibit a number of cusps (Fig. 8.16), which correspond to locations where successive electrodes cross the contact. There will be four cusps on a Wenner profile but two on a Schlumberger profile where only the potential electrodes are mobile.

Fig. 8.17 shows the results of transverse and longitudinal traversing across a series of faulted strata in Illinois, USA. Both sets of results illustrate well the strong resistivity contrasts between the

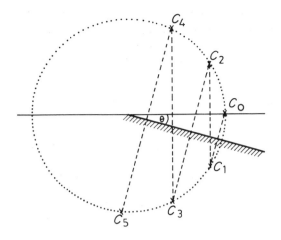

Fig. 8.14 Apparent current sources caused by a dipping interface. The sources $C_1 - C_5$ are successive images of the primary source C_0 in the interface and the surface. The sources lie on a circle centred on the outcrop of the interface, and their number is dependent upon the magnitude of the dip of the interface θ.

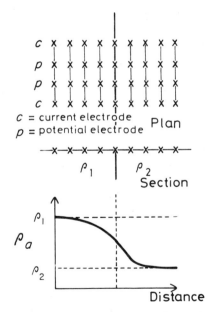

Fig. 8.15 A transverse traverse across a single vertical interface.

relatively conductive sandstone and relatively resistive limestone.

A vertical discontinuity distorts the direction of current flow and thus the overall distribution of

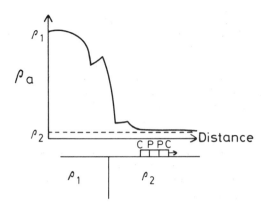

Fig. 8.16 A longitudinal traverse across a single vertical interface employing a configuration in which all four electrodes are mobile. (After Parasnis 1973.)

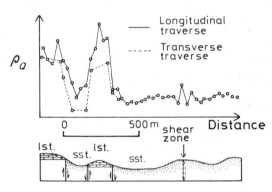

Fig. 8.17 Longitudinal and transverse traverses across a series of faulted strata in Illinois, USA. (After Hubbert 1934.)

potential in its vicinity. The potential distribution at the surface can be determined by an optical analogue in which the discontinuity is compared with a semi-transparent mirror which both reflects and transmits light. Referring to Fig. 8.18, current I is introduced at point C on the surface of a medium of resistivity ρ_1 in the vicinity of a vertical contact with a second medium of resistivity ρ_2.

In the optical analogue, a point P on the same side of the mirror as the source would receive light directly and via a single reflection. In the latter case the light would appear to originate from the image of C in the mirror C' and would be decreased in intensity with respect to the source by a factor corresponding to the reflection coefficient. Both the electric source and its image contribute to the

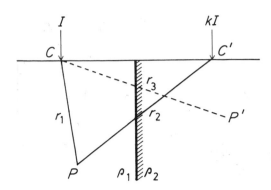

Fig. 8.18 Parameters used in the calculation of the potential due to a single surface current electrode on either side of a single vertical interface.

potential V_p at P, the latter being decreased in intensity by a factor k, the reflection coefficient. From equation (8.6)

$$V_p = \frac{I\rho_1}{2\pi}\left(\frac{1}{r_1} + \frac{k}{r_2}\right) \qquad (8.20)$$

For a point P' on the other side of the interface from the source, the optical analogue indicates that light would be received only after transmission through the mirror, resulting in a reduction in intensity by a factor corresponding to the transmission coefficient. The only contributor to the potential $V_{p'}$ at P' is the current source reduced in intensity by the factor $(1 - k)$. From equation (8.6)

$$V_{p'} = \frac{I(1 - k)\rho_2}{2\pi r_3} \qquad (8.21)$$

Equations (8.20) and (8.21) may be used to calculate the measured potential difference for any electrode spread between two points in the vicinity of the interface and thus to construct the form of an apparent resistivity profile produced by longitudinal constant separation traversing. In fact, five separate equations are required, corresponding to the five possible configurations of a four-electrode spread with respect to the discontinuity. The method can also be used to construct apparent resistivity profiles for constant separation traversing over a number of adjacent discontinuities. Albums of master curves are available for single and double vertical contacts (Logn 1954).

Three-dimensional resistivity anomalies may be obtained by contouring apparent resistivity values

from a number of CST lines. The detection of a three-dimensional body is usually only possible when its top is close to the surface, and traverses must be made directly over the body or very near to its edges if its anomaly is to be registered.

Three-dimensional anomalies may be interpreted by laboratory modelling. For example, metal cylinders, blocks or sheets may be immersed in water whose resistivity is altered by adding various salts and the model moved beneath a set of stationary electrodes. The shape of the model can then be varied until a reasonable approximation to the field curves is obtained.

The mathematical analysis of apparent resistivity variations over bodies of regular or irregular form is complex but equations are available for simple shapes such as spheres or hemispheres (Fig. 8.19), and it is also possible to compute the resistivity response of two-dimensional bodies with an irregular cross-section (Dey & Morrison 1979).

Three-dimensional anomalies may also be obtained by an extension of the CST technique known as the *mise-à-la-masse method*. This is employed when part of a conductive body, for example, an ore body, has been located either at outcrop or by drilling. One current electrode is sited within the body, the other being placed a large distance away on the surface (Fig. 8.20). A pair of potential electrodes is then moved over the surface mapping equipotential lines (lines joining the electrodes when the indicated potential difference is zero). The method provides much more information on the extent, dip, strike and continuity of the body than the normal CST techniques.

8.2.9 Limitations of the resistivity method

Resistivity surveying is an efficient method for delineating shallow layered sequences or vertical discontinuities involving changes of resistivity. It does, however, suffer from a number of limitations:

1 Interpretations are ambiguous. Consequently, independent geophysical and geological controls are necessary to discriminate between valid alternative interpretations of the resistivity data.

2 Interpretation is limited to simple structural configurations. Any deviations from these simple situations may be impossible to interpret.

3 Topography and the effects of near-surface resistivity variations can mask the effects of deeper variations.

4 The depth of penetration of the method is limited

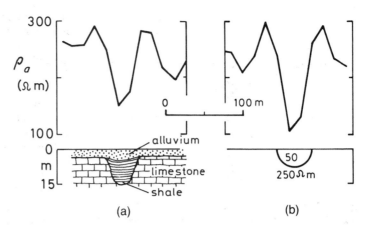

Fig. 8.19 (a) The observed Wenner resistivity profile over a shale-filled sink of known geometry in Kansas, USA. (b) The theoretical profile for a buried hemisphere. (After Cook & Van Nostrand 1954.)

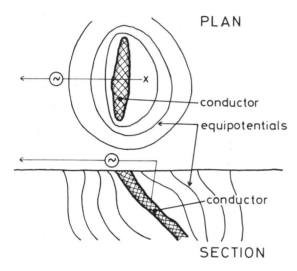

Fig. 8.20 The mise-à-la-masse method.

by the maximum electrical power that can be introduced into the ground and by the practical difficulties of laying out long lengths of cable. The practical depth limit for most surveys is about 1 km.

8.2.10 Applications of resistivity surveying

Resistivity surveys are usually restricted to relatively small-scale investigations because of the labour involved in physically planting the electrodes prior to each measurement. For this reason resistivity methods are not commonly used in reconnaissance exploration. It is probable, however, that with the increasing availability of non-contacting conductivity measuring devices (Section 9.7) this restriction will no longer apply.

Resistivity methods are widely used in engineering geological investigations of sites prior to construction. VES is a very convenient, non-destructive method of determining the depth to rockhead for foundation purposes and also provides information on the degree of saturation of subsurface materials. CST can be used to determine the variation in rockhead depth between soundings and can also indicate the presence of potentially unstable ground conditions. Fig. 8.21 shows a CST profile which has revealed the presence of a buried mineshaft from the relatively high resistivity values associated with its poorly-compacted infill. Similar techniques can be used in archaeological investigations for the location of artifacts with anomalous resistivities. For example, Fig. 8.22 shows CST profiles across an ancient buried ditch.

Probably the most widely-employed use of resistivity surveys is in hydrogeological investigations, as important information can be provided on geological structure, lithologies and subsurface water resources without the large cost of an extensive programme of drilling. The results can determine the locations of the minimum number of exploratory boreholes required for both essential aquifer tests and control of the geophysical interpretation.

The resistivity method was used by Bugg & Lloyd (1976) to delineate fresh water lenses in Grand Cayman Island of the northern Caribbean (Fig. 8.23). Because of its relatively low density, fresh water tends to float on the denser saline water which penetrates the limestone substrate of the island from the sea. Fig. 8.24 shows a fluid conductivity profile from a borehole sunk in the Central Lens compared with the results of a VES interpretation from a sounding adjacent to the borehole. It is apparent

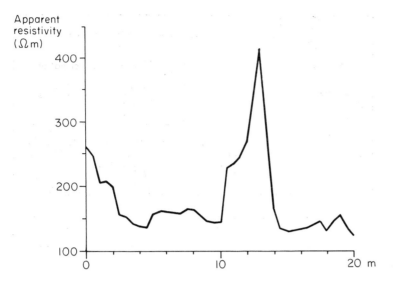

Fig. 8.21 CST resistivity profile across a buried mineshaft. (After Aspinall & Walker 1975.)

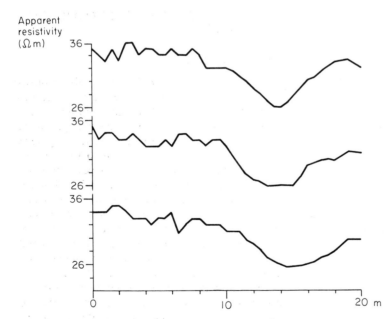

Fig. 8.22 Resistivity profiles across a buried ditch 4 m wide. (After Aspinall & Walker 1975.)

that fresh water can be distinguished from saline water by its much higher resistivity. The resistivity survey took the form of a series of VES which were interpreted using the sounding by the borehole as control. Contours on the base of the Central Lens, defined from these interpretations, are shown in Fig. 8.25.

Resistivity surveys can also be used to locate and monitor the extent of groundwater pollution. Merkel (1972) described the use of this technique in the delineation of contaminated mine drainage from old coal workings in Pennsylvania, USA. Fig. 8.26 shows a geoelectric section across part of the area, constructed from a series of VES, and its geological interpretation which indicates that no pollution is present. Fig. 8.27 shows a further geoelectric section from an adjacent area in which acid mine drainage has increased the conductivity of the groundwater, allowing its delineation as a band of low resistivity. Further VES enabled the extent of the pollution to

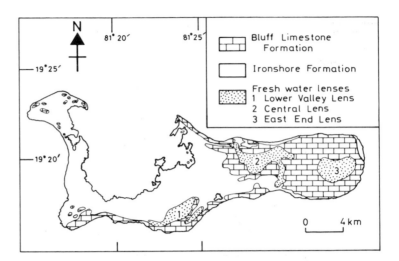

Fig. 8.23 Simplified geology and fresh water lenses of Grand Cayman. (After Bugg & Lloyd 1976.)

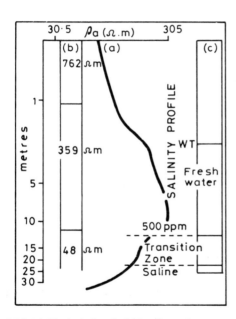

Fig. 8.24 (a) Vertical electrical sounding adjacent to a test borehole in the Central Lens, Grand Cayman. (b) Layered model interpretation of the VES. (c) Interpreted salinity profile. (After Bugg & Lloyd 1976.)

be defined. Since contamination of this type is associated with a significant change in resistivity, periodic measurements at electrodes sited in a borehole penetrating the water table could be used to monitor the onset of pollution and the degree of contamination.

8.3 INDUCED POLARIZATION (IP) METHOD

8.3.1 Principles

If, when using a standard four-electrode resistivity spread in a DC mode, the current is abruptly switched off, the voltage between the potential electrodes does not drop to zero immediately. After a large initial decrease the voltage suffers a gradual decay and can take many seconds to reach a zero value (Fig. 8.28). A similar phenomenon is observed as the current is switched on. After an initial sudden voltage increase, the voltage increases gradually over a discrete time interval to a steady state value. The ground thus acts as a capacitor and stores electrical charge, i.e. becomes electrically polarized.

If, instead of using a DC source for the measurement of resistivity, a variable low frequency AC source is used, it is found that the measured apparent resistivity of the subsurface decreases as the frequency is increased. This is because the capacitance of the ground inhibits the passage of direct currents but transmits alternating currents with increasing efficiency as the frequency rises.

The capacitative property of the ground causes both the transient decay of a residual voltage and the variation of apparent resistivity as a function of frequency. The two effects are representations of the same phenomenon in the time and frequency domains, and are linked by Fourier transformation (see Chapter 2). These two manifestations of the capacitance property of the ground provide two

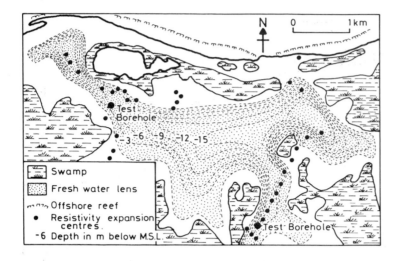

Fig. 8.25 Configuration of base of Central Lens, Grand Cayman. (After Bugg & Lloyd 1976.)

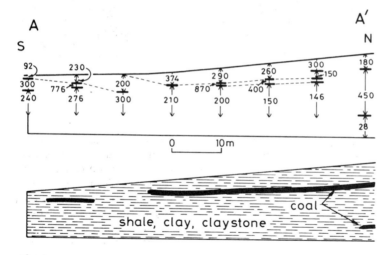

Fig. 8.26 Geoelectric section and geological interpretation of a profile near Kylertown, Pennsylvania. Numbers refer to resistivity in ohm m. (After Merkel 1972.)

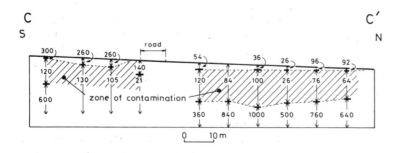

Fig. 8.27 A further geoelectric section from Kylertown, Pennsylvania. Shaded area shows zone of contamination. Numbers refer to resistivity in ohm m. (After Merkel 1972.)

different survey methods for the investigations of the effect.

The measurement of a decaying voltage over a certain time interval is known as *time domain* IP

surveying. Measurement of apparent resistivity at two or more low AC frequencies is known as *frequency domain* IP surveying.

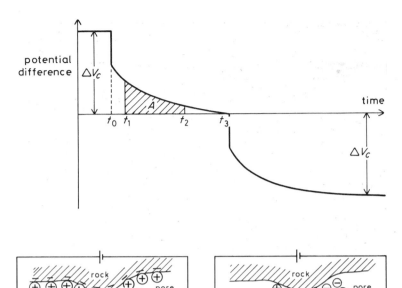

Fig. 8.28 The phenomenon of induced polarization. At time t_0 the current is switched off and the measured potential difference, after an initial large drop from the steady state value ΔV_c, decays gradually to zero. A similar sequence occurs when the current is switched on at time t_3. A represents the area under the decay curve for the time increment $t_1 - t_2$.

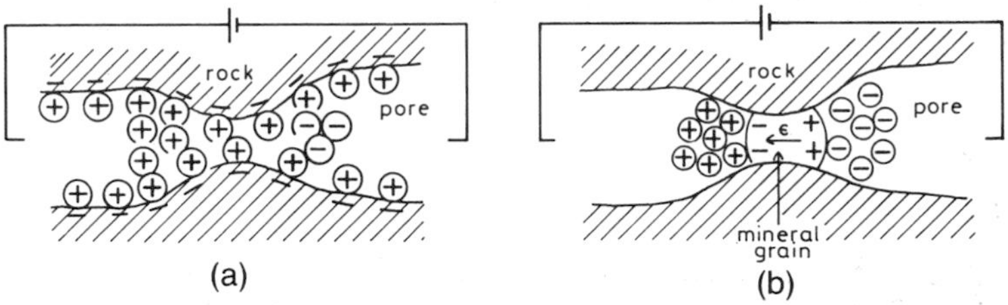

Fig. 8.29 Mechanisms of induced polarization: (a) membrane polarization (b) electrode polarization.

8.3.2 Mechanisms of induced polarization

Laboratory experiments indicate that electrical energy is stored in rocks mainly by electrochemical processes. This is achieved in two ways.

The passage of current through a rock as a result of an impressed voltage is accomplished mainly by electrolytic flow in the pore fluid. Most of the rock-forming minerals have a net negative charge on their interface with the pore fluid and attract positive ions onto this surface (Fig. 8.29(a)). The concentration of positive ions extends about $100\,\mu m$ into the pore fluid, and if this distance is of the same order as the diameter of the pore throats, the movement of ions in the fluid resulting from the impressed voltage is inhibited. Negative and positive ions thus build up on either side of the blockage and, on removal of the impressed voltage, return to their original locations over a finite period of time causing a gradually decaying voltage.

This effect is known as *membrane polarization* or *electrolytic polarization*. It is most pronounced in the presence of clay minerals where the pores are particularly small, but the effect decreases with increasing salinity of the pore fluid.

When metallic minerals are present in a rock, an alternative, electronic path is available for current flow. Fig. 8.29(b) shows a rock in which a metallic mineral grain blocks a pore. When a voltage is applied to either side of the pore space, positive and negative charges are impressed on opposite sides of the grain. Negative and positive ions then accumulate on either side of the grain which are attempting either to release electrons to the grain or to accept electrons conducted through the grain. The rate at which the electrons are conducted is slower than the rate of electron exchange with the ions. Consequently ions accumulate on either side of the grain and cause a build up of charge. When the impressed voltage is removed the ions slowly diffuse back to their original locations and cause a transitory decaying voltage.

This effect is known as *electrode polarization* or *overvoltage*. All minerals which are good conductors (e.g. metallic sulphides and oxides, graphite) contribute to this effect. The magnitude of the electrode polarization effect depends upon both the magnitude of the impressed voltage and the mineral concentration. It is most pronounced when the mineral is disseminated throughout the host rock as the surface area available for ionic-electronic interchange is then at a maximum. The effect decreases with increasing porosity as more alternative paths become available for the more efficient ionic conduction.

In prospecting for metallic ores, interest in obviously in the electrode polarization (overvoltage) effect. Membrane polarization, however, is indistinguishable from this effect during IP measurements. Membrane polarization consequently reduces the effectiveness of IP surveys and causes geological 'noise' which may be equivalent to the overvoltage effect of a rock with up to 2% metallic minerals.

8.3.3 Induced polarization measurements

Time domain IP measurements involve the monitoring of the decaying voltage after the current is switched off. The most commonly measured parameter is the *chargeability M*, defined as the area A beneath the decay curve over a certain time interval $(t_1 - t_2)$ normalized by the steady-state potential difference ΔV_c (Fig. 8.28)

$$M = \frac{A}{\Delta V_c} = \frac{1}{\Delta V_c} \int_{t_1}^{t_2} V(t)dt \qquad (8.22)$$

Chargeability is measured over a specified time interval shortly after the polarizing current is cut off (Fig. 8.28). The area A is determined within the measuring apparatus by analogue integration. Different minerals are distinguished by characteristic chargeabilities, e.g. pyrite has $M = 13.4$ ms over an interval of 1 s, and magnetite 2.2 ms over the same interval. Fig. 8.28 also shows that current polarity is reversed between successive measurements in order to destroy any remanent polarization.

Frequency domain techniques involve the measurement of apparent resistivity at two or more AC frequencies. Fig. 8.30 shows the relationship between apparent resistivity and log current frequency. Three distinct regions are apparent: region 1 is in low frequencies where resistivity is independent of frequency; region 2 is the Warberg region where resistivity is a linear function of log frequency; region 3 is the region of electromagnetic induction (Chapter 9) where current flow is by induction rather than simple conduction. Since the relationship illustrated in Fig. 8.30 varies with rock type and mineral concentration, IP measurements are usually made at frequencies at, or below, 10 Hz to remain in the non-inductive regions.

Two measurements are commonly made. The

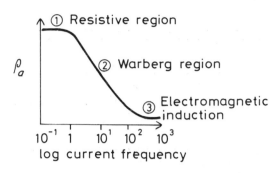

Fig. 8.30 The relationship between apparent resistivity and measuring current frequency.

percentage frequency effect PFE is defined as

$$\text{PFE} = 100(\rho_{0.1} - \rho_{10})/\rho_{10} \qquad (8.23)$$

where $\rho_{0.1}$ and ρ_{10} are apparent resistivities at measuring frequencies of 0.1 and 10 Hz. The *metal factor* MF is defined as

$$\text{MF} = 2\pi 10^5 (\rho_{0.1} - \rho_{10})/\rho_{0.1}\rho_{10} \qquad (8.24)$$

This factor normalizes the PFE with respect to the lower-frequency resistivity and consequently removes, to a certain extent, the variation of the IP effect with the effective resistivity of the host rock.

A common method of presenting IP measurements is the *pseudosection*, in which readings are plotted so as to reflect the depth of penetration. Fig. 8.31 illustrates how a pseudosection is constructed for a double-dipole array. Measured values are plotted at the intersections of lines sloping at 45° from the centres of the potential and current electrode pairs. Values are thus plotted at depths which reflect the increasing depth of penetration as the length of the array increases. The values are then contoured. VES resistivity data can also be presented in this way with the plotted depth proportional to the current electrode separation. Pseudosections give only a crude representation of the IP response distribution at depth: for example, the apparent dip of the anomalous body is not always the same as the true dip. An example of this method of presentation is shown in Fig. 8.32.

8.3.4 Field operations

IP equipment is similar to resistivity apparatus but is rather more bulky and elaborate. Theoretically, any standard electrode spread may be employed but

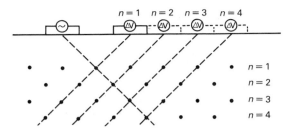

Fig. 8.31 The presentation of double-dipole IP results on a pseudosection. *n* represents the relative spacing between the current and potential electrode pairs.

in practice the double-dipole, pole-dipole and Schlumberger configurations (Fig. 8.33) are the most effective. Electrode spacings may vary from 3 to 300 m with the larger spacings used in reconnaissance surveys. To reduce the labour of moving current electrodes and generator, several pairs of current electrodes may be used, all connected via a switching device to the generator. Traverses are made over the area of interest plotting the IP reading at the mid-point of the electrode array (marked by crosses in Fig. 8.33).

Noise in an IP survey can result from several

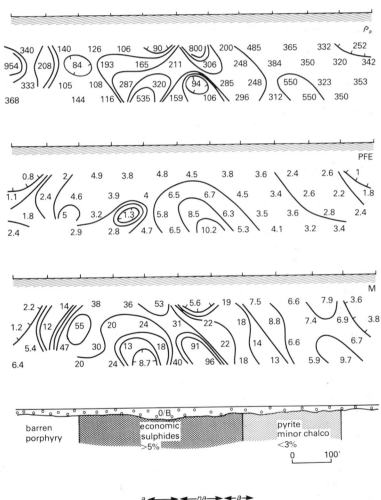

Fig. 8.32 Pseudosections of apparent resistivity (ρ_a), percentage frequency effect (PFE) and metal factor parameter (M) for a double-dipole IP traverse across a zone of massive sulphides whose shape is known from subsequent test drilling. Current and potential electrode spacing *a* was 100 feet (30.5 m). Frequencies used for the IP measurements were 0.31 and 5.0 Hz. (After Fountain 1972.)

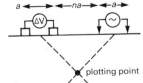

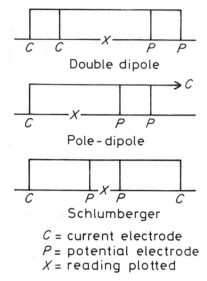

C = current electrode
P = potential electrode
X = reading plotted

Fig. 8.33 Electrode configurations used in induced polarization measurements.

phenomena. Telluric currents cause similar anomalous effects to those encountered in resistivity measurements. Noise also results from the general IP effect of barren rocks caused by membrane polarization. Noise generated by the measuring equipment results from electromagnetic coupling between adjacent wires. Such effects are common when alternating current is used since currents can be induced to flow in adjacent conductors. Consequently, cables should be at least 10 m apart and if they must cross they should do so at right angles to minimize electromagnetic induction effects.

8.3.5 Interpretation of induced polarization data

Quantitative interpretation is considerably more complex than for the resistivity method. The IP response has been computed analytically for simple features such as spheres, ellipsoids, dykes, vertical contacts and horizontal layers, enabling indirect interpretation (numerical modelling) techniques to be used.

Laboratory modelling can also be employed in indirect interpretation to simulate an observed IP anomaly. For example, apparent resistivities may be measured for various shapes and resistivities of a gelatin−copper-sulphate body immersed in water.

Much IP interpretation is, however, only qualitative. Simple parameters of the anomalies, such as sharpness, symmetry, amplitude and spatial distribution may be used to estimate the location, lateral extent, dip and depth of the anomalous zone.

The IP method suffers from the same disadvantages as resistivity surveying (see Section 8.2.9). Further, the sources of significant IP anomalies are often not of economic importance, e.g. water-filled shear zones and graphite-bearing sediments can both generate strong IP effects. Field operations are slow and the method is consequently far more expensive than most other ground geophysical techniques, survey costs being comparable with those of a gravity investigation.

8.3.6 Applications of induced polarization surveying

In spite of its drawbacks, the IP method is extensively used in base metal exploration as it has a high success rate in locating low-grade ore deposits such as disseminated sulphides (e.g. Langore *et al.* 1989). These have a strong IP effect but are non-conducting and therefore are not readily detectable by the electromagnetic methods discussed in Chapter 9. IP is by far the most effective geophysical method that can be used in the search for such targets.

Fig. 8.34 shows the chargeability profile for a time domain IP survey using a pole-dipole array across the Gortdrum copper−silver ore body in Ireland. Although the deposit is of low grade, containing less than 2% conducting minerals, the chargeability anomaly is well defined and centred over the ore body. In contrast, the corresponding apparent resistivity profile reflects the large resistivity contrast between the Old Red Sandstone and dolomitic limestone but gives no indication of the presence of the mineralization.

A further example of an IP survey is illustrated in Fig. 8.35 which shows a traverse over a copper porphyry body in British Columbia, Canada. IP and resistivity traverses were made at three different electrode spacings of a pole-dipole array. The CST results exhibit little variation over the body, but the IP (chargeability) profiles clearly show the presence of the mineralization, allow its limits to be determined and provide estimates of the depth to its upper surface.

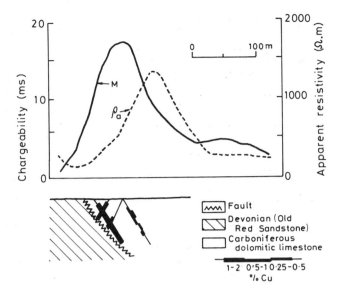

Fig. 8.34 Time domain IP profile using a pole-dipole array over the Gortdrum copper–silver body, Ireland. (After Seigel 1967.)

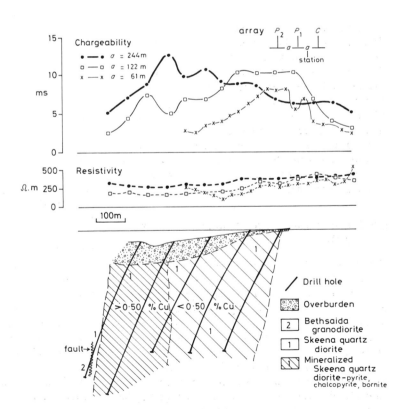

Fig. 8.35 Time domain induced polarization and resistivity profiles over a copper porphyry body in British Columbia, Canada. (After Seigel 1967.)

8.4 SELF-POTENTIAL (SP) METHOD

8.4.1 Introduction

The self-potential (or spontaneous polarization) method is based on the surface measurement of natural potential differences resulting from electrochemical reactions in the subsurface. Typical SP anomalies may have an amplitude of several hundred millivolts with respect to barren ground. They invariably exhibit a central negative anomaly and are stable over long periods of time. They are usually associated with deposits of metallic sulphides (Corry 1985), magnetite or graphite.

8.4.2 Mechanism of self potential

Field studies indicate that for a self-potential anomaly to occur its causative body must lie partially in a zone of oxidation. A widely-accepted mechanism of self potential (Sato & Mooney 1960; for a more recent analysis see Kilty 1984) requires the causative body to straddle the water table (Fig. 8.36). Below the water table electrolytes in the pore fluids undergo oxidation and release electrons which are conducted upwards through the ore body. At the top of the body the released electrons cause reduction of the electrolytes. A circuit thus exists in which current is carried electrolytically in the pore fluids and electronically in the body so that the top of the body acts as a negative terminal. This explains the negative SP anomalies that are invariably observed and, also, their stability as the ore body itself undergoes no chemical reactions and merely serves to transport electrons from depth. As a result of the subsurface currents, potential differences are produced at the surface.

8.4.3 Self-potential equipment and survey procedure

Field equipment consists simply of a pair of electrodes connected via a high-impedance millivoltmeter. The electrodes must be non-polarizing as simple metal spikes would generate their own SP effects. Non-polarizing electrodes consist of a metal immersed in a saturated solution of its own salt, such as copper in copper sulphate. The salt is contained in a porous pot which allows slow leakage of the solution into the ground.

Station spacing is generally less than 30 m. Trav-

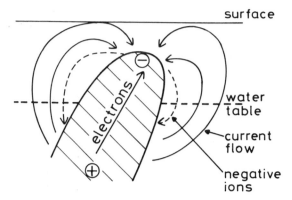

Fig. 8.36 The mechanism of self potential anomalies. (After Sato & Mooney 1960.)

erses may be performed by leapfrogging successive electrodes or, more commonly, by fixing one electrode in barren ground and moving the other over the survey area.

8.4.4 Interpretation of self-potential anomalies

The interpretation of SP anomalies is similar to magnetic interpretation because dipole fields are involved in both cases. It is thus possible to calculate the potential distributions around polarized bodies of simple shape such as spheres and ellipsoids by making assumptions about the distribution of charge over their surfaces.

Most interpretation, however, is qualitative. The anomaly minimum is assumed to occur directly over the anomalous body, although it may be displaced downhill in areas of steep topography. The anomaly half-width provides a rough estimate of depth. The symmetry or asymmetry of the anomaly provides information on the attitude of the body, the steep slope and positive tail of the anomaly lying on the downdip side.

The type of overburden can have a pronounced effect on the presence or absence of SP anomalies. Sand has little effect but a clay cover can mask the SP anomaly of an underlying body.

The SP method is only of minor importance in exploration. This is because quantitative interpretation is difficult and the depth of penetration is limited to about 30 m. It is, however, a rapid and cheap method requiring only simple field equipment. Consequently it can be useful in rapid ground

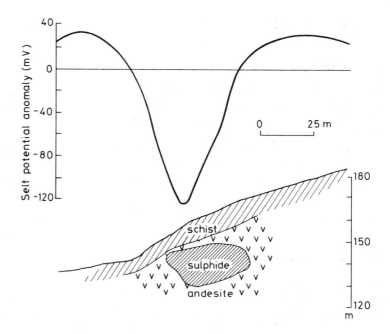

Fig. 8.37 The SP anomaly over a sulphide ore body at Sariyer, Turkey. (After Yüngül 1954.)

reconnaissance for base metal deposits when used in conjunction with magnetic, electromagnetic and geochemical techniques. It has also been used in hydrogeological investigations (e.g. Fournier 1989).

Fig. 8.37 shows the SP profile over a sulphide ore body in Turkey which contains copper concentrations of up to 14 %. The SP anomaly is negative and has an amplitude of some 140 mV. The steep topography has displaced the anomaly minimum downhill from the true location of the ore body.

8.5 PROBLEMS

1 Using the method of electrical images, derive the relationship between apparent resistivity, electrode spacing, layer thicknesses and resistivities for a VES performed with a Schlumberger spread over a single horizontal interface between media with resistivities ρ_1 and ρ_2.

2 At locations A, B, C and D along the gravity profile shown in Fig. 8.38, VES were performed with a Wenner array with the spread laid perpendicular to the profile. It

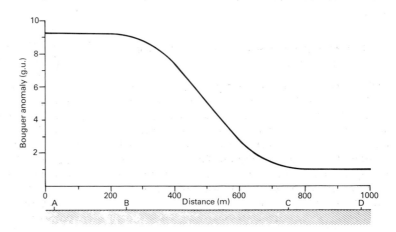

Fig. 8.38 Gravity anomaly profile pertaining to Question 2 showing also the locations of the VES at A, B, C and D.

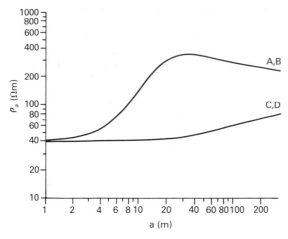

Fig. 8.39 Wenner VES sounding data for the locations shown in Fig. 8.38.

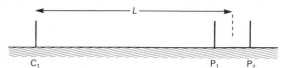

Fig. 8.40 The half-Schlumberger electrode configuration. See Question 4.

was found that the sounding curves, shown in Fig. 8.39, were similar for locations A and B and for C and D. A borehole close to A penetrated 3 m of drift, 42 m of limestone and bottomed in sandstone. Downhole geophysical surveys (Chapter 11) provided the following values of density (ρ_D) and resistivity (ρ_R) for the lithologies encountered:

Unit	ρ_R (Ω m)	ρ_D (Mg m^{-3})
Drift	40	2.00
Limestone	2000	2.75
Sandstone	200	2.40

A seismic refraction line near to D revealed 15 m of drift, although the nature of the underlying basement could not be assessed from the seismic velocity.

(a) Interpret the geophysical data so as to provide a geological section along the profile.

(b) What further techniques might be used to confirm your interpretation?

(c) If a CST were to be performed along the profile, select, giving reasons, a suitable electrode spacing to map the basement. Sketch the expected form of the CST for both longitudinal and transverse traverses.

3 Calculate the variation in apparent resistivity along a CST profile at right angles to a vertically faulted contact between sandstone and limestone, with apparent resistivities of 50 ohm m and 600 ohm m, respectively, for a Wenner configuration. What would be the effect on the profiles if the contact dipped at a shallower angle?

4 Fig. 8.40 shows a half-Schlumberger resistivity array in which the second current electrode is situated at a great distance from the other electrodes. Derive an

expression for the apparent resistivity of this array in terms of the electrode spacings and the measured resistance.

The following data represent measurements taken with a half-Schlumberger array along a profile across gneissic terrain near Kongsberg, Sweden. The potential electrode separation was kept constant at 20 m and the current electrode C_1 was fixed at the origin of the profile so that as L increased a CST was built up. R represents the resistance measured by the resistivity apparatus.

L (m)	ρ_R (Ω)
30.2	1244.818
53.8	255.598
80.9	103.812
95.1	73.846
106.0	58.820
120.0	45.502
143.8	31.416
168.4	22.786
179.6	19.993
205.1	15.290
229.3	12.209
244.0	10.785

Calculate the apparent resistivity for each reading and plot a profile illustrating the results.

In this region it is known that the gneiss can be extensively brecciated. Does the CST give any indication of brecciation?

5 The following table represents the results of a frequency domain IP survey of a Precambrian shield area. A double-dipole array was used with the separation (x) of both the current electrodes and the potential electrodes kept constant at 60 m. n refers to the number of separations between the current and potential electrode pairs and c to the distance of the centre of the array from the origin of the profile, where the results are plotted (Fig. 8.41). Measurements were taken using direct current and an alternating current of 10 Hz. These provided the apparent resistivities ρ_{dc} and ρ_{ac}, respectively.

(a) For each measurement point, calculate the percentage frequency effect (PFE) and metal factor parameter (MF).

c (m)	n = 1 ρ_{dc} (Ω m)	n = 1 ρ_{ac} (Ω m)	n = 2 ρ_{dc} (Ω m)	n = 2 ρ_{ac} (Ω m)	n = 3 ρ_{dc} (Ω m)	n = 3 ρ_{ac} (Ω m)	n = 4 ρ_{dc} (Ω m)	n = 4 ρ_{ac} (Ω m)
0	49.8	49.6			101.5	100.9		
30			72.8	72.4			99.6	98.5
60	46.0	45.8			86.2	85.2		
90			61.3	60.6			90.0	86.1
120	42.1	41.7			72.8	70.1		
150			55.5	54.4			57.5	53.5
180	44.0	43.5			49.8	46.6		
210			53.6	51.1			47.9	44.0
240	42.1	41.8			44.0	41.4		
270			65.1	64.1			47.9	44.9
300	49.8	49.6			95.8	91.7		
330			82.3	81.3			132.1	129.4
360	51.7	51.3			114.9	114.1		
390			86.2	85.9			164.7	164.0
420	49.8	49.6			120.7	120.1		
450			78.5	78.0			170.4	169.7

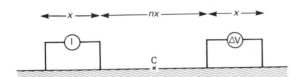

Fig. 8.41 The double-dipole electrode configuration. See Question 5.

(b) For both the PFE and MF plot four profiles for n = 1, 2, 3 and 4.

(c) Construct and contour pseudosections of the DC apparent resistivity, PFE and MF.

(d) The area is covered by highly-conductive glacial deposits 30–60 m thick. It is possible that massive sulphide mineralization is present within the bedrock. Bearing this information in mind, comment upon and interpret the profiles and pseudosections produced from (b) and (c).

6 Why are the electrical methods of exploration particularly suited to hydrogeological investigations? Describe other geophysical methods which could be used in this context, stating the reasons why they are applicable.

FURTHER READING

Bertin, J. (1976) *Experimental and Theoretical Aspects of Induced Polarisation, Vols. 1 and 2*. Gebrüder Borntraeger, Berlin.

Griffiths, D.H. & King, R.F. (1981) *Applied Geophysics for Geologists and Engineers*. Pergamon, Oxford.

Habberjam, G.M. (1979) *Apparent Resistivity and the Use of Square Array Techniques*. Gebrüder Borntraeger, Berlin.

Keller, G.V. & Frischnecht, F.C. (1966) *Electrical Methods in Geophysical Prospecting*. Pergamon, Oxford.

Koefoed, O. (1968) *The Application of the Kernel Function in Interpreting Resistivity Measurements*. Gebrüder Borntraeger, Berlin.

Koefoed, O. (1979) *Geosounding Principles. 1 — Resistivity Sounding Measurements*. Elsevier, Amsterdam.

Kunetz, G. (1966) *Principles of Direct Current Resistivity Prospecting*. Gebrüder Borntraeger, Berlin.

Marshall, D.J. & Madden, T.R. (1959) Induced polarisation: a study of its causes. *Geophysics*, **24**, 790–816.

Milsom, J. (1989) *Field Geophysics*. Open University Press, Milton Keynes.

Parasnis, D.S. (1973) *Mining Geophysics*. Elsevier, Amsterdam.

Parasnis, D.S. (1986) *Principles of Applied Geophysics* (4th edn). Chapman & Hall, London.

Parkhomenko, E.I. (1967) *Electrical Properties of Rocks*. Plenum, New York.

Sato, M. & Mooney, H.M. (1960) The electrochemical mechanism of sulphide self potentials. *Geophysics*, **25**, 226–49.

Sumner, J.S. (1976) *Principles of Induced Polarisation for Geophysical Exploration*. Elsevier, Amsterdam.

Telford, W.M., Geldart, L.P., Sheriff, R.E. & Keys, D.A. (1976) *Applied Geophysics*. Cambridge Univ. Press, Cambridge.

Ward, S.H. (1987) Electrical methods in geophysical prospecting. *In*: Samis, C.G. & Henyey, T.L. (eds.) *Methods of Experimental Physics, Vol. 24, Part B — Field Measurements*, 265–375. Academic Press, Orlando.

9 / Electromagnetic surveying

9.1 INTRODUCTION

Electromagnetic (EM) surveying methods make use of the response of the ground to the propagation of electromagnetic fields, which are composed of an alternating electric intensity and magnetizing force. Primary electromagnetic fields may be generated by passing alternating current through a small coil made up of many turns of wire or through a large loop of wire. The response of the ground is the generation of secondary electromagnetic fields and the resultant fields may be detected by the alternating currents that they induce to flow in a receiver coil by the process of electromagnetic induction.

The primary electromagnetic field travels from the transmitter coil to the receiver coil via paths both above and below the surface. Where the subsurface is homogeneous there is no difference between the fields propagated above the surface and through the ground other than a slight reduction in amplitude of the latter with respect to the former. However, in the presence of a conducting body the magnetic component of the electromagnetic field penetrating the ground induces alternating currents, or eddy currents, to flow in the conductor (Fig. 9.1). The eddy currents generate their own secondary electromagnetic field which travels to the receiver. The receiver then responds to the resultant of the arriving primary and secondary fields so that the response differs in both phase and amplitude from the response to the primary field alone. These differences between the transmitted and received electromagnetic fields reveal the presence of the conductor

and provide information on its geometry and electrical properties.

The induction of current flow results from the magnetic component of the electromagnetic field. Consequently there is no need for physical contact of either transmitter or receiver with the ground. Surface EM surveys can thus proceed much more rapidly than electrical surveys, where ground contact is required. More importantly, both transmitter and receiver can be mounted in aircraft or towed behind them. Airborne EM methods are widely used in prospecting for conductive ore bodies (see Section 9.8).

All anomalous bodies with high electrical conductivity (see Section 8.2.2) produce strong secondary electromagnetic fields. Some ore bodies containing minerals that are themselves insulators may produce secondary fields if sufficient quantities of an accessory mineral with a high conductivity are present. For example, electromagnetic anomalies observed over certain sulphide ores are due to the presence of the conducting mineral pyrrhotite distributed throughout the ore body.

9.2 DEPTH OF PENETRATION OF ELECTROMAGNETIC FIELDS

The depth of penetration of an electromagnetic field (Spies 1989) depends upon its frequency and the electrical conductivity of the medium through which it is propagating. Electromagnetic fields are attenuated during their passage through the ground, their amplitude decreasing exponentially with depth. The

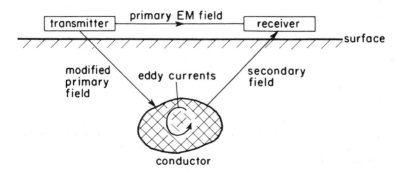

Fig. 9.1 General principle of electromagnetic surveying.

depth of penetration d can be defined as the depth at which the amplitude of the field A_d is decreased by a factor e^{-1} compared with its surface amplitude A_0

$$A_d = A_0 e^{-1} \qquad (9.1)$$

In this case

$$d = 503.8 \, (\sigma f)^{-1/2} \qquad (9.2)$$

where d is in metres, the conductivity of the ground σ is in $S\,m^{-1}$ and the frequency f of the field is in Hz.

The depth of penetration thus increases as both the frequency of the electromagnetic field and the conductivity of the ground decrease. Consequently, the frequency used in an EM survey can be tuned to a desired depth range in any particular medium. For example, in relatively dry glacial clays with a conductivity of $5 \times 10^{-4} \, S\,m^{-1}$, d is about 225 m at a frequency of 10 kHz.

Equation (9.2) represents a theoretical relationship. Practically, an effective depth of penetration z_e can be defined which represents the maximum depth at which a conductor may lie and still produce a recognizable electromagnetic anomaly

$$z_e \approx 100(\sigma f)^{-1/2} \qquad (9.3)$$

The relationship is approximate as the penetration depends upon such factors as the nature and magnitude of the effects of near-surface variations in conductivity, the geometry of the subsurface conductor and instrumental noise. The frequency dependence of depth penetration places constraints on the EM method. Normally, very low frequencies are difficult to generate and measure and the maximum penetration is of the order of 500 m.

9.3 DETECTION OF ELECTROMAGNETIC FIELDS

Electromagnetic fields may be mapped in a number of ways, the simplest of which employs a small search coil consisting of several hundred turns of copper wire wound on a circular or rectangular frame typically between 0.5 m and 1 m across. The ends of the coil are connected via an amplifier to earphones. The amplitude of the alternating voltage induced in the coil by an electromagnetic field is proportional to the component of the field perpendicular to the plane of the coil. Consequently, the strength of the signal in the earphones is at a maximum when the plane of the coil is at right angles to the direction of the arriving field. Since the ear is more sensitive to sound minima than maxima, the coil is usually turned until a null position is reached. The plane of the coil then lies in the direction of the arriving field.

9.4 TILT-ANGLE METHODS

When only a primary electromagnetic field H_p is present at a receiver coil, a null reading is obtained when the plane of the coil lies parallel to the field direction. There are an infinite number of such null positions as the coil is rotated about a horizontal axis in the direction of the field (Fig. 9.2).

In many EM systems the induced secondary field H_s lies in a vertical plane. Since the primary and secondary fields are both alternating, the total field vector describes an ellipse in the vertical plane with time (Fig. 9.3). The resultant field is then said to be *elliptically polarized* in the vertical plane. In this case there is only one null position of the search coil, namely where the plane of the coil coincides with the plane of polarization.

For good conductors it can be shown that the direction of the major axis of the ellipse of polarization corresponds reasonably accurately to that of the resultant of the primary and secondary electromagnetic field directions. The angular deviation of this axis from the horizontal is known as the *tilt-angle* θ of the resultant field (Fig. 9.3). There are a number of EM techniques (known as *tilt-angle* or *dip-angle* methods) which simply measure spatial variations in this angle. The primary field may be generated by a fixed transmitter, which usually consists of a large horizontal or vertical coil, or by a small mobile transmitter. Traverses are made across the survey area normal to the geological strike. At each station the search coil is rotated about three orthogonal axes until a null signal is obtained so that the plane of the coil lies in the plane of the polarization ellipse. The tilt-angle may then be determined

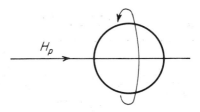

Fig. 9.2 The rotation of a search coil about an axis corresponding to the direction of arriving electromagnetic radiation H_p producing an infinite number of null positions.

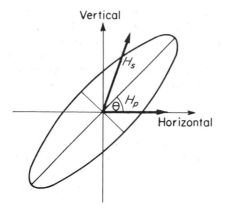

Fig. 9.3 The polarization ellipse and tilt-angle θ. H_p and H_s represent the primary and secondary electromagnetic fields.

by rotating the coil about a horizontal axis at right angles to this plane until a further minimum is encountered.

9.4.1 Tilt-angle methods employing local transmitters

In the case of a fixed, vertical transmitter coil, the primary field is horizontal. Eddy currents within a subsurface conductor then induce a magnetic field whose lines of force describe concentric circles around the eddy current source, which is assumed to lie along its upper edge (Fig. 9.4(a)). On the side of the body nearest the transmitter the resultant field dips upwards. The tilt decreases towards the body and dips downwards on the side of the body remote from the transmitter. The body is located directly below the crossover point where the tilt-angle is zero, as here both primary and secondary fields are horizontal. When the fixed transmitter is horizontal the primary field is vertical (Fig. 9.4(b)) and the body is located where the tilt is at a minimum. An example of the use of tilt-angle methods (vertical transmitter) in the location of a massive sulphide body is presented in Fig. 9.5.

If the conductor is near the surface both the

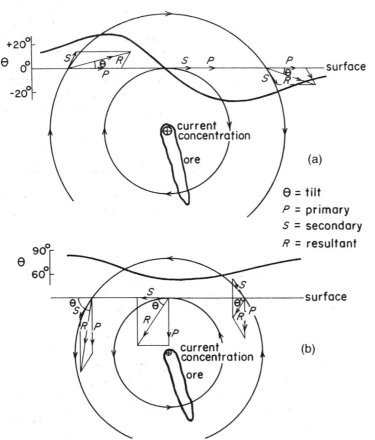

θ = tilt
P = primary
S = secondary
R = resultant

Fig. 9.4 Tilt-angle profiles resulting from (a) vertical and (b) horizontal transmitter loops. (After Parasnis 1973.)

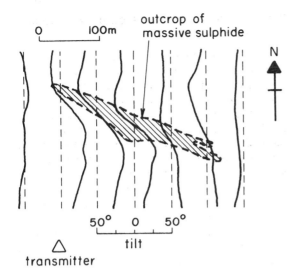

outcrop of massive sulphide

Fig. 9.5 Example of tilt-angle survey using a vertical loop transmitter. (After Parasnis 1973.)

mobile and which can provide much more quantitative information on subsurface conductors. However, two tilt-angle methods still in common use are the very low frequency (VLF) and audio frequency magnetic field (AFMAG) methods, neither of which requires the erection of a special transmitter.

9.4.2 The VLF method

The source utilized by the VLF method is electromagnetic radiation generated in the low frequency band of 15–25 kHz by the powerful radio transmitters used in long-range communications and navigational systems. Several stations using this frequency range are available around the world and transmit continuously either an unmodulated carrier wave or a wave with superimposed morse code. Such signals may be used for surveying up to distances of several thousand kilometres from the transmitter.

At large distances from source the electromagnetic field is essentially planar and horizontal (Fig. 9.6). The electric component E lies in a vertical plane and the magnetic component H lies at right angles to the direction of propagation in a horizontal plane. A conductor that strikes in the direction of the transmitter is cut by the magnetic vector and the induced eddy currents produce a secondary electromagnetic field. Conductors striking at right angles to the direction of propagation are not cut effectively by the magnetic vector.

The VLF receiver is a small hand-held device incorporating two orthogonal aerials which can be tuned to the particular frequencies of the transmitters. The direction of a transmitter is found by rotating the horizontal instrument around a vertical axis until a null position is found. Traverses are then

amplitude and gradients of the tilt-angle profile are large. These quantities decrease as the depth to the conductor increases and may consequently be used to derive semi-quantitative estimates of the conductor depth. A vertical conductor would provide a symmetrical tilt-angle profile with equal gradients on either side of the body. As the inclination of the conductor decreases, the gradients on either side become progressively less similar. The asymmetry of the tilt-angle profile can thus be used to obtain an estimate of the dip of the conductor.

Tilt-angle methods employing fixed transmitters have been largely superseded by survey arrangements in which both transmitter and receiver are

Fig. 9.6 Principle of VLF method. Dashed lines show a tabular conductor striking towards the antenna which is cut by the magnetic vector of the electromagnetic field.

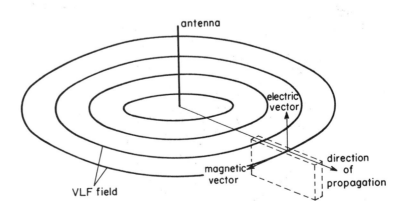

performed over the survey area at right angles to this direction. The instrument is rotated about a horizontal axis orthogonal to the traverse and the tilt recorded at the null position. Profiles are similar in form to Fig. 9.4(a), with the conductor lying beneath locations of zero tilt. See Hjelt *et al.* (1985) for a discussion of the interpretation of VLF data.

The VLF method has the advantages that the field equipment is small and light, being conveniently operated by one person, and that there is no need to install a transmitter. However, for a particular survey area, there may be no suitable transmitter providing a magnetic vector across the geological strike. A further disadvantage is that the depth of penetration is somewhat less than that attainable by tilt-angle methods using a local transmitter. The VLF method can be used in airborne EM surveying.

9.4.3 The AFMAG method

The AFMAG method (Labson *et al.* 1985) can similarly be used on land or in the air. The source in this case is the natural electromagnetic fields generated by thunderstorms and known as *sferics*. Sferics propagate around the Earth between the ground surface and the ionosphere. This space constitutes an efficient electromagnetic waveguide and the low attenuation means that thunderstorms anywhere in the world make an effective contribution to the field at any given point. The field also penetrates the subsurface where, in the absence of electrically-conducting bodies, it is practically horizontal. The sferic sources are random so that the signal is generally quite broadband between 1 and 1000 Hz.

The AFMAG receiver differs from conventional tilt-angle coils since random variations in the direction and intensity of the primary field make the identification of minima impossible with a single coil. The receiver consists of two orthogonal coils each inclined at 45 degrees to the horizontal (Fig. 9.7). In the absence of a secondary field the components of the horizontal primary field perpendicular to the coils are equal and their subtracted output is zero (Fig. 9.7(a)). The presence of a conductor gives rise to a secondary field which causes deflection of the resultant field from the horizontal (Fig. 9.7(b)). The field components orthogonal to the two coils are then unequal, so that the combined output is no longer zero and the presence of a conductor is indicated. The output provides a measure of the tilt.

On land both the azimuths and tilts of the resultant

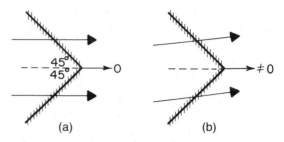

Fig. 9.7 Principle of AFMAG receiver: (a) conductor absent, (b) conductor present.

electromagnetic field can be determined by rotating the coils about a vertical axis until a maximum signal is obtained. These are conventionally plotted as dip vectors. In the air, azimuths cannot be determined as the coils are attached to the aircraft so that their orientation is controlled by the flight direction. Consequently, only perturbations from the horizontal are monitored along the flight lines. The output signal is normally fed into an amplifier tuned to two frequencies of about 140 and 500 Hz. Comparison of the amplitude of the signals at the two frequencies provides an indication of the conductivity of the anomalous structure as it can be shown that the ratio of low-frequency response to high-frequency response is greater than unity for a good conductor and less than unity for a poor conductor.

The AFMAG method has the advantage that the frequency range of the natural electromagnetic fields used extends to an order of magnitude lower than can be produced artificially so depths of investigation of several hundred metres are feasible.

9.5 PHASE MEASURING SYSTEMS

Tilt-angle methods such as VLF and AFMAG are widely used since the equipment is simple, relatively cheap and the technique is rapid to employ. However, they provide little quantitative information on the conductor. More sophisticated EM surveying systems measure the phase and amplitude relationships between primary, secondary and resultant electromagnetic fields. The various types of system available are discussed in McCracken *et al.* (1986).

An alternating electromagnetic field can be represented by a sine wave with a wavelength of 2π (360°) (Fig. 9.8(a)). When one such wave lags behind another the waves are said to be out-of-phase. The phase difference can be represented by a phase angle θ corresponding to the angular separation of the waveforms. The phase relationships of electromagnetic waves can be represented on special vector diagrams in which vector length is proportional to field amplitude and the angle measured counterclockwise from the primary vector to the secondary vector represents the angular phase lag of the secondary field behind the primary.

The primary field P travels directly from transmitter to receiver above the ground and suffers no modification other than a small reduction in amplitude caused by geometric spreading. As the primary field penetrates the ground it is reduced in amplitude to a greater extent but remains in phase with the surface primary. The primary field induces an alternating voltage in a subsurface conductor with the same frequency as the primary but with a phase lag of $\pi/2$ (90°) according to the laws of electromagnetic induction. This may be represented on the vector diagram (Fig. 9.8(b)) by a vector $\pi/2$ counterclockwise to P.

The electrical properties of the conductor cause a further phase lag ϕ,

$$\phi = \tan^{-1}(2\pi f L/r) \qquad (9.4)$$

where f is the frequency of the electromagnetic field, L the inductance of the conductor (its tendency to oppose a change in the applied field) and r the resistance of the conductor. For a good conductor ϕ will approach $\pi/2$ while for a poor conductor ϕ will be almost zero.

The net effect is that the secondary field S produced by the conductor lags behind the primary with a phase angle of $(\pi/2 + \phi)$. The resultant field R can now be constructed (Fig. 9.8(b)).

The projection of S on the horizontal (primary field) axis is $S \sin \phi$ and is an angle π out of phase with P. It is known as the *in-phase* or *real component* of S. The vertical projection is $S \cos \phi$, $\pi/2$ out of phase with P, and is known as the *out-of-phase*, *imaginary* or *quadrature* component.

Modern instruments are capable of splitting the secondary electromagnetic field into its real (Re) and imaginary (Im) components. The larger the

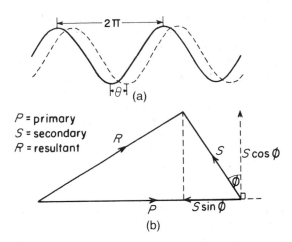

Fig. 9.8 (a) The phase difference θ between two waveforms. (b) Vector diagram illustrating the phase and amplitude relationships between primary, secondary and resultant electromagnetic fields.

ratio Re/Im, the better the conductor. Some systems, mainly airborne, simply measure the phase angle ϕ.

Classical phase-measuring systems employed a fixed source, usually a very large loop of wire laid on the ground. These systems include the *Two-frame*, *Compensator* and *Turam* systems. They are still in use but are more cumbersome than modern systems in which both transmitter and receiver are mobile.

A typical field set is shown in Fig. 9.9. The transmitter and receiver coils are about one metre in diameter and are usually carried horizontally, although different orientations may be used. The coils are linked by a cable which carries a reference signal and also allows the coil separation to be accurately maintained at, normally, between 30 m and 100 m. The transmitter is powered by a portable AC generator. Output from the receiver coil passes through a compensator and decomposer (see below). The equipment is first read on barren ground and the compensator adjusted to produce zero output. By this means, the primary field is compensated so

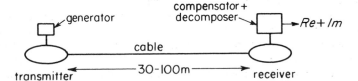

Fig. 9.9 Mobile transmitter–receiver EM field equipment.

that the system subsequently responds only to secondary fields. Consequently, such EM methods reveal the presence of bodies of anomalous conductivity without providing information on absolute conductivity values. Over the survey area the decomposer splits the secondary field into real and imaginary components which are usually displayed as a percentage of the primary field whose magnitude is relayed via the interconnecting cable. Traverses are generally made perpendicular to geological strike and readings plotted at the mid-point of the system. The maximum detection depth is about half the transmitter−receiver separation.

Fieldwork is simple and requires a crew of only two or three operators. The spacing and orientation of the coils is critical as a small percentage error in spacing can produce appreciable error in phase measurement. The coils must also be kept accurately horizontal and coplanar as small relative tilts can produce substantial errors. The required accuracy of spacing and orientation is difficult to maintain with large spacings and over uneven terrain.

Fig. 9.10 shows a mobile transmitter−receiver EM profile across a sheet-like conductor in the Kankberg area of northern Sweden. A consequence of the coplanar horizontal coil system employed is that conducting bodies produce negative anomalies in both real and imaginary components with maximum amplitudes immediately above the conductor. The asymmetry of the anomalies is diagnostic of the inclination of the body, with the maximum gradient lying on the downdip side. In this case the large ratio of real to imaginary components over the ore body indicates the presence of a very good conductor, while a lesser ratio is observed over a sequence of graphite-bearing phyllites to the north.

9.6 TIME-DOMAIN ELECTROMAGNETIC SURVEYING

A significant problem with many EM surveying techniques is that a small secondary field must be measured in the presence of a much larger primary field, with a consequent decrease in accuracy. This problem is overcome in *time-domain electromagnetic surveying* (TDEM), sometimes called *pulsed* or *transient-field EM*, by using a primary field which is not continuous but consists of a series of pulses separated by periods when it is inactive. The secondary field induced by the primary is only measured during the interval when the primary is absent. The eddy currents induced in a subsurface conductor tend to diffuse inwards towards its centre when the inducing field is removed and gradually dissipate by resistive heat loss. Within highly conductive bodies, however, eddy currents circulate around the boundary of the body and decay more slowly. Measurement of the rate of decay of the waning eddy currents thus provides a means of locating anomalously conducting bodies and estimating their conductivity. The analysis of the decaying secondary field is equivalent to analysing the response to a continuous EM wave at a number of frequencies. TDEM consequently bears the same relationship to continuous-wave EM as, for example, time-domain IP does to frequency-domain IP. INPUT® (Section 9.8.1) is an example of an air-

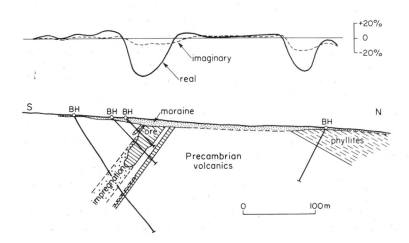

Fig. 9.10 Mobile transmitter− receiver profile, employing horizontal coplanar coils with a separation of 60 m and an operating frequency of 3.6 kHz, in the Kankberg area, north Sweden, Real and imaginary components are expressed as a percentage of the primary field. (After Parasnis 1973.)

borne version of the method.

In ground surveys, the primary pulsed EM field is generated by a transmitter that usually consists of a large loop of wire, several tens of metres in diameter, which is laid on the ground. The transmitter loop can also be utilized as the receiver, or a second coil can be used for this purpose, either on the ground surface or down a borehole (Dyck & West 1984). The transient secondary field produced by the decaying eddy currents can last from less than a millisecond for poor conductors to more than 20 ms for good conductors. The decaying secondary field is quantified by measuring the temporal variation of the amplitude of the secondary at a number of fixed times (channels) after primary cut-off (Fig. 9.11). In good conductors the secondary field is of long duration and will register in most of the channels; in poor conductors the secondary field will only register in the channels recorded soon after the primary field becomes inactive. Repeated measurements can be stacked in a manner analogous to seismic waves (Section 4.5) to improve the signal-to-noise ratio. The position and attitude of the conductor can be estimated from the change in amplitude from place to place of the secondary field in selected channels, while depth estimates can be made from the anomaly half-width. More quantitative interpretations can be made by simulation of the anomaly in terms of the computed response of simple geometric shapes such as spheres, cylinders or plates, or more simply by using the concept of equivalent current filaments (Barnett 1984) which models the distribution of eddy currents in the conductor. Limited two-dimensional modelling (Oristaglio & Hohmann 1984) is also possible using a finite-difference approach.

A form of depth sounding can be made utilizing TDEM (Frischnecht & Raab 1984). Only short offsets of transmitter and receiver are necessary and the array therefore crosses a minimum of geological boundaries such as faults and lithological contacts. By contrast, VES or continuous-wave EM methods are much more affected by near-surface conductivity inhomogeneities since long arrays are required. It is claimed that penetration of up to about 10 km can be achieved by TDEM sounding.

An example of a surface application of TDEM is presented in Fig. 9.12, which shows the results of a survey undertaken near the Rum Jungle Mine, Northern Territory, Australia (Spies 1976). The target, which had been revealed by other geophysical methods (Fig. 9.13), was a band of highly-conductive

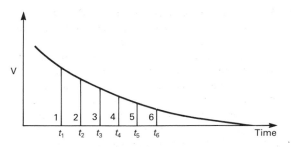

Fig. 9.11 The quantification of a decaying TDEM response by measurement of its amplitude in a number of channels (1–6) at increasing times (t_{1-6}) after primary field cut-off. The amplitudes of the responses in the different channels are recorded along a profile.

graphitic black shale, which has a conductivity in excess of $0.1\,\mathrm{S\,m^{-1}}$ in its pristine condition. In Fig. 9.12 the TDEM response is expressed in terms of the induced voltage in the loop $e(t)$ normalized with respect to the current in the transmitter loop I. The response is shown for a number of different times after primary cut-off. The response persists into the latest channels, indicating the presence of a good conductor which corresponds to the graphitic shale. The asymmetry of the response curves and their variation from channel to channel allows the dip of the conductor to be estimated. The first channel, which logs the response to relatively shallow depths, peaks to the right. The maximum moves to the left in later channels, which give the response to progressively greater depths, indicating that the conductor dips in that direction.

An example of a survey using a borehole TDEM system is presented in Fig. 9.14, which shows results from the Single Tree Hill area, N.S.W., Australia (Boyd & Wiles 1984). Here semi-massive sulphides (pyrite and pyrrhotite), which occur in intensely sericitized tuffs with shale bands, have been penetrated by three drillholes. The TDEM responses at a suite of times after primary field cut-off, recorded as the receiver was lowered down the three drillholes, are shown. In hole PDS1, the response at early times indicates the presence of a conductor at a depth of 145 m. The negative response at later times at this depth is caused by the diffusion of eddy currents into the conductor past the receiver and indicate that the hole is near the edge of the conductor. In holes DS1 and DS2 the negative responses at 185 m and 225 m, respectively, indicate that the receiver passed outside, but near the edges of, the conductor at these depths. Also shown in the section

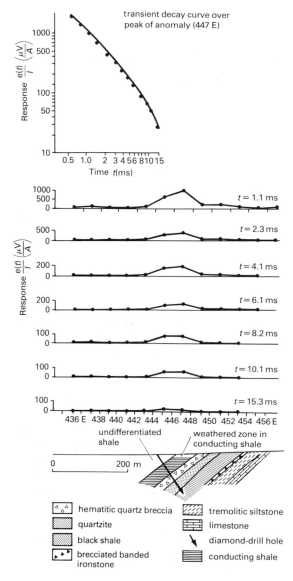

is an interpretation of the TDEM data in terms of a model consisting of a rectangular current-carrying loop.

9.7 NON-CONTACTING CONDUCTIVITY MEASUREMENT

It is possible to obtain readings of ground conductivity by EM measurements (McNeill 1980).

Measurements of this type can be made using standard resistivity methods (Section 8.2), but, since these require the introduction of current into the ground via electrodes, they are labour intensive, slow and therefore costly. Moreover, resistivity measurements are influenced by geological noise arising from near-surface resistivity variations which limit the resolution that can be achieved. The more recently developed non-contacting conductivity meters utilize EM fields and do not suffer from these drawbacks. No ground contact is required so that measurements can be made at walking pace and the subsurface volume sampled is averaged in such a way that resolution is considerably improved (Zalasiewicz *et al.* 1985).

The secondary EM field measured in a mobile transmitter-receiver survey (Section 9.5) is generally a complex function of the coil spacing s, the operating frequency f and the conductivity of the subsurface σ. However, it can be shown that if the product of s and the skin depth d (Section 9.2), known as the *induction number*, is much less than unity, the following relationship results:

$$H_s/H_p \approx i\omega\mu_0\sigma s^2/4 \tag{9.5}$$

where H_s and H_p are the amplitudes of the secondary and primary EM fields, respectively, $\omega = 2\pi f$, μ_0 is the magnetic permeability of vacuum and $i = \sqrt{(-1)}$, its presence indicating that the quadrature component is measured. Thus the ratio H_s/H_p is proportional to the ground conductivity. Since d depends on the product σf, estimation of the maximum probable value of σ allows the selection of f such that the above condition of low induction number is satisfied. The depth of penetration depends upon s and is independent of the conductivity distribution of the subsurface. Measurements taken at low induction number thus provide an apparent conductivity σ_a given by

$$\sigma_a = \frac{4}{\omega\,\mu_0 s^2}\frac{(H_s)}{(H_p)} \tag{9.6}$$

This relationship allows the construction of electromagnetic instruments which provide a direct reading of ground conductivity down to a predetermined depth. In one application the transmitter and receiver are horizontal dipoles mounted on a boom 3.7 m apart, providing a fixed depth of investigation of about 6 m. The instrument provides a rapid means of performing constant separation traversing (Section 8.2.3) to a depth suitable for engineering and archaeological investigations. Where a greater depth of penetration is required, an instrument is

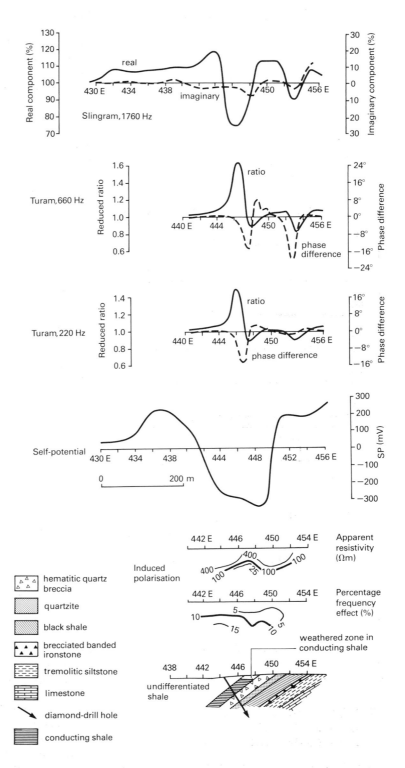

Fig. 9.13 Comparison of various geophysical methods over the same profile as shown in Fig. 9.12 near the Rum Jungle Mine, Northern Territory, Australia. (After Spies 1976.)

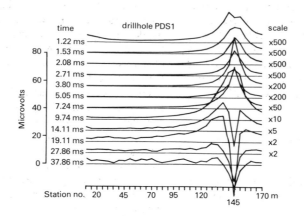

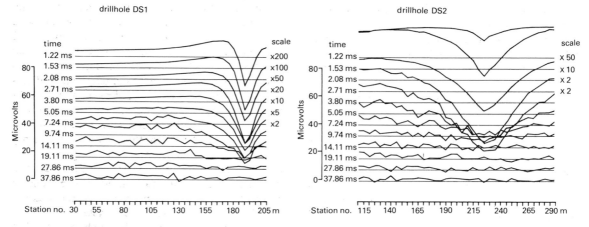

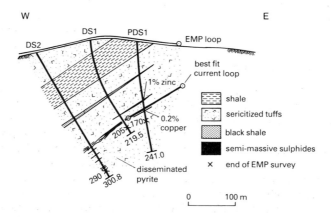

Fig. 9.14 Drillhole TDEM profiles and geological section over Single Tree Hill, N.S.W., Australia. (Redrawn from Boyd & Wiles 1984.)

used in which the transmitter and receiver, which usually take the form of vertical coplanar coils, are separate, so that their spacing is variable. CST can be performed with the subsurface energized to a desired depth, while vertical electrical sounding (Section 8.2.3) can be undertaken by progressively increasing the transmitter–receiver separation.

9.8 AIRBORNE ELECTROMAGNETIC SURVEYING

Airborne EM techniques are widely used because of their speed and cost-effectiveness and a large number of systems are available.

There is a broad division into *passive systems*, where only the receiver is airborne, and *active systems*, where both transmitter and receiver are mobile. Passive systems include airborne versions of the VLF and AFMAG methods. Independent transmitter methods can also be used with an airborne receiver, but are not very attractive as prior ground access to the survey area is required.

Active systems are more commonly used, as surveys can be performed in areas where ground access is difficult and provide more information than the passive tilt-angle methods. They are, basically, ground mobile transmitter–receiver systems lifted into the air and interfaced with a continuous recording device. Certain specialized methods, described later, have been adopted to overcome the specific difficulties encountered in airborne work. Active systems comprise two main types, *fixed separation* and *quadrature*.

9.8.1 Fixed separation systems

In fixed separation systems the transmitter and receiver are maintained at a fixed separation, and real and imaginary components are monitored as in ground surveys. The coils are generally arranged to be vertical and either coplanar or coaxial. Accurate maintenance of separation and height is essential, and this is usually accomplished by mounting the transmitter and receiver either on the wings of an aircraft or on a beam carried beneath a helicopter. Compensating methods have to be employed to correct for minute changes in the relative positions of transmitter and receiver resulting from such factors as flexure of the wing mountings, vibration and temperature changes. Since only a small transmitter–receiver separation is used to generate and

detect an electromagnetic field over a relatively large distance, such minute changes in separation would cause significant distortion of the signal. Fixed-wing systems are generally flown at a ground clearance of 100–200 m, while helicopters can survey at elevations as low as 20 m.

Greater depth of penetration can be achieved by the use of two planes flying in tandem (Fig. 9.15), the rear plane carrying the transmitter and the forward plane towing the receiver mounted in a bird. Although the aircraft have to fly at a strictly regulated speed, altitude and separation, the use of a rotating primary field compensates for relative rotation of the receiver and transmitter. The rotating primary field is generated by a transmitter consisting of two orthogonal coils in the plane perpendicular to the flight direction. The coils are powered by the same AC source with the current to one coil shifted $\pi/2$ (90°) out-of-phase with respect to the other. The resulting field rotates about the flight line and is detected by a receiver with a similar coil configuration which passes the signals through a phase-shift network so that the output over a barren area is zero. The presence of a conductor is then indicated by non-zero output and the measured secondary field decomposed into real and imaginary components. Although penetration is increased and orientation errors are minimized, the method is relatively expensive and the interpretation of data is complicated by the complex coil system.

Airborne TDEM methods, such as INPUT® (INduced PUlse Transient) (Barringer 1962), may be used to enhance the secondary field measurement. The discontinuous primary field shown in Fig. 9.16 is generated by passing pulses of current through a transmitter coil strung about an aircraft. The transient primary field induces currents within a subsurface conductor. These currents persist during the period when the primary field is shut off and the receiver becomes active. The exponential decay curve is sampled at several points and the signals displayed on a strip chart. The signal amplitude in successive sampling channels is, to a certain extent, diagnostic of the type of conductor present. Poor conductors produce a rapidly decaying voltage and only register on those channels sampling the voltage shortly after primary cut-off. Good conductors appear on all channels.

INPUT® is more expensive than other airborne EM methods but provides greater depth penetration, possibly in excess of 100 m, because the secondary signal can be monitored more accurately in the

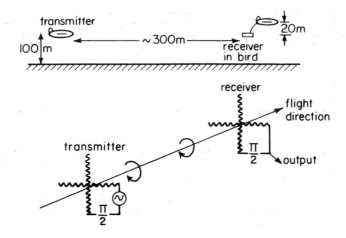

Fig. 9.15 The two-plane, rotary field, EM system.

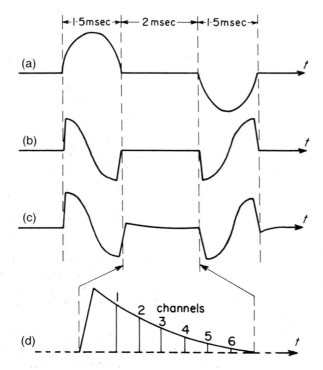

Fig. 9.16 Principle of the INPUT® system. (a) Primary field. (b) Receiver response to primary alone. (c) Receiver response in the presence of a secondary field. (d) Enlargement of the receiver signal during primary field cut-off. The amplitude of the decaying induced voltage is here sampled on six channels.

absence of the primary field. It also provides a direct indication of the type of conductor present from the duration of the induced secondary field.

As well as being employed in the location of conducting ore bodies, airborne EM surveys can also be used as an aid to geological mapping. In humid and sub-tropical areas a weathered surface layer develops whose thickness and conductivity depend upon the local rock type. Fig. 9.17 shows an INPUT® profile across part of the Itapicuru Greenstone Belt in Brazil, with sampling times increasing from 0.3 ms at channel 1 to 2.1 ms at channel 6. The transient response over mafic volcanic rocks and Mesozoic sediments is developed in all six

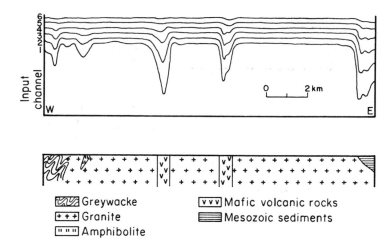

Fig. 9.17 INPUT® profile across part of the Itapicuru Greenstone Belt, Brazil. (After Palacky 1981.)

▨ Greywacke
+++ Granite
¹ ¹ ¹ Amphibolite

v v v Mafic volcanic rocks
≡ Mesozoic sediments

channels, indicating that their weathered layer is highly conductive, while the response over greywacke is only apparent in channels 1 to 4, indicating a comparatively less conductive layer.

9.8.2 Quadrature systems

Quadrature systems were the first airborne EM methods devised. The transmitter is usually a large aerial slung between the tail and wingtips of a fixed-wing aircraft and a nominally-horizontal receiver is towed behind the aircraft on a cable some 150 m long.

In quadrature systems the orientation and height of the receiver cannot be rigorously controlled as the receiver 'bird' oscillates in the slipstream. Consequently, the measurement of real and imaginary components is not possible as the strength of the field varies irregularly with movement of the receiver coil. However, the phase difference between the primary field and the resultant field caused by a conductor is independent of variation in the receiver orientation. A disadvantage of the method is that a given phase shift φ may be caused by either a good or a poor conductor (Fig. 9.18). This problem is overcome by measuring the phase shift at two different primary frequencies, usually of the order of 400 and 2300 Hz. It can be shown that if the ratio of low-frequency to high-frequency response exceeds unity, a good conductor is present.

Fig. 9.19 shows a contour map of real component anomalies (in ppm of the primary field) over the

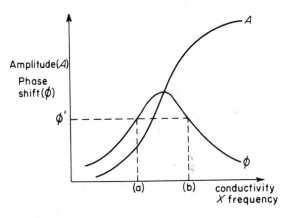

Fig. 9.18 The relationship between the phase/amplitude of a secondary electromagnetic field and the product of conductivity and frequency. A given phase shift φ′ could result from a poor conductor (a) or a good conductor (b).

Skellefteå orefield, northern Sweden. A fixed separation system was used, with vertical, coplanar coils mounted perpendicular to the flight direction on the wingtips of a small aircraft. Only contours above the noise level of some 100 ppm are presented. The pair of continuous anomaly belts in the southwest, with amplitudes exceeding 1000 ppm, corresponds to graphitic shales, which serve as guiding horizons in this orefield. The belt to the north of these is not continuous, and although in part related to sulphide ores, also results from a power cable. In the northern part of the area the three distinct anomaly centres all correspond to strong sulphide mineralization.

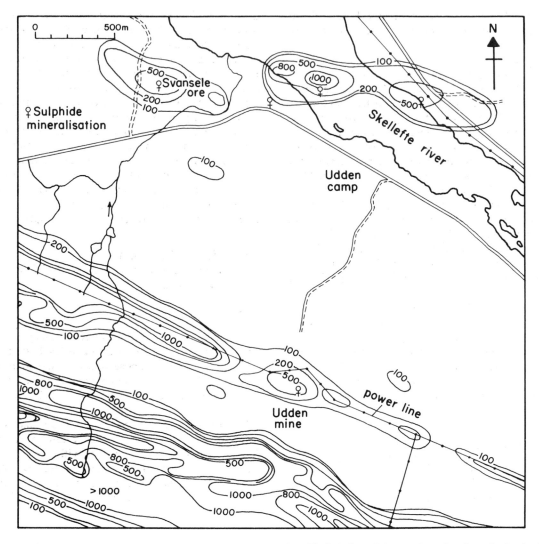

Fig. 9.19 Contour map of real component anomalies over part of the Skellefteå orefield, northern Sweden, obtained using an airborne system with vertical coplanar coils. Mean ground clearance 30 m, operating frequency 3.5 kHz. Contours in ppm of the primary field. (After Parasnis 1973.)

9.9 INTERPRETATION OF ELECTROMAGNETIC DATA

As with other types of geophysical data an indirect approach can be adopted in the interpretation of electromagnetic anomalies. The observed electromagnetic response is compared with the theoretical response, for the type of equipment used, to conductors of various shapes and conductivities. Theoretical computations of this type are quite complex and limited to simple geometric shapes such as spheres, cylinders, thin sheets and horizontal layers.

If the causative body is of complex geometry and variable conductivity, laboratory modelling may be used (Chakridi & Chouteau 1988). Because of the complexity of theoretical computations, this technique is used far more extensively in electromagnetic interpretation than in other types of geophysical interpretation. For example, to model a massive sulphide body in a well-conducting host rock, an aluminium model immersed in salt water may be used.

Master curves are available for simple interpretation of moving source–receiver data in cases where it may be assumed that the conductor has a simple geometric form. Fig. 9.20 shows such a set of curves for a simple sheet-like dipping conductor of thickness t and depth d where the distance between horizontal, coplanar coils is a. The point corresponding to the maximum real and imaginary values, expressed as a percentage of the primary field, is plotted on the curves. From the curves coinciding with this point, the corresponding λ/a and d/a values are determined. The latter ratio is readily converted into conductor depth. λ corresponds to $10^7(\sigma f t)^{-1}$, where σ is the conductivity of the sheet and f the frequency of the field. Since a and f are known, the product σt can be determined. By performing measurements at more than one frequency, σ and t can be computed separately.

Much electromagnetic interpretation is, however, only qualitative, particularly for airborne data. Contour maps of real or imaginary components provide information on the length and conductivity of conductors while the asymmetry of the profiles provides an estimate of the inclination of sheet-like bodies.

9.10 LIMITATIONS OF THE ELECTROMAGNETIC METHOD

The electromagnetic method is a versatile and ef-ficient survey technique, but it suffers from several drawbacks. As well as being caused by economic sources with a high conductivity such as ore bodies, electromagnetic anomalies can also result from non-economic sources such as graphite, water-filled shear zones, bodies of water and man-made features. Superficial layers with a high conductivity such as wet clays and graphite-bearing rocks may screen the effects of deeper conductors. Penetration is not very great, being limited by the frequency range that can be generated and detected. Unless natural fields are used, maximum penetration in ground surveys is limited to about 500 m, and is only about 50 m in airborne work. Finally, the quantitative interpretation of electromagnetic anomalies is complex.

9.11 TELLURIC AND MAGNETOTELLURIC FIELD METHODS

9.11.1 Introduction

Within and around the Earth there exist large scale, low frequency, natural magnetic fields known as *magnetotelluric fields*. These induce natural alternating electric fields to flow within the Earth, known as *telluric currents*. Both of these natural fields can be used in prospecting.

Magnetotelluric fields are believed to result from the flow of charged particles in the ionosphere, as fluctuations in the fields correlate with diurnal variations in the geomagnetic field caused by solar emissions. Magnetotelluric fields penetrate the ground and there induce telluric currents to flow.

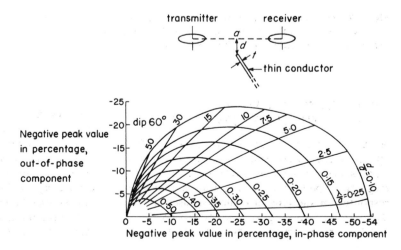

Fig. 9.20 Example of a vector diagram used in estimating the parameters of a thin dipping conductor from the peak real and imaginary component values. (Redrawn from Nair *et al.* 1968.)

The fields are of variable frequency, ranging from 10^{-5} Hz up to the audio range, and overlap the frequency range utilized in the AFMAG method (Section 9.4.3).

9.11.2 Surveying with telluric currents

Telluric currents flow within the Earth in large circular patterns that stay fixed with respect to the Sun. They normally flow in sheets parallel to the surface and extend to depths of several kilometres in the low frequencies. The telluric method is, in fact, the only electrical technique capable of penetrating to the depths of interest to the oil industry. Although variable in both their direction and intensity, telluric currents cause a mean potential gradient at the Earth's surface of about 10 mV km^{-1}.

Telluric currents are used in prospecting by measuring the potential differences they cause between points at the surface. Obviously no current electrodes are required and potential differences are monitored using non-polarizing electrodes or plates made of a chemically inert substance such as lead. Electrode spacing is typically 300–600 m in oil exploration and 30 m or less in mineral surveys. The potential electrodes are connected to an amplifier which drives a strip chart recorder or tape recorder.

If the electrical conductivity of the subsurface were uniform the potential gradient at the surface would be constant (Fig. 9.21(a)). Zones of differing conductivity deflect the current flow from the horizontal and cause distortion of the potential gradients measured at the surface. Fig. 9.21(b) shows the distortion of current flow lines caused by a salt dome which, since it is a poor conductor, deflects the current lines into the overlying layers. Similar effects may be produced by anticlinal structures. Interpretation of anomalous potential gradients measured at the surface permits the location of subsurface zones of distinctive conductivity.

Telluric potential gradients are measured using orthogonal electrode pairs (Fig. 9.22(a)). In practice, the survey technique is complicated by temporal variation in direction and intensity of the telluric currents. To overcome this problem, one orthogonal electrode pair is read at a fixed base located on nearby barren ground and another moved over the survey area. At each observation point the potential differences between the pairs of electrodes at the base and at the mobile station are recorded simultaneously over a period of about ten minutes. From the magnitude of the two horizontal components of

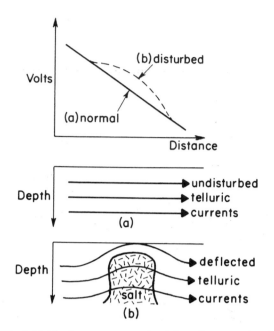

Fig. 9.21 The instantaneous potential gradient associated with telluric currents. (a) Normal, undisturbed gradient. (b) Disturbed gradient resulting from deflection of current flow by a salt dome.

the electrical field it is simple to find the variation in direction and magnitude of the resultant field at the two locations over the recording interval. The assumption is made that the ground is uniform beneath the base electrodes so that the conductivity is the same in all directions. The resultant electrical field should also be constant in all directions and would describe a circle with time (Fig. 9.22(b)). To correct for variations in intensity of the telluric currents, a function is determined which, when applied to the base electrode results, constrains the resultant electric vector to describe a circle of unit radius. The same function is then applied to the mobile electrode data. Over an anomalous structure the conductivity of the ground is not the same in all directions and the magnitude of the corrected resultant electric field varies with direction. The resultant field vector traces an ellipse whose major axis lies in the direction of maximum conductivity. The relative disturbance at this point is conveniently measured by the ratio of the area of the ellipse to the area of the corresponding base circle. The results of a survey of this type over the Haynesville Salt Dome, Texas, USA

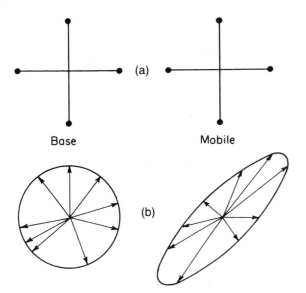

Fig. 9.22 (a) Base and mobile potential electrode sets used in telluric surveys. (b) The figure traced by the horizontal component of the telluric field over an undisturbed area (circle) and in the presence of a subsurface conductor (ellipse) after correction for temporal variations in telluric current intensity.

are presented in Fig. 1.4. The solid circles represent locations where ellipse areas relative to a unit base circle have been computed. Contours of these values outline the known location of the dome with reasonable accuracy.

The telluric method is applicable to oil exploration as it is capable of detecting salt domes and anticlinal structures, both of which constitute potential hydrocarbon traps. As such, the method has been used in Europe, North Africa and the USSR. It is not widely used in the USA where oil traps tend to be too small in area to cause a significant distortion of telluric current flow. The telluric method can also be adapted to mineral exploration.

9.11.3 Magnetotelluric surveying

Prospecting using magnetotelluric fields is more complex than the telluric method as both the electric and magnetic fields must be measured. The technique does, however, provide more information on subsurface structure. The method is, for example, used in investigations of the crust and upper mantle (e.g. Hutton *et al.* 1980).

Telluric currents are monitored as before, although no base station is required. The magnetotelluric field is measured by its inductive effect on a coil about a metre in diameter or by use of a sensitive fluxgate magnetometer. Two orthogonal components are measured at each station.

The depth z to which a magnetotelluric field penetrates is dependent on its frequency f and the resistivity ρ of the substrate, according to equations of the form of (9.2) and (9.3), i.e.

$$z = k(\rho/f)^{1/2} \qquad (9.7)$$

where k is a constant. Consequently, depth penetration increases as frequency decreases. It can be shown that the amplitudes of the electric and magnetic fields, E and B, are related

$$\rho_a = \frac{0.2}{f}\left(\frac{E}{B}\right)^2 \qquad (9.8)$$

where f is in Hz, E in mV km^{-1} and B in nT. The apparent resistivity ρ_a thus varies inversely with frequency. The calculation of ρ_a for a number of decreasing frequencies thus provides resistivity information at progressively increasing depths and is essentially a form of vertical electrical sounding (Section 8.2.3).

Interpretation of magnetotelluric data is most reliable in the case of horizontal layering. Master curves of apparent resistivity against period are available for two and three horizontal layers, vertical contacts and dykes, and interpretation may proceed in a similar manner to curve-matching techniques in the resistivity method (Section 8.2.7). Routines are now available, however, which allow the modelling of two-dimensional structures.

9.12 GROUND-PENETRATING RADAR

Ground-penetrating radar (GPR) (Davis & Annan 1989) is a technique of imaging shallow soil and rock structure at high resolution. Although analogous in some ways to the seismic methods, it is included in this chapter as the propagation of radar waves through a medium is controlled by its electrical properties at high frequencies.

GPR is similar in its principles to seismic reflection profiling (Chapter 4) and sonar (Section 4.10) surveying. A short radar pulse in the frequency band 10–1000 MHz is introduced into the ground. The propagation of the pulse is controlled by the *dielectric constant* (*relative permittivity*), which is dimensionless, and conductivity of the subsurface. Dielectric conduction takes place in poor conductors and insulators, which have no free carriers, by the slight displacement of electrons with respect to their nuclei. Water has a dielectric constant of 80, whereas in most dry geological materials the dielectric constant is in the range 4–8. Consequently the water content of materials exerts a strong influence on the propagation of a radar pulse.

Contrasts in the conductivity and dielectric properties across an interface cause part of an impinging radar pulse to be reflected. Within the frequency range utilized, the velocity and attenuation of the radar pulses are essentially constant. Penetration is of the order of 20 m, although this may increase to 50 m under ideal conditions of low conductivity. As with seismic waves, there is a trade-off between depth of penetration and resolution. The returned radar signals are amplified, digitized and recorded, and the reflections can subsequently be enhanced by digital data-processing techniques very similar to those used in reflection seismology (Section 4.7).

9.13 APPLICATIONS OF ELECTROMAGNETIC SURVEYING

The principle use of EM surveys is in the exploration for metalliferous mineral deposits, which differ significantly in their electrical properties from their host rocks. In spite of the limited depth of penetration, airborne techniques are frequently used in reconnaissance surveys, with aeromagnetic surveys often run in conjunction. EM methods are also used in the follow-up ground surveys which provide more precise information on the target area. Standard moving source–receiver methods (Section 9.5) may be used for this purpose, although in rugged or forested terrain the VLF (Section 9.4.2) or AFMAG (Section 9.4.3) methods may be preferred as no heavy equipment is required and there is no need to cut tracks for survey lines.

On a small scale, EM methods can be used in geotechnical and archaeological surveys to locate buried objects such as mine workings, pipes or treasure trove. The instruments used can take a form similar to the mine detectors used by army engineers, which have a depth of penetration of only a few centimetres and respond only to metal, or may be of the non-contacting conductivity meter-type described in Section 9.7, which have greater pen-

etration and also respond to non-metallic resistivity anomalies.

9.14 PROBLEMS

1 Calculate the depth of penetration of electromagnetic fields with frequencies of 10, 500 and 2000 kHz in:
(a) wet sandstone with a conductivity of 10^{-2} S m^{-1},
(b) massive limestone with a conductivity of 2.5×10^{-4} S m^{-1},
(c) granite with a conductivity of 10^{-6} S m^{-1}.

2 Figure 9.23(a) shows four profiles obtained during a tilt-angle EM survey near Uchi Lake, Ontario, the horizontal axes being displayed in their correct relative geographical positions. The survey was performed using transmitter and receiver in the form of vertical loops kept at a fixed separation of 120 m. Sketch in the location of the sub-

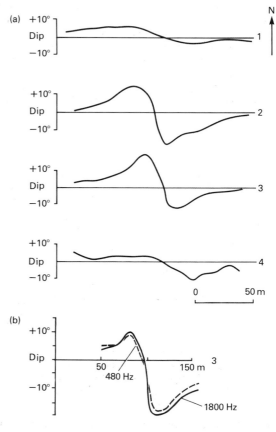

Fig. 9.23 (a) Tilt angle profiles from an EM survey near Uchi Lake, Ontario. (b) Profile 3 repeated with dual frequency EM equipment. See Question 2. (After Telford *et al.* 1976).

surface conductor and comment on its geometry. Fig. 9.23(b) shows a repeat of profile 2 using a fixed transmitter and a mobile receiver operated at frequencies of 480 and 1800 Hz. Where was the transmitter located and what form did it take? What additional information is provided by this profile?

3 During a phase-measuring EM survey, the resultant EM field was observed to have an amplitude 78% of that of the primary field and lagged behind it with an angular phase difference of 22°. Determine the amplitude of the secondary field of the subsurface conductor and of its real and imaginary components, all expressed as a percentage of the primary. What do these results reveal about the nature of the conductor?

4 Figure 9.24 shows various ground geophysical measurements taken over volcanic terrain in Bahia, Brazil. The EM survey was conducted with a system using horizontal, coplanar coils 100 m apart and a frequency of 444 Hz. The time-domain IP survey used a double-dipole array with a basic electrode separation of 25 m. Interpret these data as fully as possible. What further information would be necessary before an exploratory borehole were sunk?

5 Fig. 9.25 shows the results of airborne and ground geophysical surveys over an area of the Canadian Shield. The airborne EM survey used a quadrature system with measurements of phase angle taken at 2300 and 400 Hz. The ground tilt-angle EM survey was undertaken with a vertical-loop system using a local transmitter. Interpret and comment upon these results. Fig. 9.26 can be used

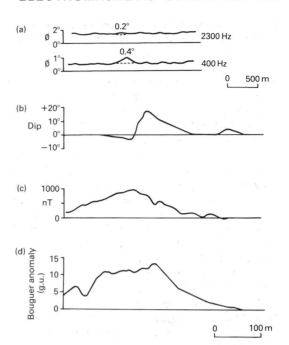

Fig. 9.25 (a) Dual frequency airborne EM, (b) ground tilt angle EM, (c) magnetic and (d) gravity profiles from the Canadian Shield. See Question 5. (After Paterson 1967.)

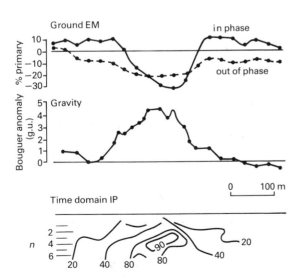

Fig. 9.24 Ground EM profile, Bouguer gravity profile and chargeability pseudosection representing results from a double-dipole IP electrode spread, all from a survey in Bahia, Brazil. See Question 4. (After Palacky & Sena 1979.)

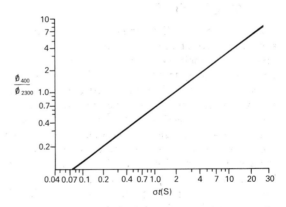

Fig. 9.26 Characteristic curve for an airborne EM system over a half-plane. ϕ_{400}/ϕ_{2300} is the ratio of peak responses at 400 Hz and 2300 Hz respectively, σ and t are the conductivity in $S\,m^{-1}$ and thickness of the conductor in metres, respectively. See Question 5. (After Paterson 1967.)

to estimate the product of conductivity and conductor thickness from the airborne data.

6 Which geophysical methods are particularly suitable for archaeological applications?

FURTHER READING

Boissonas, E. & Leonardon, E.G. (1948) Geophysical exploration by telluric currents with special reference to a survey of the Haynesville Salt Dome, Wood County, Texas. *Geophysics*, **13**, 387–403.

Cagniard, L. (1953) Basic theory of the magnetotelluric method of geophysical prospecting. *Geophysics*, **18**, 605–35.

Davis, J.L. & Annan, A.P. (1989) Ground-penetrating radar for high-resolution mapping of soil and rock stratigraphy. *Geophys. Prosp.*, **37**, 531–51.

Dobrin, M.B. & Savit, C.H. (1988) *Introduction to Geophysical Prospecting* (4th edn) McGraw-Hill, New York.

Jewell, T.R. & Ward, S.H. (1963) The influence of con-ductivity inhomogeneities upon audiofrequency magnetic fields. *Geophysics*, **28**, 201–21.

Keller, G.V. & Frischnecht, F.C. (1966) *Electrical Methods in Geophysical Prospecting*. Pergamon, Oxford.

Milsom, J. (1989) *Field Geophysics*. Open University Press, Milton Keynes.

Parasnis, D.S. (1973) *Mining Geophysics*. Elsevier, Amsterdam.

Parasnis, D.S. (1986) *Principles of Applied Geophysics*. Chapman & Hall, London.

Telford, W.M. Geldart, L.P., Sheriff, R.E. & Keys, D.A. (1976) *Applied Geophysics*. Cambridge University Press, Cambridge.

Wait, J.R. (1982) *Geo-Electromagnetism*. Academic Press, New York.

10 / Radiometric surveying

10.1 INTRODUCTION

Surveying for radioactive minerals has become important over the last few decades because of the demand for nuclear fuels. Radiometric surveying is employed in the search for deposits necessary for this application, and also for deposits associated with radioactive elements such as titanium and zirconium. Radiometric surveys are of use in geological mapping as different rock types can be recognized from their distinctive radioactive signature (Moxham 1963, Pires & Harthill 1989). There are in excess of fifty naturally occurring radioactive isotopes, but the majority are rare or only very weakly radioactive. The elements of principal concern in radiometric exploration are uranium (U^{238}), thorium (Th^{232}) and potassium (K^{40}). The latter isotope is widespread in potassium-rich rocks which may not be associated with concentrations of U and Th. Potassium can thus obscure the presence of economically important deposits and constitutes a form of geological 'noise' in this type of surveying. Fig. 10.1 shows a ternary

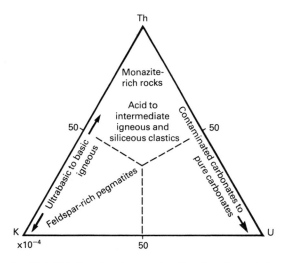

Fig. 10.1 Relative abundances of radioactive elements in different rock types. Also shown are the relative radioactivities of the radioelements. (After Wollenberg 1977.)

diagram illustrating the relative abundances of U^{238}, Th^{232} and K^{40} in different rock types.

Radiometric surveys are less widely used than the other geophysical methods as they seek a very specific target. Probably the most common application of radiometric techniques is in geophysical borehole logging (Section 11.7).

10.2 RADIOACTIVE DECAY

Elements whose atomic nuclei contain the same number of protons but different numbers of neutrons are termed isotopes. They are forms of the same element with different atomic weights. A conventional notation for describing an element A in terms of its atomic number n and atomic weight w is $_nA^w$. Certain isotopes are unstable and may disintegrate spontaneously to form other elements. The disintegration is accompanied by the emission of radioactivity of three possible types.

Alpha particles are helium nuclei $_2He^4$ which are emitted from the nucleus during certain disintegrations:

$$_nA^w \rightarrow _{n-2}B^{w-4} + _2He^4$$

Beta particles are electrons which may be emitted when a neutron splits into a proton and an electron during certain disintegrations. The proton remains within the nucleus so that the atomic weight remains the same but the atomic number increases by one to form a new element:

$$_nA^w \rightarrow _{n+1}B^w + e^-$$

Gamma rays are pure electromagnetic radiation released from excited nuclei during disintegrations. They are characterized by frequencies in excess of about 10^{16} Hz and differ from X-rays only in being of higher energy.

In addition to these emissions, a further process occurs in some radioactive elements which also releases energy in the form of gamma rays. This is known as k-capture and takes place when an electron from the innermost(k) shell enters the nucleus. The atomic number decreases and a new element is formed:

$$_nA^w + e^- \rightarrow _{n-1}B^w$$

219

Radioactive decay may lead to the formation of a stable element or a further radioactive product which itself undergoes decay. The rate of decay is exponential so that

$$N = N_0 e^{-\lambda t}$$

where N is the number of atoms remaining after time t from an initial number N_0 at time $t = 0$. λ is a decay constant characteristic of the particular element. The half-life of an element is defined as the time taken for N_0 to decrease by a half. Half-lives vary from 10^{-7} s for $_{84}Po^{212}$ to about 10^{13} Ma for $_{82}Pb^{204}$. The fact that decay constants are accurately known and unaffected by external conditions such as temperature, pressure and chemical composition forms the basis of radiometric dating.

The radioactive emissions have very different penetrating properties. Alpha particles are effectively stopped by a sheet of paper, beta particles are stopped by a few millimetres of aluminium and gamma rays are only stopped by several centimetres of lead. In air, alpha particles can travel no more than a few centimetres, beta particles only a few decimetres and gamma rays several hundreds of metres. Alpha particles thus cannot be detected in radiometric surveying and beta particles only in ground surveys. Only gamma rays can be detected in airborne surveys.

There are three radioactive series of uranium and thorium whose parents are $_{92}U^{235}$, $_{92}U^{238}$ and $_{90}Th^{232}$. These all decay eventually to stable isotopes of lead via intermediate, daughter radioisotopes. About 42% of K^{40} decays by beta emission to Ca^{40} and 58% to Ar^{40} by K-capture.

10.3 RADIOACTIVE MINERALS

There are a large number of radioactive minerals (for a full list see Durrance, 1986), but the more common are given in Table 10.1 with their modes of occurrence.

The nature of the mineral in which the radioisotope is found is irrelevant for detection purposes as the prospecting techniques locate the element itself.

10.4 INSTRUMENTS FOR MEASURING RADIOACTIVITY

Several types of detector are available for radiometric surveys, results being conventionally displayed as the number of counts of emissions over a fixed period of time. Radioactive decay is a random process following a Poisson distribution with time so that adequate count times are important if the statistical error in counting decay events is to be kept at an acceptable level.

The standard unit of gamma radiation is the roentgen (R). This corresponds to the quantity of radiation that would produce 2.083×10^{15} pairs of ions per cubic metre at standard temperature and pressure. Radiation anomalies are usually expressed in μR per hour.

10.4.1 Geiger counter

The *Geiger* (or *Geiger–Müller*) *counter* responds primarily to beta particles. The detecting element is a sealed glass tube containing an inert gas, such as argon, at low pressure plus a trace of a quenching agent such as water vapour, alcohol or methane. Within the tube a cylindrical cathode surrounds a thin axial anode and a power source maintains a potential difference of several hundred volts between them. Incoming beta particles ionize the gas and the positive ions and electrons formed are accelerated towards the electrodes, ionizing more gas en route. These cause discharge pulses across an anode resistor which, after amplification, may be registered as clicks, while an integrating circuit displays the number of counts per minute. The quenching agent suppresses the secondary emission of electrons resulting from bombardment of the cathode by positive ions.

The Geiger counter is cheap and easy to use. However, since it only responds to beta particles, its use is limited to ground surveys over terrain with little soil cover.

10.4.2 Scintillation counter

The *scintillation counter* is based on the phenomenon that certain substances such as thallium-treated sodium iodide and lithium-drifted germanium convert gamma rays to light, i.e. they *scintillate*. Photons of light impinging upon a semi-transparent cathode of a photomultiplier cause the emission of electrons. The photomultiplier amplifies the electron pulse before its arrival at the anode where it is further amplified and integrated to provide a display in counts per minute.

The scintillation counter is more expensive than the Geiger counter and less easy to transport, but it is almost 100% efficient in detecting gamma rays.

Table 10.1. Radioactive minerals. (From Telford *et al.* 1976.)

Potassium

Mineral	(i) Orthoclase and microcline feldspars [$KAlSi_3O_8$]
	(ii) Muscovite [$H_2KAl(SiO_4)_3$]
	(iii) Alunite [$K_2Al_6(OH)_{12}Si\,O_4$]
	(iv) Sylvite, carnallite [$KCl, MgCl_2, 6H_2O$]
Occurrence	(i) Main constituents in acid igneous rocks and pegmatites
	(ii) Same
	(iii) Alteration in acid volcanics
	(iv) Saline deposits in sediments

Thorium

Mineral	(i) Monazite [ThO_2 + Rare earth phosphate]
	(ii) Thorianite [$(Th,U)O_2$]
	(iii) Thorite, uranothorite [$ThSiO_4$ + U]
Occurrence	(i) Granites, pegmatites, gneiss
	(ii), (iii) Granites, pegmatites, placers

Uranium

Mineral	(i) Uraninite [Oxide of U, Pb, Ra + Th, Rare earths]
	(ii) Carnotite [$K_2O.2UO_3.V_2O_5.2H_2O$]
	(iii) Gummite [Uraninite alteration]
Occurrence	(i) Granites, pegmatites and with vein deposits of Ag, Pb, Cu, etc.
	(ii) Sandstones
	(iii) Associated with uraninite

Versions are available which can be mounted in ground transport or aircraft.

10.4.3 Gamma-ray spectrometer

The *gamma-ray spectrometer* is an extension of the scintillation counter that enables the source element to be identified. This is possible as the spectra of gamma rays from K^{40}, Th and U contain peaks which represent stages in the decay series. Since the higher the frequency of gamma radiation, the higher its contained energy, it is customary to express the spectrum in terms of energy levels. A form of windowing whereby the energy levels between predetermined upper and lower levels are monitored then provides a diagnostic means of discriminating between different sources. Fig. 10.2 shows the gamma ray spectra of U^{238}, Th^{232} and K^{40} and it is apparent that measurements at 1.76, 2.62 and 1.46 MeV, respectively, provide a discrimination of the source (1 Mev = 10^6 electron volts, one electron volt being the energy acquired by a particle of unit charge falling through a potential of 1 volt). These devices are sometimes termed *pulse-height*

analysers as the intensity of the scintillation pulses is approximately proportional to the original gamma ray energy.

Gamma ray spectrometers for airborne use are often calibrated by flying over an area of known radioisotope concentration or by positioning the aircraft on a concrete slab fabricated with a known proportion of radioisotopes. The actual concentrations of U^{238}, Th^{232} and K^{40} in the field can then be estimated from survey data.

10.4.4 Radon emanometer

Radon is the only gaseous radioactive element. Being a noble gas it does not form compounds with other elements and moves freely through pores, joints and faults in the subsurface either as a gas or dissolved in groundwater. It is one of the products of the U^{238} decay series, with a half-life of 3.8 days, and the presence of Rn^{222} at the surface is often an indication of buried uranium concentrations.

The *radon emanometer* samples air drawn from a shallow drillhole. The sample is filtered, dried and passed to an ionization chamber or zinc sulphide

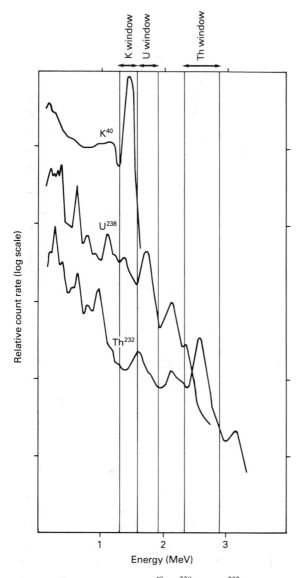

Fig. 10.2 Energy spectra of K^{40}, U^{238} and Th^{232} and their measurement windows.

scintillometer where alpha particle activity is immediately monitored to provide a count rate.

The emanometer is relatively slow to use in the field. It does, however, represent a means of detecting deeper deposits of uranium than the other methods described above, since spectrometers will only register gamma rays originating in the top metre or so of the subsurface (Telford 1982). Because of its high mobility, radon can have travelled a

considerable distance from the source of uranium before being detected. The emanometer has also been used to map faults, which provide channels for the transport of radon generated at depth (Abdoh & Pilkington 1989). This technique is advantageous when there is no great difference in rock properties across the fault that could be detected by other geophysical methods.

10.5 FIELD SURVEYS

As previously stated, Geiger counter investigations are limited to ground surveys. Count rates are noted and their significance assessed with respect to background effects resulting from the potassium content of the local rocks, nuclear fallout and cosmic radiation. An appreciable anomaly would usually be in excess of three times the background count rate.

Scintillation counters may also be used in ground surveys and are usually sited on rock exposures. The ground surface should be relatively flat so that radioactive emissions originate from the half-space below the instrument. If this condition does not obtain, a lead collimator can be used to ensure that radioactive emissions do not arrive from elevated areas flanking the instrument.

Most radiometric surveying is carried out from the air, employing larger scintillation sensors than in ground instruments, with a consequent increase in measurement sensitivity. Instruments are interfaced with strip recorders and position fixing is by means of the methods discussed in Section 7.8. Radiometric measurements are normally taken in conjunction with magnetic and electromagnetic readings, so providing additional datasets at minimal extra cost. In surveying for relatively small deposits the slow speed of helicopters is often advantageous and provides greater discrimination and amplitude of response. Flight altitude is usually less than 100 m and, because of the weak penetrative powers of radioactive emissions, the information obtained relates only to the top metre or so of the ground.

The interpretation of radiometric data is mainly qualitative, although characteristic curves are available for certain elementary shapes which provide the parameter: (surface area) × (source intensity).

10.6 EXAMPLE OF RADIOMETRIC SURVEYING

Fig. 10.3 shows a ground magnetic and gamma-ray profile across a zone of uranium mineralization in

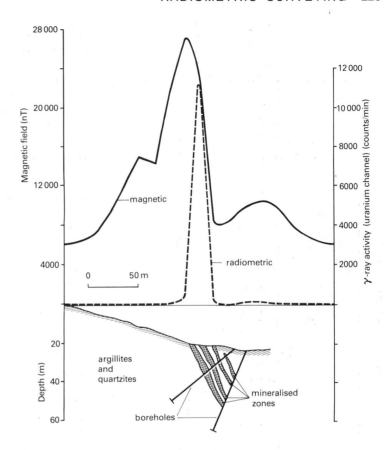

Fig. 10.3 Radiometric and magnetic profiles over pitchblende-magnetite mineralization in Labrador. (After Telford *et al.* 1976.)

Labrador. This was obtained from contour maps of a small area identified from a regional airborne survey. There are strong coincident magnetic and radiometric anomalies, the source of which was investigated by two boreholes. The anomalies arise from magnetite and pitchblende, located immediately beneath the anomaly maxima, in an argillaceous and quartzitic host. Pitchblende is a variety of massive, botryoidal or colloform uraninite.

FURTHER READING

Durrance, E.M. (1986) *Radioactivity in Geology*. Ellis Horwood, Chichester.

Milsom, J. (1989) *Field Geophysics*. Open University Press, Milton Keynes.

Telford, W.M. (1982) Radon mapping in the search for uranium. *In*: Fitch, A.A. (ed.) *Developments in Geophysical Exploration Methods*. Applied Science, London, 155–94.

Telford, W.M., Geldart, L.P., Sheriff, R.E. & Keys, D.A. (1976) *Applied Geophysics*. Cambridge University Press, Cambridge.

Wollenberg, H.A. (1977) Radiometric methods. *In*: Morse, J.G. (ed.) *Nuclear Methods in Mineral Exploration and Production*. Elsevier, Amsterdam, 5–36.

11 / Geophysical borehole logging

11.1 INTRODUCTION TO DRILLING

Shallow boreholes may be excavated by percussion drilling, in which rock fragments are blown out of the hole by air pressure. Most boreholes, however, are sunk by rotary drilling, in which the detritus produced by rotating teeth on a rock bit drilling head is flushed to the surface by a drilling fluid (or 'mud'), which holds it in suspension. The drilling fluid also lubricates and cools the bit and its density is carefully controlled so that the pressure it exerts is sufficient to exceed that of any pore fluids encountered so as to prevent blowouts. The deposition of particles held in suspension in the drilling fluid seals porous wall rocks to form a *mudcake* (Fig. 11.1). Mudcakes up to several millimetres thick can build up on the borehole wall and since the character of the mudcake is determined by the porosity and permeability of the wallrock in which it is developed, investigation of the mudcake properties indirectly provides insight into these poroperm properties. The drilling fluid filtrate penetrates the wallrock and completely displaces indigenous fluids in a 'flushed zone' which can be several centimetres thick (Fig. 11.1). Beyond lies an *annulus of invasion* where the

proportion of filtrate gradually decreases to zero. This zone of invasion is a few centimetres thick in rock such as shale, but can be up to a few metres wide in more permeable and porous rocks.

Casing may be introduced into borehole sections immediately after drilling to prevent collapse of the wallrock into the hole. Cased holes are lined with piping, the voids between wallrock and pipe being filled with cement so as to prevent invasion and collapse of the hole. Boreholes with no casing are termed *open holes*.

11.2 PRINCIPLES OF WELL LOGGING

The production of cores during drilling, which provides a full sample of the rocks penetrated, is very expensive. The fragments of rock flushed to the surface during other types of drilling are often difficult to interpret as they have been mixed and leached by the drilling fluid and often provide little information on the intrinsic physical properties of the formations from which they derive. Geophysical borehole logging, also known as *downhole geophysical surveying* or *wire-line logging*, is used to derive further information about the sequence of

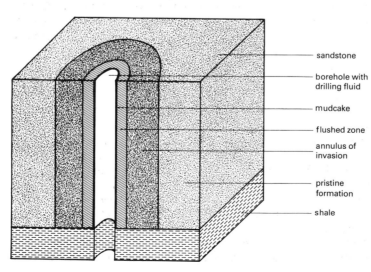

sandstone

borehole with drilling fluid

mudcake

flushed zone

annulus of invasion

pristine formation

shale

Fig. 11.1 The borehole environment.

rocks penetrated by a borehole. Of particular value is the ability to define the depth to geological interfaces or beds that have a characteristic geophysical signature, to provide a means of correlating geological information between boreholes and to obtain information on the *in situ* properties of the wallrock. Potentially, any of the normal geophysical surveying techniques described in previous chapters may be adapted for use in borehole logging, but in practice the most useful and widely-applied methods are based on electrical resistivity, electromagnetic induction, self potential, natural and induced radioactivity, sonic velocity and temperature.

These methods and some other specialized logging techniques, such as gravity and magnetic logging, are described below. In addition, several other types of subsurface geophysical measurements may be taken in a borehole environment. Of these, perhaps the most important and widely used is vertical seismic profiling, as discussed in Section 4.13.

The instrumentation necessary for borehole logging is housed in a cylindrical metal tube known as a *sonde*. Sondes are suspended in the borehole from an armoured multicore cable. They are lowered to the base of the section of the hole to be logged, and logging is carried out as the sonde is winched back up through the section. Logging data are commonly recorded on a paper strip chart and also on magnetic tape in analogue or digital form for subsequent computer processing. The surface instrumentation, including recorders, cable drums and winches, is usually installed in a special recording truck located near the wellhead. Sondes normally contain combinations of logging tools that do not mutually interfere, so that a wide suite of geophysical logs may be obtained from a limited number of logging runs.

Several techniques of borehole logging are used together to overcome the problems of mudcake and drilling fluid filtrate invasion so as to investigate the properties of the pristine wallrock. Open holes can be surveyed with the full complement of logging tools. Casing prevents the use of logging methods based on electrical resistivity and distorts measurement of seismic velocities. Consequently only a few of the logging methods, such as those based on radioactivity, can be used in cased holes.

Logging techniques are very widely used in the investigation of boreholes drilled for hydrocarbon exploration, as they provide important *in situ* properties of possible reservoir rocks. They are also used in hydrogeological exploration for similar reasons.

11.3 FORMATION EVALUATION

The geological properties obtainable from borehole logging are: formation thickness and lithology, porosity, permeability, proportion of water and/or hydrocarbon saturation, stratal dip and temperature.

Formation thickness and lithology are normally determined by comparison of borehole logs with the log of a cored hole. The most useful logs are those based on resistivity (Section 11.4), self potential (Section 11.6), radioactivity (Section 11.7) and sonic velocity (Section 11.8), and these are often used in combination to obtain an unambiguous section. The caliper log, which measures changes in borehole diameter, also provides information on the lithologies present. In general, larger diameters reflect the presence of less cohesive wallrocks which are easily eroded during drilling.

Porosity estimates are usually based on measurements of resistivity, sonic velocity and radioactivity. In addition, porosity estimates may be obtained by gamma-ray density logging (Section 11.7.2), neutron−gamma-ray logging (Section 11.7.3) and nuclear magnetic resonance logging (Section 11.10). The methodology is described in the relevant sections which follow. Permeability and water and hydrocarbon saturation are derived from resistivity measurements. Stratal dip and temperature are determined by their own specialized logs.

11.4 RESISTIVITY LOGGING

In this chapter the symbol R is used for resistivity to avoid confusion with the symbol ρ used for density.

The general equation for computing apparent resistivity R_a for any downhole electrode configuration is

$$R_a = \frac{4\pi\Delta V}{I\left\{\left(\frac{1}{C_1P_1} - \frac{1}{C_2P_1}\right) - \left(\frac{1}{C_1P_2} - \frac{1}{C_2P_2}\right)\right\}} \quad (11.1)$$

where C_1, C_2 are the current electrodes, P_1, P_2 the potential electrodes between which there is a potential difference ΔV, and I is the current flowing in the circuit (Fig. 11.2). This is similar to equation (8.9) but with a factor of four instead of two, as the current is flowing in a full space rather than the half-space associated with surface surveying.

Different electrode configurations are used to give information on different zones around the borehole.

Switching devices allow the connection of different sets of electrodes so that several types of resistivity log can be measured during a single passage of the sonde.

The region energized by any particular current electrode configuration can be estimated by considering the equipotential surfaces on which the potential electrodes lie. In an homogeneous medium, the potential difference between the electrodes reflects the current density and resistivity in that region. The same potential difference would be obtained no matter what the position of the potential electrode pair. The zone energized is consequently the region between the equipotential surfaces on which the potential electrodes lie. Fig. 11.2 shows the energized zone in an homogeneous medium.

11.4.1 Normal log

In the *normal log*, only one potential and current electrode are mounted on the sonde, the other pair being grounded some distance from the borehole (Fig. 11.3). By substitution in equation (11.1)

$$R_a = 4\pi C_1 P_1 \, \Delta V/I \qquad (11.2)$$

Since $C_1 P_1$ and I are constant, R_a varies with ΔV, and the output can be calibrated directly in ohm m. The zone energized by this configuration is a thick shell with an inner radius $C_1 P_1$ and a large outer radius. However, the current density decreases rapidly as the separation of C_1 and P_2 increases, so that measurements of resistivity correspond to those in a relatively thin spherical shell. The presence of drilling fluid and resistivity contrasts across lithological boundaries cause current refractions so that the zone tested changes in shape with position in the hole.

It is possible to correct for the invasion of drilling fluid by using the results of investigations with different electrode separations (*short normal log* 16 in (406 mm), *long normal log* 64 in (1626 mm)) which give different penetration into the wallrock. Comparison of these logs with standard correction charts (known as departure curves) allows removal of drilling-fluid effects.

The normal log is characterized by smooth changes in resistivity as lithological boundaries are traversed by the sonde because the zone of testing precedes

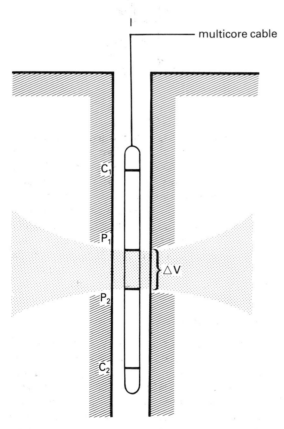

Fig. 11.2 The general form of electrode configuration in resistivity logging. The shaded area represents the effective region energized by the system.

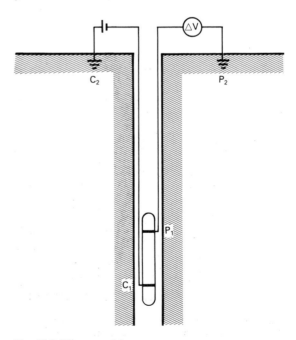

Fig. 11.3 The normal log.

the sonde and the adjacent bed controls the apparent resistivity. Examples of short and long normal logs are given in Fig. 11.4.

11.4.2 Lateral log

In the *lateral log* the in-hole current electrode C_1 is a considerable distance above the potential electrode pair, and is usually mounted on the wire about 6 m above a short sonde containing P_1 and P_2 about 800 mm apart (Fig. 11.5). For this electrode configuration

$$R_a = \frac{4\pi\Delta V}{I\left(\dfrac{1}{C_1P_1} - \dfrac{1}{C_1P_2}\right)} \qquad (11.3)$$

An alternative configuration uses C_1 mounted below the potential electrode pair.

The measured potential difference varies in proportion to the resistivity, so the output can be calibrated directly in ohm m. The zone energized extends much farther into the wallrock than with normal logs, and the apparent resistivity thus approaches the pristine wallrock value more closely.

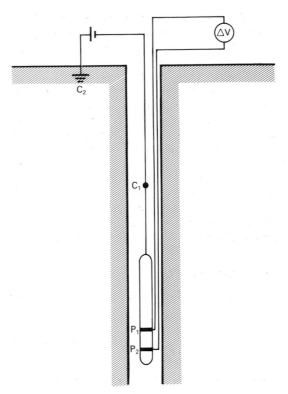

Fig. 11.5 The lateral log.

The electrode configuration causes asymmetry in the apparent resistivity signature as the potential electrode pair descends through one bed while the current electrode may be moving through another. Thin beds produce spurious peaks below them. The lateral log does, however, give a clear indication of the lower boundary of a formation. An example of a lateral log and comparison with normal and self-potential logs is given in Fig. 11.6. As with the normal logs, corrections for the effects of invasion can be applied by making use of standard charts.

11.4.3 Laterolog

The normal and lateral logs described above have no control on the direction of current flow through the wallrock. By contrast, the *laterolog* (or *guard log*) is a focussed log in which the current is directed horizontally so that the zone tested has the form of a circular disc. This may be achieved by the use of a short electrode 75–300 mm long between two long (guard) electrodes about 1.5 m long (Fig. 11.7). The current supply to the electrodes is automatically ad-

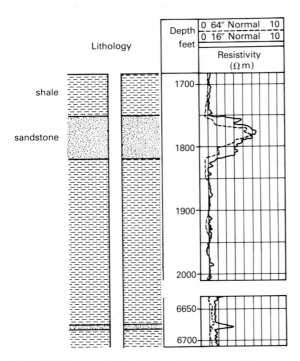

Fig. 11.4 A comparison of short and long normal logs through a sequence of sandstone and shale. (After Robinson & Çoruh 1988.)

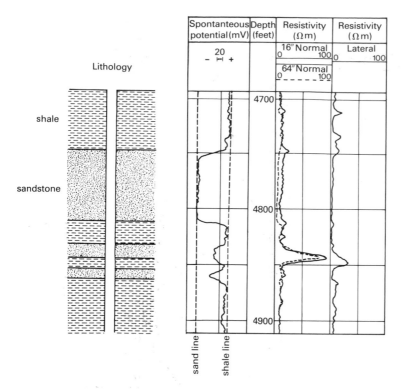

Lithology

shale

sandstone

sand line

shale line

Fig. 11.6 The lateral log compared with normal and self potential logs. (After Guyod 1974.)

justed so as to maintain them all at the same potential. Since no potential difference exists between the electrodes, the current flows outwards horizontally, effectively energizing the wallrock to a depth of about three times the length of the guard electrodes. The use of a fixed potential has the consequence that the current in the central electrode varies in proportion to the apparent resistivity so that the output can be calibrated in ohm m.

The focussing of the log makes it sensitive to thin beds down to the same thickness as the length of the central electrode. The zone of invasion has a pronounced effect which can be estimated from the results of normal and lateral logging and corrected using standard charts.

11.4.4 **Microlog**

The *microlog* (or *wall-resistivity log*) makes measurements at very small electrode spacings by using small, button-shaped electrodes 25–50 mm apart mounted on an insulating pad pressed firmly against the wallrock by a power-driven expansion device (Fig. 11.8). The depth of penetration is typically about 100 mm. Different electrode arrangements

allow the measurement of micronormal, microlateral and microlaterolog apparent resistivities that are equivalent to normal, lateral and laterolog measurements with much smaller electrode spacings. The log has to be moved very slowly and it is normally used only in short borehole sections which are of particular interest.

As the electrode spacing is so small, the effects of the borehole diameter, drilling fluid and adjacent beds are negligible. Very thin beds register sharply, but the main use of the microlog is to measure the resistivities of the mudcake and zone of invasion, which are needed to convert log measurements into true resistivities.

11.4.5 **Porosity estimation**

Porosity is defined as the fractional volume of pore spaces in a rock. The method of *porosity estimation* is based on the relationship between *formation factor* F and porosity ϕ discovered by Archie (1942). F is a function of rock texture and defined as

$$F = R_f/R_w \qquad (11.4)$$

where R_f and R_w are the resistivities of the saturated

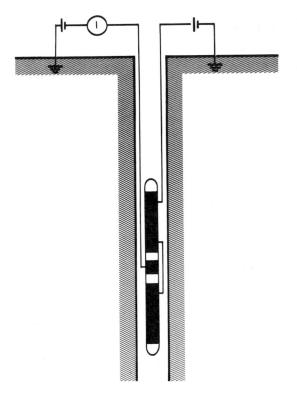

Fig. 11.7 The laterolog.

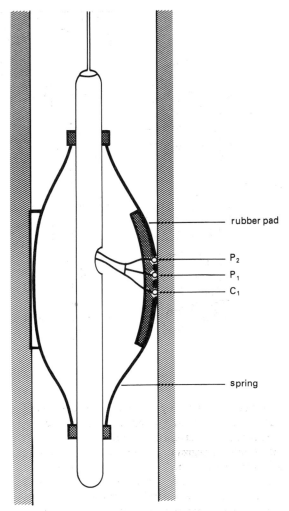

Fig. 11.8 The microlog.

formation and pore fluid, respectively (Section 8.22). Porosity and formation factor are related by

$$\phi = aF^{-m} \tag{11.5}$$

where a is an empirical constant specific to the rocks of the area of interest, and m a constant known as the *cementation factor* which depends on the grain size and complexity of the paths between pores (*tortuosity*). Normal limits on a and m, derived experimentally, are given by

$$0.62 < a < 1.0, \ 2.0 < m < 3.0$$

11.4.6 Water and hydrocarbon saturation estimation

Natural pore water is generally a good conductor of electricity because of the presence of dissolved salts. Hydrocarbons, however, are poor conductors and cause an increase in the measured resistivity of a rock relative to that in which water is the pore fluid. Hydrocarbons displace pore water and cause it to be reduced to an irreducible minimum level. Archie

(1942) described a method of estimating the proportion of pore water present (the *water saturation S*) based on laboratory measurements of the resistivities of sandstone cores containing varying proportions of hydrocarbons and pore water of fixed salinity. If R_f and R_h are the resistivities of (matrix + pore water) and (matrix + pore water + hydrocarbons), respectively

$$S = (R_f/R_h)^{1/n} \tag{11.6}$$

where n is the *saturation exponent*. The experimentally determined limits of n are $1.5 < n < 3.0$, although n is usually assumed to be two where there is no evidence to the contrary.

Combining equations (11.4) and (11.6) gives an alternative expression for S

$$S = (FR_w/R_h)^{1/n} \tag{11.7}$$

R_f is determined in parts of the borehole which are known to be saturated with water.

11.4.7 Permeability estimation

Permeability (k) is a measure of the capacity of a formation to transmit fluid under the influence of a pressure gradient. It is dependent upon the degree of interconnection of the pores, the size of the pore throats and the active capillary forces. It is estimated from the minimum pore water remaining after displacement of the rest by hydrocarbons (the *irreducible water saturation* S_{irr}), which in turn is estimated from resistivity measurements in parts of the formation where irreducible saturation obtains.

$$k = (c\phi^3/S_{irr})^2 \tag{11.8}$$

where ϕ is determined as in Section 11.4.5 and c is a constant dependent on the lithology and grain size of the formation. Large errors in determining the parameters from which k is derived render permeability the most difficult reservoir property to estimate.

k is commonly expressed in darcies, a unit corresponding to a permeability which allows a flow of $1\,\mathrm{mm\,s^{-1}}$ of a fluid of viscosity $10^{-3}\,\mathrm{Pa\,s}$ through an area of $100\,\mathrm{mm^2}$ under a pressure gradient of $0.1\,\mathrm{atm\,mm^{-1}}$. Reservoirs commonly exhibit values of permeability from a few millidarcies to 1 darcy.

11.4.8 Resistivity dipmeter log

The sonde of the *dipmeter log* contains four equally-spaced microresistivity electrodes at the same horizontal level, which allow the formation dip and strike to be estimated. The orientation of the sonde is determined by reference to a magnetic compass and its deviation from the vertical by reference to a spirit level or pendulum. The four electrodes are mounted at right angles to each other round the sonde. If the beds are horizontal, identical readings are obtained at each electrode. Non-identical readings can be used to determine dip and strike. In fact the four electrodes can be used to make four three-point dip calculations as a control on data quality. Dipmeter results are commonly displayed on *tadpole plots* (Fig. 11.9).

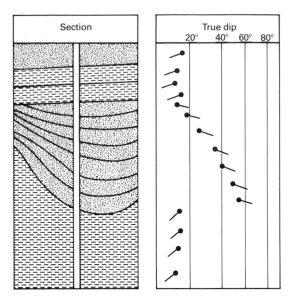

Fig. 11.9 A typical tadpole plot obtained from a dipmeter log.

11.5 INDUCTION LOGGING

The *induction log* is used in dry holes or boreholes that contain non-conductive drilling fluid which electrically insulates the sonde. The wallrock is energized by an electromagnetic field, typically of about 20 kHz, which generates eddy currents in the wallrock by electromagnetic induction. The secondary EM field created is registered at a receiver which is compensated for direct coupling with the primary field and which allows a direct estimate of apparent resistivity to be made. The set-up is thus similar to the surface moving coil-receiver EM system described in Section 9.5.

The two-coil system shown in Fig. 11.10(a) is unfocussed and the induced EM field flows in circular paths around the borehole, with a depth of investigation of about 75% of the transmitter−receiver separation. Lithological boundaries show up as gradual changes in apparent resistivity as they are traversed. When combined with information from other logs, corrections for invasion can be made from standard charts.

Clearer indications of lithological contacts can be obtained using a focussed log such as that shown in Fig. 11.10(b), in which two extra coils are mounted near the receiver and transmitter and wired in series with them. Such an arrangement provides a depth

transmitter   receiver

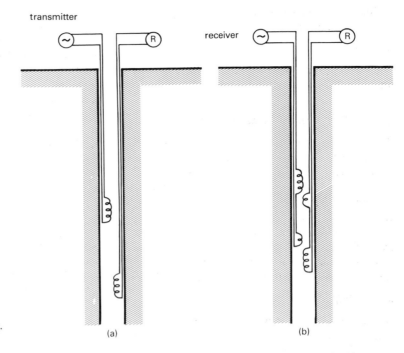

(a)   (b)

Fig. 11.10 (a) A simple induction log.
(b) A focussed induction log.

of penetration of about twice the transmitter-receiver
separation. This particular focussed system has the
disadvantage that spurious apparent resistivities are
produced at boundaries, but this effect may be
compensated by employing additional coils.

See Section 9.6 for the application of time–
domain electromagnetic techniques in borehole
surveys.

11.6 SELF-POTENTIAL LOGGING

In the *self-potential (SP) log*, measurements of
potential difference are made in boreholes filled with
conductive drilling fluid between an electrode on
the sonde and a grounded electrode at the surface
(Fig. 11.11).

The SP effect (Section 8.4.2) originates from the
movement of ions at different speeds between two
fluids of differing concentration. The effect is pronounced across the boundary between sandstone
and shale, as the invasion of drilling mud filtrate is
greater into the sandstone. Near the borehole there
is a contact between mud filtrate in the sandstone
and pore fluid of different salinity in the shale. The

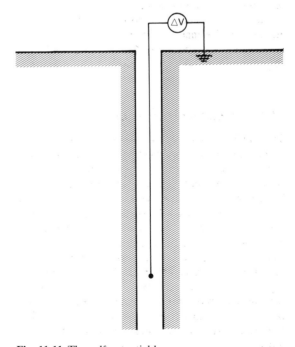

Fig. 11.11 The self-potential log.

movement of ions necessary to nullify this difference is impeded by the membrane polarization effect (Section 8.3.2) of the clay minerals in the shale. This causes an imbalance of charge across the boundary and generates a potential difference of a few tens to a few hundreds of millivolts.

In sequences of sandstone and shale, the sandstone anomaly is negative with respect to the shale. This SP effect provides a sharper indication of the boundary than resistivity logs. In such sequences it is possible to draw a 'shale line' through the anomaly maxima and a 'sand line' through the minima (Fig. 11.6). The proportion of sand to shale at intermediate anomalies can then be estimated by interpolation.

The main applications of SP logging are the identification of boundaries between shale horizons and more porous beds, their correlation between boreholes, and the determination of the volume of shale in porous beds. They have also been used to locate coal seams. In hydrocarbon-bearing zones the SP log has less deflection than normal and this 'hydrocarbon suppression' can be an indicator of their presence.

11.7 RADIOMETRIC LOGGING

Radiometric logs make use of either the natural radioactivity produced by the unstable elements U^{238}, Th^{232} and K^{40} (Section 10.3), or radioactivity induced by the bombardment of stable nucleii with gamma rays or neutrons. Gamma rays are detected by a scintillation counter (Section 10.4.2) or occasionally by a Geiger–Müller counter (Section 10.4.1) or an ionization chamber. Radioactivity in borehole measurements is usually expressed in API (American Petroleum Institute) units, which are defined according to reference levels in a test pit at the University of Houston.

11.7.1 Natural gamma radiation log

Shales usually contain small quantities of radioactive elements, in particular K^{40} which occurs in micas, alkali feldspars and clay minerals, and trace amounts of U^{238} and Th^{232}. These produce detectable gamma radiation from which the source can be distinguished by spectrometry, i.e. measurements in selected energy bands (Section 10.4.3). The *natural gamma radiation log* consequently detects shale horizons and can provide an estimate of the clay content of other sedimentary rocks. Potassium-rich evaporites are also distinguished. An example of this type of log is shown in Fig. 11.12.

The *natural gamma radiation log* (or *gamma log*) measures radioactivity originating within a few decimetres of the borehole. Because of the statistical nature of gamma-ray emissions, a recording time of several seconds is necessary to obtain a reasonable count, so the sensitivity of the log depends on the count time and the speed with which the hole is logged. Reasonable results are obtained with a count time of 2 s and a speed of $150\,\text{mm s}^{-1}$. Measurements can be made in cased wells, but the intensity of the radiation is reduced by about 30%.

11.7.2 Gamma-ray density log

In the *gamma-ray density* (or *gamma–gamma*) *log*, artificial gamma rays from a Co^{60} or Cs^{137} source are utilized. Gamma-ray photons collide 'elastically' with electrons and are reduced in energy, a phenomenon known as *Compton scattering*. The number of collisions over any particular interval of time depends upon the abundance of electrons present (the *electron density index*), which in turn is a function of the density of the formation. Density is thus estimated by measuring the proportion of gamma radiation returned to the detector by Compton scattering.

The relationship between the formation density ρ_f and electron density index ρ_e depends upon the elements present

$$\rho_f = \rho_e w / 2\Sigma N \qquad (11.9)$$

where w is the molecular weight of the constituents of the formation and N is the atomic number of the elements present, which specifies the number of electrons.

The sonde has a plough-shaped leading edge which cuts through the mudcake, and is pressed against the wallrock by a spring. Most of the scattering takes place within about 75 mm of the sonde. A modern version of the sonde uses long and short spacings for the detectors which are sensitive to material far from and near to the sonde, respectively.

Porosity ϕ may be estimated from the density measurements. For a rock of formation density ρ_f, matrix density ρ_m and pore fluid density ρ_w

$$\rho_f = \phi \rho_w + (1 - \phi) \rho_m \qquad (11.10)$$

Thus

$$\phi = (\rho_m - \rho_f)/(\rho_m - \rho_w) \qquad (11.11)$$

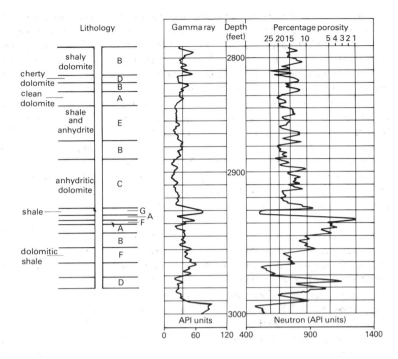

Fig. 11.12 Natural gamma and neutron logs over the same sequence of dolomite and shale. (After Wood *et al*. 1974.)

11.7.3 **Neutron–gamma-ray log**

In the *neutron–gamma-ray* (or *neutron*) *log*, non-radioactive elements are bombarded with neutrons and, as a result of neutron capture by the nuclei, they are stimulated to emit gamma rays which provide information on porosity. The sonde contains a neutron source, consisting of a small quantity of a radioactive substance such as Pu–Be, and a scintillation counter (Section 10.4.2) a fixed distance apart.

The neutrons collide with atomic nuclei in the wallrock. Most nuclei are much more massive than neutrons, which rebound 'elastically' with very little loss of kinetic energy. However, a hydrogen ion has almost the same mass as a neutron, so collision transfers considerable kinetic energy and slows the neutron to the point at which it can be absorbed by a larger nucleus. This neutron capture, which normally occurs within 600 mm of the borehole, gives rise to gamma radiation, a proportion of which impinges on the scintillation counter. The intensity of the radiation is controlled by how far it has travelled from the point of neutron capture. This distance depends mainly on the hydrogen-ion concentration: the higher the concentration, the closer the neutron

capture to the borehole and the higher the level of radiation.

In sandstone and limestone all hydrogen ions are present in pore fluids or hydrocarbons, so the hydrogen-ion concentration is entirely dependent upon the porosity. In shales, however, hydrogen can also derive from micas and clay minerals. Consequently the lithology must be determined by other logs (e.g. gamma log) before porosity estimates can be made in this way. Similar count times and logging speeds to other radiometric methods are used. The method is suitable for use in both cased and uncased boreholes. An example is given in Fig. 11.12.

11.8 **SONIC LOGGING**

The *sonic log*, also known as the *continuous velocity* or *acoustic log*, determines the seismic velocities of the formations traversed. The sonde normally contains two receivers about 300 mm apart and an acoustic source some 900–1500 mm from the nearest receiver (Fig. 11.13(a)). The source generates ultrasonic pulses at a frequency of 20–40 kHz.

Since the wallrock invariably has a greater velocity than the drilling fluid, part of the sonic pulse is critically refracted in the wallrock and part of its

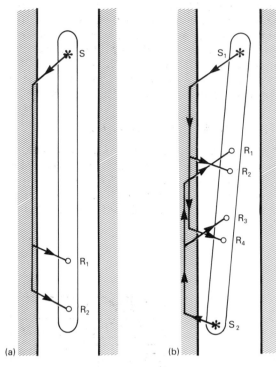

Fig. 11.13 (a) A simple sonic log. (b) A borehole-compensated sonic log.

energy returns to the sonde as a head wave. Each sonic pulse activates a timer so that the differential travel time between the receivers can be measured. If the sonde is tilted in the well, or if the well diameter varies, different path lengths result. This problem is overcome, in a borehole-compensated log, by using a second source on the other side of the receivers (Fig. 11.13(b)) so that the tilt effect is self-cancelling when all four travel paths are considered.

Porosity φ may be estimated from the sonic measurements. For a rock whose matrix velocity (the velocity of its solid components) is V_m and pore fluid velocity is V_w, the formation velocity V_f is given by

$$\frac{1}{V_f} = \frac{\phi}{V_w} + \frac{1-\phi}{V_m}. \tag{11.12}$$

The velocity of the matrix can be determined from cuttings and that of the fluid from standard values.

Sondes of the dimensions described above have transmission path lengths that lead to penetrations of only a few centimetres into the wallrock and allow the discrimination of beds only a few decimetres in thickness. However, they are greatly affected by drilling damage to the wallrock and to overcome this, longer sondes with source-geophone spacings of 2.1–3.7 m may be used. In addition to providing porosity estimates, sonic logs may be used for correlation between boreholes and are also used in the interpretation of seismic reflection data by providing velocities for the conversion of reflection times into depths. An example is given in Fig. 11.14.

Sonic logs can also provide useful attenuation information, usually from the first P-wave arrival. Attenuation (Section 3.5) is a function of many variables including wavelength, wavetype, rock texture, type and nature of pore fluid and the presence of fractures and fissures. However, in a cased well, the attenuation is at a minimum when the casing is held in a thick annulus of cement and at a maximum when the casing is free. This forms the basis of the cement bond log (or cement evaluation probe) which is used to investigate the effectiveness of the casing. Other techniques make use of both P- and S-travel times to estimate the *in situ* elastic moduli (Section 5.11). See also the description of vertical seismic profiling in Section 4.13.

11.9 TEMPERATURE LOGGING

Temperature gradients may be measured through a borehole section using a sonde on which a number of closely-spaced thermistor probes are mounted. The vertical *heat flux H* is estimated by

$$H = k_z d\theta/dz \tag{11.13}$$

where $d\theta/dz$ is the vertical temperature gradient and k_z is the *thermal conductivity* of the relevant wallrock, which is usually determined by laboratory measurement.

Temperature gradients within about 20 m of the Earth's surface are strongly affected by diurnal and seasonal changes in solar heating and do not provide reliable estimates of heat flux. Porous strata can also strongly influence temperature gradients by the ingress of connate water and because their contained pore fluids act as a thermal sink. Heat flux measurements are commonly made to assess the potential of an area for geothermal energy utilization.

11.10 MAGNETIC LOGGING

11.10.1 Magnetic log

The normal *magnetic log* has only limited application. The magnetic field is measured either with a from the Earth's field. A receiver measures the amplitude and decay rate of the precession of the protons as they realign in the geomagnetic field direction when the polarizing field is inactive. The amplitude measurements provide an estimate of the amount of fluid in the pore spaces and the rate of decay is diagnostic of the type of fluid present.

11.11 GRAVITY LOGGING

In situations where density is a function of depth only, the strata being substantially horizontal, step-wise measurement of the vertical gravity gradient with a *gravity log* can be used to estimate mean densities according to the calculation given in Section 6.9.

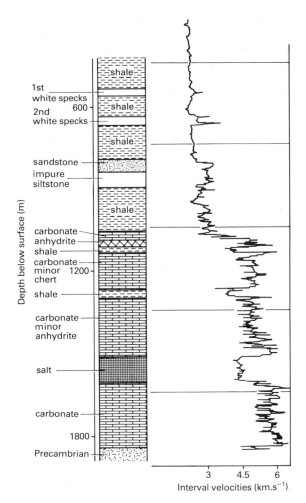

Fig. 11.14 A continuous velocity log. (After Grant & West 1965.)

downhole fluxgate or proton magnetometer (Section 7.6) or a susceptibility meter is utilized. Anomalous readings indicate the presence of magnetic minerals.

11.10.2 Nuclear magnetic resonance log

The *nuclear magnetic resonance* (or *free fluid index*) *log* is used to estimate the hydrogen-ion concentration in formation fluids and, hence, to obtain a measure of porosity. The method of measurement resembles that of the proton magnetometer, but with the formation fluid taking the place of the sensor. A pulsed magnetic field causes the alignment of some of the hydrogen ions in a direction different

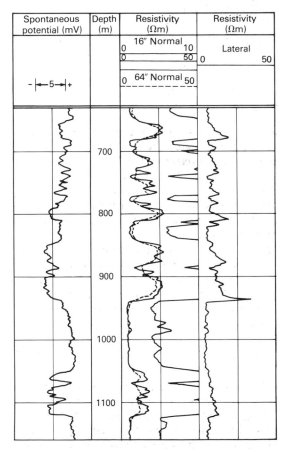

Fig. 11.15 SP and resistivity logs pertaining to Question 4. (After Desbrandes 1985.)

A specialized borehole gravimeter of LaCoste & Romberg type (Section 6.4) is used for gravity logging. The instrument has a diameter of about 100 mm, ±5 microgal accuracy, and is capable of operation in temperatures up to 120°C and pressures up to 80 MPa. The normal vertical spacing of observations is about 6 m and if depths are determined to ±50 mm, densities can be estimated to ±0.01 Mg m^{-3}, which corresponds to an accuracy of porosity estimation of about ±1%. The density applies to the part of the formation lying within about five times the spacing between observations. This is more accurate than other methods of measuring density in boreholes and can be used in cased holes. It is, however, time-consuming as each reading can take 10–20 minutes and the meter is so costly that it can only be risked in boreholes in excellent condition.

11.12 PROBLEMS

1 A sandstone, when saturated with water of resistivity 5 ohm m, has a resistivity of 40 ohm m. Calculate the probable range of porosity for this rock.

2 On a sonic log, the travel time observed in a sandstone was 568 μs over a source−receiver distance of 2.5 m. Given that the seismic velocities of quartz and pore fluid are 5.95 and 1.46 km s^{-1}, respectively, calculate the porosity of the sandstone. What would be the effect on the observed travel time and velocity of the sandstone if the pore fluid were methane with a velocity of 0.49 km s^{-1}?

3 During the drilling of an exploratory borehole, the rock chippings flushed to the surface indicated the presence of a sandstone−shale sequence. The lateral log revealed a discontinuity at 10 m depth below which the resistivity decreased markedly. The SP log showed no deflection at this depth and recorded consistently low values. The

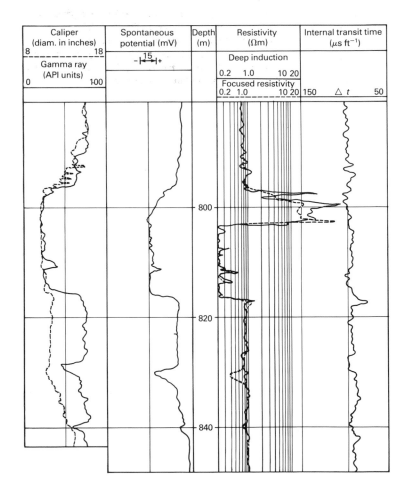

Fig. 11.16 SP, induction, resistivity, sonic, caliper and gamma-ray logs pertaining to Question 5. (After Ellis 1987.)

gamma-ray density log indicated an increase in density with depth across the discontinuity from 2.24 to $2.35\,Mg\,m^{-3}$.

(a) Infer, giving your reasons, the nature of the discontinuity.

(b) What porosity information is provided by the data?

4 Fig. 11.15 shows the SP and short normal (including a partial expanded scale version), long normal and lateral resistivity logs of a borehole penetrating a sedimentary sequence. Interpret the logs as fully as possible.

5 Fig. 11.16 shows the SP, induction, laterolog, sonic, caliper and gamma logs of a borehole in a sequence of shale and sandstone. Interpret the logs as fully as possible.

6 Two gravity readings in a borehole, 100 m apart vertically, reveal a measured gravity difference of 107.5 gu. What is the average density of the rocks between the two observation levels?

FURTHER READING

Asquith, G.B. & Gibson, C.R. (1982) *Basic Well Log Analysis for Geologists*. Am. Assoc. Petroleum Geologists, Tulsa.

Desbrandes, R. (1985) *Encyclopedia of Well Logging*. Graham & Trotman, London.

Dyck, A.V. & Young, R.P. (1985) Physical characterization of rock masses using borehole methods. *Geophysics*, **50**, 2530−41.

Ellis, D.V. (1987) *Well Logging for Earth Scientists*. Elsevier, Amsterdam.

Hearst, J.R. & Nelson, P.H. (1985) *Well Logging for Physical Properties*. McGraw-Hill, New York.

Labo, J. (1986) *A Practical Introduction to Borehole Geophysics*. Soc. Econ. Geologists, Tulsa.

Pirson, S.J. (1977) *Geologic Well Log Analysis* (2nd edn). Gulf, Houston.

Rider, M.H. (1986) *The Geological Interpretation of Well Logs*. Blackie, London.

Robinson, E.S. & Çoruh, C. (1988) *Basic Exploration Geophysics*. John Wiley & Sons, New York.

Segesman, F.F. (1980) Well logging method. *Geophysics*, **45**, 1667−84.

Serra, O. (1984) *Fundamentals of Well-log Interpretation, 1. The Acquisition of Logging Data*. Elsevier, Amsterdam.

Serra, O. (1986) *Fundamentals of Well-log Interpretation, 2. The Interpretation of Logging Data*. Elsevier, Amsterdam.

Snyder, D.D. & Fleming, D.B. (1985) Well logging − a 25 year perspective. *Geophysics*, **50**, 2504−29.

Tittman, J. (1987) Geophysical well logging. *In*: Samis, C.G. & Henyey, T.L. (eds) *Methods of Experimental Physics, Vol. 24, Part B − Field Measurements*, Academic Press, Orlando, 441−615.

Appendix

SI, c.g.s. and Imperial (customary U.S.A.) units and conversion factors

Quantity	SI name	SI Symbol	c.g.s. equivalent	Imperial (U.S.A.) equivalent
Mass	kilogram	kg	10^3 g	2.205 lb
Time	second	s	s	s
Length	metre	m	10^2 cm	39.37 in
				3.281 ft
Acceleration	metre s^{-2}	$m\,s^{-2}$	10^2 cm $s^{-2} = 10^2$ gal	39.37 in s^{-2}
Gravity	gravity unit	$gu = \mu m\,s^{-2}$	10^{-1} milligal (mgal)	3.937×10^{-5} in s^{-2}
Density	megagram m^{-3}	$Mg\,m^{-3}$	$g\,cm^{-3}$	3.613×10^{-2} lb in^{-3}
				62.421 lb ft^{-3}
Force	newton	N	10^5 dyne	0.2248 lb (force)
Pressure	pascal	$Pa = N\,m^{-2}$	10 dyne cm$^{-2} = 10^{-5}$ bar	1.45×10^{-4} lb in^{-2}
Energy	joule	J	10^7 erg	0.7375 ft lb
Power	watt	$W = J\,s^{-1}$	10^7 erg s^{-1}	0.7375 ft lb s^{-1}
				1.341×10^{-3} hp
Temperature	T	°C*	°C	$(1.8T + 32)$ °F
Current	ampere	A	A	A
Potential	volt	V	V	V
Resistance	ohm	$\Omega = V\,A^{-1}$	Ω	Ω
Resistivity	ohm m	$\Omega\,m$	$10^2\,\Omega$ cm	3.281 ohm ft
Conductance	siemen	$S = \Omega^{-1}$	mho	mho
Conductivity	siemen m^{-1}	$S\,m^{-1}$	10^{-2} mho cm^{-1}	0.3048 mho ft^{-1}
Dielectric constant	dimensionless			
Magnetic flux	weber	$Wb = V\,s$	10^8 maxwell	
Magnetic flux density (B)	tesla	$T = Wb\,m^{-2}$	10^4 gauss (G)	
Magnetic anomaly	nanotesla	$nT = 10^{-9}\,T$	gamma $(\gamma) = 10^{-5}\,G$	
Magnetizing field (H)	ampere m^{-1}	$A\,m^{-1}$	$4\pi 10^{-3}$ oersted (Oe)	
Inductance	henry	$H = Wb\,A^{-1}$	10^9 emu (electromagnetic unit)	
Permeability of vacuum (μ_0)	henry m^{-1}	$4\pi 10^{-7} H\,m^{-1}$	1	
Susceptibility	dimensionless	k	4π emu	
Magnetic pole strength	ampere m	$A\,m$	10 emu	
Magnetic moment	ampere m^2	$A\,m^2$	10^3 emu	
Magnetization (J)	ampere m^{-1}	$A\,m^{-1}$	10^{-3} emu cm^{-3}	

* Strictly, SI temperatures should be stated in Kelvin ($K = 273.15 + °C$). In this book, however, temperatures are given in the more familiar Centigrade (Celsius) scale.

References

Abdoh, A. & Pilkington, M. (1989) Radon emanation studies of the Ile Bizard Fault, Montreal. *Geoexploration*, **25**, 341–54.

Al-Chalabi, M. (1972) Interpretation of gravity anomalies by non-linear optimisation. *Geophys. Prosp.*, **20**, 1–16.

Al-Sadi, H.N. (1980) *Seismic Exploration*. Birkhauser Verlag, Basel.

Anstey, N.A. (1965) Wiggles. *J. Can. Soc. Exploration Geophysicists*, **1**, 13–43.

Anstey, N.A. (1966) Correlation techniques – a review. *J. Can. Soc. Exploration Geophysicists*, **2**, 55–82.

Anstey, N.A. (1977) *Seismic Interpretation: The Physical Aspects*. IHRDC, Boston.

Anstey, N.A. (1981) *Seismic Prospecting Instruments, Vol. 1: Signal Characteristics and Instrument Specifications*. Gerbrüder Borntraeger, Berlin.

Anstey, N.A. (1982) *Simple Seismics*. IHRDC, Boston.

Archie, G.E. (1942) The electrical resistivity log as an aid in determining some reservoir characteristics. *Trans. Am. Inst. Mining Met. Eng.*, **146**, 54–62.

Arnaud Gerkens, J.C.d'. (1989) *Foundations of Exploration Geophysics*. Elsevier, Amsterdam.

Arzi, A.A. (1975) Microgravimetry for engineering applications. *Geophys. Prosp.*, **23**, 408–25.

Aspinall, A. & Walker, A.R. (1975) The earth resistivity instrument and its application to shallow earth surveys. *Underground Services*, **3**, 12–5.

Asquith, G.B. & Gibson, C.R. (1982) *Basic Well Log Analysis for Geologists*. Am. Assoc. Petroleum Geologists, Tulsa.

Baeten, G., Fokkema, J. & Ziolkowski, A. (1988) The marine vibrator source. *First Break*, **6**(9), 285–94.

Balch, A.H., Lee, M.W., Miller, J.J. & Taylor, R.T. (1982) The use of vertical seismic profiles in seismic investigations of the earth. *Geophysics*, **47**, 906–18.

Bally, A.W. (ed.) (1983) *Seismic Expression of Structural Styles* (a picture and work atlas: 3 vols). AAPG Studies in Geology No 15, American Association of Petroleum Geologists, Tulsa.

Bally, A.W. (ed.) (1987) *Atlas of Seismic Stratigraphy* (3 vols). AAPG Studies in Geology No **27**, American Association of Petroleum Geologists, Tulsa.

Bamford, D., Nunn, K., Prodehl, C. & Jacob, B. (1978) LISPB-IV. Crustal structure of northern Britain. *Geophys. J. R. astr. Soc.*, **54**, 43–60.

Baranov, W. (1975) *Potential Fields and Their Transformations in Applied Geophysics*. Gebrüder Borntraeger, Berlin.

Baranov, V. & Naudy, H. (1964) Numerical calculation of the formula of reduction to the magnetic pole (airborne). *Geophysics*, **29**, 67–79.

Barazangi, M. & Brown L. (eds.) (1986) *Reflection Seismology: The Continental Crust*. AGU Geodynamics Series, **14**. American Geophysical Union, Washington.

Barker, R.D. (1981) The offset system of electrical resistivity sounding and its use with a multicore cable. *Geophys. Prosp.*, **29**, 128–43.

Barker, R.D. & Worthington, P.F. (1972) Location of disused mineshafts by geophysical methods. *Civil Engineering and Public Works Review*, **67**, No. 788, 275–6.

Barnett, C.T. (1984) Simple inversion of time-domain electromagnetic data. *Geophysics*, **49**, 925–33.

Barraclough, D.R. & Malin, S.R.C. (1971) *Synthesis of International Geomagnetic Reference Field Values*. Inst. Geol. Sci. Rep. No. 71/1.

Barringer, A.R. (1962) A new approach to exploration – the INPUT airborne electrical pulse prospecting system. *Mining Congress J.*, **48**, 49–52.

Barton, P.J. (1986) Comparison of deep reflection and refraction structures in the North Sea. *In*: Barazangi, M. & Brown, L. (eds) *Reflection Seismology: a Global Perspective*. Geodynamics Series, **13**, 297–300, American Geophysical Union, Washington DC.

Bayerly, M. & Brooks, M. (1980) A seismic study of deep structure in South Wales using quarry blasts. *Geophys. J. R. astr. Soc.*, **60**, 1–19.

Bell, R.E. & Watts, A.B. (1986) Evaluation of the BGM-3 sea gravity meter system onboard R/V Conrad. *Geophysics*, **51**, 1480–93.

Berg, O.R. & Woolverton, D.G. (eds) (1985) *Seismic Stratigraphy II: An Integrated Approach to Hydrocarbon Exploration*. AAPG Memoir 39, American Association of Petroleum Geologists, Tulsa.

Berry, M.J. & West, G.F. (1966) An interpretation of the first-arrival data of the Lake Superior experiment by the time term method. *Bull. Seismol. Soc. Am.*, **56**, 141–71.

Bertin, J. (1976) *Experimental and Theoretical Aspects of Induced Polarisation, Vols 1 and 2*. Gebrüder Borntraeger, Berlin.

Bhattacharya, P.K. & Patra, H.P. (1968) *Direct Current Electrical Sounding*. Elsevier, Amsterdam.

Birch, F. (1960) The velocity of compressional waves in rocks to ten kilobars, Part I. *J. Geophys. Res.*, **65**, 1083–102.

Birch, F. (1961) The velocity of compressional waves in rocks to ten kilobars, Part 2. *J. Geophys. Res.*, **66**, 2199–224.

Boissonnas, E. & Leonardon, E.G. (1948) Geophysical exploration by telluric currents with special reference to a survey of the Haynesville Salt Dome, Wood County, Texas. *Geophysics*, **13**, 387–403.

Bolt, B.A. (1976) *Nuclear Explosions and Earthquakes: The Parted Veil*. Freeman, San Francisco.

Bolt, B.A. (1982) *Inside the Earth*. Freeman, San Francisco.

Bott, M.H.P. (1973) Inverse methods in the interpretation of magnetic and gravity anomalies. *In*: Alder, B., Fernbach, S. & Bolt, B.A. (eds.), *Methods in Computational Physics*, **13**, 133–62.

Bott, M.H.P. (1982) *The Interior of the Earth*. Edward Arnold, London.

Bott, M.H.P., Day, A.A. & Masson-Smith, D. (1958) The geological interpretation of gravity and magnetic surveys in Devon and Cornwall. *Phil. Trans. R. Soc.*, **215A**, 161–91.

Bott, M.H.P. & Scott, P. (1964) Recent geophysical studies in southwest England. *In*: Hosking, K.F.G. & Shrimpton, G.H. (eds), *Present Views of Some Aspects of the Geology of Devon and Cornwall*. Royal Geological Society of Cornwall.

Bowin, C., Aldrich, T.C. & Folinsbee, R.A. (1972) VSA gravity meter system: Tests and recent developments. *J. geophys. Res.*, **77**, 2018–33.

Boyd, G.W. & Wiles, C.J. (1984) The Newmont drill-hole EMP system – Examples from eastern Australia. *Geophysics*, **49**, 949–56.

Brewer, J.A. (1983) Profiling continental basement: the key to understanding structures in the sedimentary cover. *First Break*, **1**, 25–31.

Brewer, J.A. & Oliver, J.E. (1980) Seismic reflection studies of deep crustal structure. *Ann. Rev. Earth planet. Sci.*, **8**, 205–30.

Brigham, E.O. (1974) *The Fast Fourier Transform*. Prentice-Hall, New Jersey.

Brooks, M., Doody, J.J. & Al-Rawi, F.R.J. (1984) Major crustal reflectors beneath SW England. *J. Geol. Soc. Lond.*, **141**, 97–103.

Brooks, M. & Ferentinos, G. (1984) Tectonics and sedimentation in the Gulf of Corinth and the Zakynthos and Kefallinia Channels, western Greece. *Tectonophysics*, **101**, 25–54.

Brooks, M., Mechie, J. & Llewellyn, D.J. (1983) Geophysical investigations in the Variscides of southwest Britain. *In*: Hancock, P.L. (ed.), *The Variscan Fold Belt in the British Isles*. Hilger, Bristol, Ch. 10, 186–97.

Brown, A.R. (1986) *Interpretation of three-dimensional seismic data*. AAPG Memoir **42**, American Association of Petroleum Geologists, Tulsa.

Brown, L.F. & Fisher, W.L. (1980) *Seismic Stratigraphic Interpretation and Petroleum Exploration*. AAPG Continuing Education Course Note Series No. 16.

Brozena, J.M. & Peters, M.F. (1988) An airborne gravity study of eastern North Carolina. *Geophysics*, **53**, 245–53.

Bugg, S.F. & Lloyd, J.W. (1976) A study of freshwater lens configuration in the Cayman Islands using resistivity methods. *Q. J. Eng. Geol*, **9**, 291–302.

Cady, J.W. (1980) Calculation of gravity and magnetic anomalies of finite-length right polygonal prisms. *Geophysics*, **45**, 1507–12.

Cagniard, L. (1953) Basic theory of the magnetotelluric method of geophysical prospecting. *Geophysics*, **18**, 605–35.

Camina, A.R. & Janacek, G.J. (1984) *Mathematics for Seismic Data Processing and Interpretation*. Graham & Trotman, London.

Cassell, B. (1984) Vertical seismic profiles – an introduction. *First Break*, **2**(11), 9–19.

Červeny, V., Langer, J. & Pšenčik, I. (1974) Computation of geometric spreading of seismic body waves in laterally inhomogeneous media with curved interfaces. *Geophys. J. R. astr. Soc.*, **38**, 9–19.

Červený, V. & Ravindra, R. (1971) *Theory of Seismic Head Waves*. University of Toronto Press, Toronto.

Chakridi, R. & Chouteau, M. (1988) Design of models for electromagnetic scale modelling. *Geophys. Prosp.*, **36**, 537–50.

Christensen, N.I. & Fountain, D.M. (1975) Constitution of the lower continental crust based on experimental studies of seismic velocities in granulites. *Bull. Geol.Soc. Am.*, **86**, 227–36.

Claerbout, J.F. (1985) *Fundamentals of Geophysical Data Processing*. McGraw Hill, New York.

Clark, A.J. (1986) Archaeological geophysics in Britain. *Geophysics*, **51**, 1404–13.

Cook, K.L. & Van Nostrand, R.G. (1954) Interpretation of resistivity data over filled sinks. *Geophysics*, **19**, 761–90.

Corry, C.E. (1985) Spontaneous polarization associated with porphyry sulfide mineralization. *Geophysics*, **50**, 1020–34.

Cunningham, A.B. (1974) Refraction data from single-ended refraction profiles. *Geophysics*, **39**, 292–301.

Daniels, J. (1988) Locating caves, tunnels and mines. *Geophysics: The Leading Edge of Exploration*, 32–7.

Davis, J.L. & Annan, A.P. (1989) Ground-penetrating radar for high-resolution mapping of soil and rock stratigraphy. *Geophys. Prosp.*, **37**, 531–51.

Davis, J.L., Prescott, W.H., Svarc, J.L. & Wendt, K.J. (1989) Assessment of Global Positioning System measurements for studies of crustal deformation. *J. geophys. Res.*, **94**, 13635–50.

Day, G.A., Cooper, B.A., Anderson, C., Burgers, W.F.J., Rønnevik, H.C. & Schöneich, H. (1981) Regional structural maps of the North Sea. *In*: Illing, L.V. & Hobson, G.D. (eds.), *Petroleum Geology of the Continental Shelf of NW Europe*. Heyden & Son, London, Ch. 5, 76–84.

Dehlinger, P. (1978) *Marine Gravity*. Elsevier, Amsterdam.

Desbrandes, R. (1985) *Encyclopedia of Well Logging*. Graham & Trotman, London.

Dey, A. & Morrison, H.F. (1979) Resistivity modelling for arbitrarily shaped two-dimensional structures. *Geophys. Prosp.*, **27**, 106–36.

Dix, C.H. (1955) Seismic velocities from surface measurements. *Geophysics*, **20**, 68–86.

Dix, C.H. (1981) *Seismic Prospecting for Oil*. IHRDC, Boston.

Dobrin, M.B. & Savit, C.H. (1988) *Introduction to Geophysical Prospecting* (4th edn). McGraw Hill, New York.

Duncan, P.M. & Garland, G.D. (1977) A gravity study of the Saguenay area, Quebec. *Can. J. Earth Sci.*, **14**, 145–52.

Durrance, E.M. (1986) *Radioactivity in Geology*. Ellis Horwood, Chichester.

Dyck, A.V. & West, G.F. (1984) The role of simple computer models in interpretations of wide-band, drill-hole electromagnetic surveys in mineral exploration. *Geophysics*, **49**, 957–80.

Dyck, A.V. & Young, R.P. (1985) Physical characterization of rock masses using borehole methods. *Geophysics*, **50**, 2530–41.

Ellis, D.V. (1987) *Well Logging for Earth Scientists*. Elsevier, New York.

Fitch, A.A. (1976) *Seismic Reflection Interpretation*. Gebrüder Borntraeger, Berlin.

Fitch, A.A. (1981) *Developments in Geophysical Exploration Methods, Vol. 2*. Applied Science Publishers, London.

Fountain D.K. (1972) Geophysical case-histories of disseminated sulfide. *Geophysics*, **37**, 142–59.

Fournier, C. (1989) Spontaneous potentials and resistivity surveys applied to hydrogeology in a volcanic area: case history of the Chaîne des Puys (Puy de Dôme, France). *Geophys. Prosp.*, **37**, 647–68.

Frischknecht, F.C. & Raab, P.V. (1984) Time-domain electromagnetic soundings at the Nevada Test Site, Nevada. *Geophysics*, **49**, 981–92.

Gardner, G.H.F., Gardner, L.W. & Gregory, A.R. (1974) Formation velocity and density – the diagnostic basics for stratigraphic traps. *Geophysics*, **39**, 770–80.

Garland, G.D. (1951) Combined analysis of gravity and magnetic anomalies *Geophysics*, **16**, 51–62.

Garland, G.D. (1965) *The Earth's Shape and Gravity*. Pergamon, Oxford.

Giese, P., Prodehl, C. & Stein, A. (eds.) (1976) *Explosion Seismology in Central Europe*. Springer-Verlag, Berlin.

Gilbert, D. & Galdeano, A. (1985) A computer program to perform transformations of gravimetric and aeromagnetic surveys. *Computers & Geosciences*, **11**, 553–88.

Götze, H-J. & Lahmeyer, B. (1988) Application of three-dimensional interactive modelling in gravity and magnetics. *Geophysics*, **53**, 1096–108.

Grant, F.S. & West, G.F. (1965) *Interpretation Theory in Applied Geophysics*. McGraw-Hill, New York.

Gregory, A.R. (1977) Aspects of rock physics from laboratory and log data that are important to seismic interpretation. *In*: Payton, C.E. (ed.), *Seismic Stratigraphy – Applications to Hydrocarbon Exploration*. Memoir 26, American Association of Petroleum Geologists, Tulsa, 15–46.

Griffiths, D.H. & King, R.F. (1981) *Applied Geophysics for Geologists and Engineers*. Pergamon, Oxford.

Gunn, P.J. (1975) Linear transformations of gravity and magnetic fields. *Geophys. Prosp.*, **23**, 300–12.

Guyod, H. (1974) Electrolog. *In*: *Log Review 1*. Dresser Industries, Houston.

Habberjam, G.M. (1979) *Apparent Resistivity and the Use of Square Array Techniques*. Gebrüder Borntraeger, Berlin.

Hagedoorn, J.G. (1959) The plus-minus method of interpreting seismic refraction sections. *Geophys. Prosp.*, **7**, 158–82.

Hatton, L., Worthington, M.H. & Makin, J. (1986) *Seismic Data Processing*. Blackwell Scientific Publications, Oxford.

Hearst, J.R. & Nelson, P.H. (1985) *Well Logging for Physical Properties*. McGraw-Hill, New York.

Heirtzler, J.R., Dickson, G.O., Herron, E.M., Pitman, W.C. & Le Pichon, X. (1968) Marine magnetic anomalies, geomagnetic field reversals, and motions of the ocean floor and continents. *J. Geophys. Res.*, **73**, 2119–36.

Hjelt, S.E. Kaikkonen, P. & Pietilä, R. (1985) On the interpretation of VLF resistivity measurements. *Geoexploration*, **23**, 171–81.

Hood, P.J. & Teskey, D.J. (1989) Aeromagnetic gradiometer program of the Geological Survey of Canada. *Geophysics*, **54**, 1012–22.

Hooper, W. & McDowell, P. (1977) Magnetic surveying for buried mineshafts and wells. *Ground Engineering*, **10**, 21–3.

Hubbert, M.K. (1934) Results of Earth-resistivity survey on various geological structures in Illinois. *Trans. Am. Inst. Mining Met. Eng.*, **110**, 9–29.

Hubral, P. & Krey, T. (1980) *Interval Velocities from Seismic Reflection Time Measurements*. Society of Exploration Geophysicists, Tulsa.

Hutton, V.R.S., Ingham, M.R. & Mbipom, E.W. (1980) An electrical model of the crust and upper mantle in Scotland. *Nature, Lond.*, **287**, 30–3.

I.A.G. (International Association of Geodesy). (1971) *Geodetic Reference System 1967*. Pub. Spec. No. 3 du Bulletin Géodésique.

Jewel, T.R. & Ward, S.H. (1963) The influence of conductivity inhomogeneities upon audio-frequency magnetic fields. *Geophysics*, **28**, 201–21.

Johnson, S.H. (1976) Interpretation of split-spread refraction data in terms of plane dipping layers. *Geophysics*, **41**, 418–24.

Kanasewich, E.R. (1981) *Time Sequence Analysis in Geophysics* (3rd edn.). Univ. of Alberta.

Kanasewich, E.R. & Agarwal, R.G. (1970) Analysis of combined gravity and magnetic fields in wave-number domain. *J. Geophys. Res.*, **75**, 5702–12.

Kearey, P. & Allison, J.R. (1980) A geomagnetic investigation of Carboniferous igneous rocks at Tickenham, County of Avon. *Geol. Mag.*, **117**, 587–93.

Kearey, P. & Vine, F.J. (1990) *Global Tectonics*. Blackwell Scientific Publications, Oxford.

Keller, G.V. & Frischnecht, F.C. (1966) *Electrical Methods in Geophysical Prospecting*. Pergamon, Oxford.

Kilty, K.T. (1984) On the origin and interpretation of self-potential anomalies. *Geophys. Prosp.*, **32**, 51–62.

Klemperer, S.L. and the BIRPS group (1987) Reflectivity of the crystalline crust: hypotheses and tests. *Geophys. J. R. Astr. Soc.*, **89**, 217–22.

Kleyn, A.H. (1983) *Seismic Reflection Interpretation*. Applied Science Publishers, London.

Knopoff, L. (1983) The thickness of the lithosphere from the dispersion of surface waves. *Geophys. J. R. Astr. Soc.*, **74**, 55–81.

Koefoed, O. (1968) *The Application of the Kernel Function in Interpreting Resistivity Measurements*. Gebrüder Borntraeger, Berlin.

Koefoed, O. (1979) *Geosounding Principles, 1 — Resistivity Sounding Measurements*. Elsevier, Amsterdam.

Kulhánek, O. (1976) *Introduction to Digital Filtering in Geophysics*. Elsevier, Amsterdam.

Kunetz, G. (1966) *Principles of Direct Current Resistivity Prospecting*. Gebrüder Borntraeger, Berlin.

Labo, J. (1986) *A Practical Introduction to Borehole Geophysics*. Soc. Econ. Geologists, Tulsa.

Labson, V.F., Becker, A., Morrison, H.F. & Conti, U. (1985) Geophysical exploration with audiofrequency natural magnetic fields. *Geophysics*, **50**, 656–64.

LaCoste, L.J.B. (1967) Measurement of gravity at sea and in the air. *Rev. Geophysics*, **5**, 477–526.

LaCoste, L.J.B., Ford, J., Bowles, R. & Archer, K. (1982) Gravity measurements in an airplane using state-of-the-art navigation and altimetry. *Geophysics*, **47**, 832–7.

Langore, L., Alikaj, P. & Gjovreku, D. (1989) Achievements in copper sulphide exploration in Albania with IP and EM methods. *Geophys. Prosp.*, **37**, 975–91.

Lavergne, M. (1989) *Seismic Methods*. Editions Technip, Paris.

Lee, M.K., Pharaoh, T.C. & Soper, N.J. (1990) Structural trends in central Britain from images of gravity and aeromagnetic fields. *J. Geol. Soc. Lond.*, **147**, 241–58.

Le Tirant, P. (1979) *Seabed Reconnaissance and Offshore Soil Mechanics*. Editions Technip, Paris.

Leggo, P.J. (1982) Geological implications of ground impulse radar. *Trans. Instn. Min. Metall. (Sect. B: Appl. earth sci.)*, **91**, B1–6.

Lines, L.R., Schultz, A.K. & Treitel, S. (1988) Cooperative inversion of geophysical data. *Geophysics*, **53**, 8–20.

Logn, O. (1954) Mapping nearly vertical discontinuities by Earth resistivities. *Geophysics*, **19**, 739–60.

March, D.W. & Bailey, A.D. (1983) Two-dimensional transform and seismic processing. *First Break*, **1**(1), 9–21.

Marshall, D.J. & Madden, T.R. (1959) Induced polarisation: a study of its causes. *Geophysics*, **24**, 790–816.

Mason, R.G. & Raff, R.D. (1961) Magnetic survey off the west coast of North America, 32°N to 42°N. *Bull. Geol. Soc. Am.*, **72**, 1259–66.

Mayne, W.H. (1967) Practical considerations of the use of common reflection point techniques. *Geophysics*, **32**, 225–9.

McCracken, K.G., Oristaglio, M.L. & Hohmann, G.W. (1986) A comparison of electromagnetic exploration systems. *Geophysics*, **51**, 810–8.

McNeill, J.D. (1980) *Electromagnetic Terrain Conductivity Measurement at Low Induction Numbers*. Technical Note TN-6, Geonics, Mississauga.

McQuillin, R. & Ardus, D.A. (1977) *Exploring the Geology of Shelf Seas*. Graham & Trotman, London.

McQuillin, R., Bacon, M. & Barclay, W. (1979) *An Introduction to Seismic Interpretation*. Graham & Trotman, London.

Meckel, L.D. & Nath, A.K. (1977) Geologic considerations for stratigraphic modelling and interpretation. *In*: Payton, C.E. (ed.), *Seismic Stratigraphy — Applications to Hydrocarbon Exploration*. AAPG Memoir **26**, 417–38.

Menke, W. (1989) *Geophysical Data Analysis: Discrete Inverse Theory*. Academic Press, London.

Merkel, R.H. (1972) The use of resistivity techniques to delineate acid mine drainage in groundwater. *Groundwater*, **10**, No. 5, 38–42.

Milsom, J. (1989) *Field Geophysics*. Open University Press, Milton Keynes.

Mitchum, R.M., Vail, P.R. & Thompson, S. (1977) Seismic stratigraphy and global changes of sea level, Part 2: The depositional sequence as a basic unit for stratigraphic analysis. *In*: Payton, C.E. (ed.), *Seismic Stratigraphy — Applications to Hydrocarbon Exploration*. Memoir **26**, American Association of Petroleum Geologists, Tulsa, 53–62.

Mittermayer, E. (1969) Numerical formulas for the Geodetic Reference System 1967. *Bolletino di Geofisca Teorica ed Applicata*, **11**, 96–107.

Morelli, C., Gantor, C., Honkasalo, T., McConnell, R.K., Tanner, J.G., Szabo, B., Votila, V. & Whalen, C.T. (1971) *The International Gravity Standardisation Net*. Pub. Spec. No. 4 du Bulletin Géodésique.

Moxham, R.M. (1963) Natural radioactivity in Washington County, Maryland. *Geophysics*, **28**, 262–72.

Musgrave, A.W. (ed.) (1967) *Seismic Refraction Prospecting*. Society of Exploration Geophysicists, Tulsa.

Nafe, J.E. & Drake, C.L. (1963) Physical properties of marine sediments. *In*: Hill, M.N. (ed.), *The Sea. Vol. 3*. Interscience Publishers, New York, 794–815.

Nair, M.R., Biswas, S.K. & Mazumdar, K. (1968) Experimental studies on the electromagnetic response of tilted conducting half-planes to a horizontal-loop prospective system. *Geoexploration*, **6**, 207–44.

Neidell, N.S. & Poggiagliolmi, E. (1977) Stratigraphic modelling and interpretation — geophysical principles.

In: Payton, C.E. (ed.), *Seismic Stratigraphy – Applications to Hydrocarbon Exploration*. Memoir **26**, American Association of Petroleum Geologists, Tulsa, 389–416.

Nettleton, L.L. (1971) *Elementary Gravity and Magnetics for Geologists and Seismologists*. Society of Exploration Geophysicists, Tulsa, Monograph Series, No. 1.

Nettleton, L.L. (1976) *Gravity and Magnetics in Oil Exploration*. McGraw-Hill, New York.

O'Brien, P.N.S. (1974) Aspects of seismic research in the oil industry. *Geoexploration*, **12**, 75–96.

Orellana, E. & Mooney, H.M. (1966) *Master Tables and Curves for Vertical Electrical Sounding Over Layered Structures*. Interciencia, Madrid.

Orellana, E. & Mooney, H.M. (1972) *Two and Three Layer Master Curves and Auxilliary Point Diagrams for Vertical Electrical Sounding Using Wenner Arrangement*. Interciencia, Madrid.

Oristaglio, M.L. & Hohmann, G.W. (1984) Diffusion of electromagnetic fields into a two-dimensional earth: A finite-difference approach. *Geophysics*, **49**, 870–94.

Palacky, G.J. (1981) The airborne electromagnetic method as a tool of geological mapping. *Geophys. Prosp.*, **29**, 60–88.

Palacky, G.J. & Sena, F.O. (1979) Conductor identification in tropical terrains – case histories from the Itapicuru greenstone belt, Bahia, Brazil. *Geophysics*, **44**, 1941–62.

Palmer, D. (1980) *The Generalised Reciprocal Method of Seismic Refraction Interpretation*. Society of Exploration Geophysicists, Tulsa.

Palmer, D. (1986) *Handbook of Geophysical Exploration: Section 1, Seismic Exploration. Vol. 13: Refraction Seismics*. Expro Science Publications, Amsterdam.

Parasnis, D.S. (1966, 1973) *Mining Geophysics*. Elsevier, Amsterdam.

Parasnis, D.S. (1986) *Principles of Applied Geophysics*. Chapman & Hall, London.

Parker, R.L. (1977) Understanding inverse theory. *Ann. Rev. Earth Planet. Sci.*, **5**, 35–64.

Parkhomenko, E.I. (1967) *Electrical Properties of Rocks*. Plenum, New York.

Paterson, N.R. & Reeves, C.V. (1985) Applications of gravity and magnetic surveys: the state-of-the-art in 1985. *Geophysics*, **50**, 2558–94.

Paterson, N.R. (1967) Exploration for massive sulphides in the Canadian Shield. *In*: Morley, L.W. (ed.), *Mining and Groundwater Geophysics*. Econ. Geol. Report No. **26**, Geol. Surv. Canada, 275–89.

Payton, C.E. (ed.) (1977) *Seismic Stratigraphy – Applications to Hydrocarbon Exploration*. Memoir **26**, American Association of Petroleum Geologists, Tulsa.

Peddie, N.W. (1983) International geomagnetic reference field – its evolution and the difference in total field intensity between new and old models for 1965–1980. *Geophysics*, **48**, 1691–6.

Peters, J.W. & Dugan, A.F. (1945) Gravity and magnetic investigations at the Grand Saline Salt Dome, Van Zandt Co., Texas. *Geophysics*, **10**, 376–93.

Pires, A.C.B. & Harthill, N. (1989) Statistical analysis of airborne gamma-ray data for geologic mapping purposes: Crixas–Itapaci area, Goias, Brazil. *Geophysics*, **54**, 1326–32.

Pirson, S.J. (1977) *Geologic Well Log Analysis* (2nd edn). Gulf, Houston.

Pritchett, W.C. (1990) *Acquiring Better Seismic Data*. Chapman & Hall, London.

Ramsey, A.S. (1964) *An Introduction to the Theory of Newtonian Attraction*. Cambridge University Press, Cambridge.

Rayner, J.N. (1971) *An Introduction to Spectral Analysis*. Pion, England.

Reilly, W.I. (1972) Use of the International System of Units (SI) in geophysical publications. *N.Z.J. Geol. Geophys.*, **15**, 148–58.

Rider, M.H. (1986) *The Geological Interpretation of Well Logs*. Blackie, London.

Robinson, E.A. (1983) *Migration of Geophysical Data*. IHRDC, Boston.

Robinson, E.A. (1983) *Seismic Velocity Analysis and the Convolutional Model*. IHRDC, Boston.

Robinson, E.A. & Treitel, S. (1967) Principles of digital Wiener filtering. *Geophys. Prosp.*, **15**, 311–33.

Robinson, E.A. & Treitel, S. (1980) *Geophysical Signal Analysis*. Prentice-Hall, London.

Robinson, E.S. & Çoruh, C. (1988) *Basic Exploration Geophysics*. Wiley, New York.

Sato, M. & Mooney, H.H. (1960) The electrochemical mechanism of sulphide self potentials. *Geophysics*, **25**, 226–49.

Schramm, M.W., Dedman, E.V. & Lindsey, J.P. (1977) Practical stratigraphic modelling and interpretation. *In*: Payton, C.E. (ed.), *Seismic Stratigraphy – Applications to Hydrocarbon Exploration*. Memoir **26**, American Association of Petroleum Geologists, Tulsa, 477–502.

Segesman, F.F. (1980) Well logging method. *Geophysics*, **45**, 1667–84.

Seigel, H.O. (1967) The induced polarisation method. *In*: Morley, L.W. (ed.) *Mining and Groundwater Geophysics*. Econ. Geol. Report No. **26**, Geol. Survey of Canada, 123–37.

Sengbush, R.L. (1983) *Seismic Exploration Methods*. IHRDC, Boston.

Serra, O. (1984) *Fundamentals of Well-log Interpretation, 1. The Acquisition of Logging Data*. Elsevier, Amsterdam.

Serra, O. (1986) *Fundamentals of Well-log Interpretation, 2. The Interpretation of Logging Data*. Elsevier, Amsterdam.

Sharma, P. (1976) *Geophysical Methods in Geology*. Elsevier, Amsterdam.

Sheriff, R.E. (1973) *Encyclopedic Dictionary of Exploration Geophysics*. Society of Exploration Geophysicists, Tulsa.

Sheriff, R.E. (1978) *A First Course in Geophysical Exploration and Interpretation*. IHRDC, Boston.

Sheriff, R.E. (1980) *Seismic Stratigraphy*. IHRDC, Boston.

Sheriff, R.E. (1982) *Structural Interpretation of Seismic Data*. American Association of Petroleum Geologists Continuing Education Course Note Series No. 23.

Sheriff, R.E. & Geldart, L.P. (1982) *Exploration Seismology Vol. 1: History, Theory and Data Acquisition*. Cambridge University Press, Cambridge.

Sheriff, R.E. & Geldart, L.P. (1983) *Exploration Seismology Vol. 2: Data-processing and Interpretation*. Cambridge University Press, Cambridge.

Sjögren, B. (1984) *Shallow Refraction Seismics*. Chapman & Hall, London.

Smith, R.A. (1959) Some depth formulae for local magnetic and gravity anomalies. *Geophys. Prosp.*, **7**, 55–63.

Smythe, D.K., Dobinson, A., McQuillin, R., Brewer, J.A., Matthews, D.H., Blundell, D.J. & Kelk, B. (1982) Deep structure of the Scottish Caledonides revealed by the MOIST profile. *Nature, Lond.*, **299**, 338–40.

Snyder, D.B. & Flack, C.A. (1990) A Caledonian age for reflections within the mantle lithosphere north and west of Scotland. *Tectonics*.

Snyder, D.D. & Fleming, D.B. (1985) Well logging – a 25 year perspective. *Geophysics*, **50**, 2504–29.

Spector, A. & Grant, F.S. (1970) Statistical models for interpreting aeromagnetic data. *Geophysics*, **35**, 293–302.

Spies, B.R. (1976) The transient electromagnetic method in Australia. *B.M.R. J. Austral. Geol. & Geophys.*, **1**, 23–32.

Spies, B.R. (1989) Depth of investigation in electromagnetic sounding methods. *Geophysics*, **54**, 872–88.

Stacey, F.D. & Banerjee, S.K. (1974) *The Physical Principles of Rock Magnetism*. Elsevier, Amsterdam.

Stacey, R.A. (1971) Interpretation of the gravity anomaly at Darnley Bay, N.W.T. *Can. J. Earth Sci.*, **8**, 1037–42.

Stoffa, P.L. & Buhl, P. (1979) Two-ship multichannel seismic experiments for deep crustal studies: expanded spread and constant offset profiles. *J. Geophys. Res.*, **84**, 7645–60.

Sumner, J.S. (1976) *Principles of Induced Polarisation for Geophysical Exploration*. Elsevier, Amsterdam.

Talwani, M. (1965) Comparison with the help of a digital computer of magnetic anomalies caused by bodies of arbitrary shape. *Geophysics*, **30**, 797–817.

Talwani, M. & Ewing, M. (1960) Rapid computation of gravitational attraction of three-dimensional bodies of arbitrary shape. *Geophysics*, **25**, 203–25.

Talwani, M., Le Pichon, X. & Ewing, M. (1965) Crustal structure of the mid-ocean ridges 2. Computed model from gravity and seismic refraction data. *J. Geophys. Res.*, **70**, 341–52.

Talwani, M., Worzel, J.L. & Landisman, M. (1959) Rapid gravity computations for two-dimensional bodies with applications to the Mendocino submarine fracture zones. *J. Geophys. Res.*, **64**, 49–59.

Taner, M.T. & Koehler, F. (1969) Velocity spectra – digital computer derivation and applications of velocity functions. *Geophysics*, **34**, 859–81.

Tarling, D.H. (1983) *Palaeomagnetism*. Chapman & Hall, London.

Telford, W.M. (1982) Radon mapping in the search for uranium. *In*: Fitch, A.A. (ed.), *Developments in Geophysical Exploration Methods*. Applied Science, London, 155–94.

Telford, W.M., Geldart, L.P., Sheriff, R.E. & Keys, D.A. (1976) *Applied Geophysics*. Cambridge University Press, Cambridge.

Thomas, M.D. & Kearey, P. (1980) Gravity anomalies, block-faulting and Andean-type tectonism in the eastern Churchill Province. *Nature, Lond.*, **283**, 61–3.

Thornburgh, H.R. (1930) Wave-front diagrams in seismic interpretation. *Bull. Am. Assoc. Petrol. Geol.*, **14**, 185–200.

Torge, W. (1989) *Gravimetry*. Walter de Gruyter, Berlin.

Tittman, J. (1987) Geophysical well logging. *In*: Samis, C.G. & Henyey, T.L. (eds), *Methods of Experimental Physics, Vol. 24, Part B – Field Measurements*. Academic Press, Orlando, 441–615.

Vacquier, V., Steenland, N.C., Henderson, R.G. & Zeitz, I. (1951) Interpretation of aeromagnetic maps. *Geol. Soc. Am. Mem.*, **47**.

Vail, P.R., Mitchum, R.M. & Thompson, S. (1977) Seismic stratigraphy and global changes of sea level, Part 3: Relative changes of sea level from coastal onlap. *In*: Payton, C.E. (ed.), *Seismic Stratigraphy – Applications to Hydrocarbon Exploration*. Memoir **26**, American Association of Petroleum Geologists, Tulsa, 63–81.

Vail, P.R., Mitchum, R.M. & Thompson, S. (1977) Seismic stratigraphy and global changes of sea-level, Part 4: Global cycles of relative changes of sea level. *In*: Payton, C.E. (ed.), *Seismic Stratigraphy – Applications to Hydrocarbon Exploration*. Memoir **26**, American Association of Petroleum Geologists, Tulsa, 83–97.

Van Overmeeren, R.A. (1975) A combination of gravity and seismic refraction measurements, applied to groundwater explorations near Taltal, Province of Antofagasta, Chile. *Geophys. Prosp.*, **23**, 248–58.

Van Overmeeren, R.A. (1989) Aquifer boundaries explored by geoelectrical measurements in the coastal plain of Yemen: a case of equivalence. *Geophysics*, **54**, 38–48.

Vine, F.J. & Matthews, D.H. (1963) Magnetic anomalies over oceanic ridges. *Nature, Lond.*, **199**, 947–9.

Wait, J.R. (1982) *Geo-Electromagnetism*. Academic Press, New York.

Ward, S.H. (1967) *In: Mining Geophysics, vol. 2*, Society of Exploration Geophysicists, Tulsa, 10–196, 224–372.

Ward, S.H. (1987) Electrical methods in geophysical prospecting. In: Samis, C.G. & Henyey, T.L. (eds.), *Methods of Experimental Physics, Vol. 24, Part B – Field Measurements*. Academic Press, Orlando, 265–375.

Waters, K.H. (1978) *Reflection Seismology — a Tool for Energy Resource Exploration*. Wiley, New York.

Webb, J.E. (1966) The search for iron ore, Eyre Peninsula, South Australia. *In: Mining Geophysics, Vol. 1*, Society of Exploration Geophysicists, Tulsa, 379–90.

Westbrook, G.K. (1975) The structure of the crust and upper mantle in the region of Barbados and the Lesser Antilles. *Geophys. J. R. Astr. Soc.*, **43**, 201–42.

Whitcomb, J.H. (1987) Surface measurements of the Earth's gravity field. *In*: Samis, C.G. & Henyey, T.L. (eds.), *Methods of Experimental Physics, Vol. 24, Part B — Field Measurements*. Academic Press, Orlando, 127–61.

White, P.S. (1966) Airborne electromagnetic survey and ground follow-up in northwestern Quebec. *In: Mining Geophysics, Vol. 1*, Society of Exploration Geophysicists, Tulsa, 252–61.

Willmore, P.L. & Bancroft, A.M. (1960) The time-term approach to refraction seismology. *Geophys. J. R. Astr. Soc.*, **3**, 419–32.

Wold, R.J. & Cooper, A.K. (1989) Marine magnetic gradiometer — a tool for the seismic interpreter. *Geophysics: The Leading Edge of Exploration*, 22–7.

Wollenberg, H.A. (1977) Radiometric methods. *In*: Morse, J.G. (ed.), *Nuclear Methods in Mineral Exploration and Production*, Elsevier Science, Amsterdam, 5–36.

Wood, R.D., Wichmann, P.A. & Watt, H.B. (1974) Gamma ray — neutron log. *In*: *Log Review 1*. Dresser Industries, Houston.

Wong, J., Bregman, N., West, G. & Hurley P. (1987) Cross-hole seismic scanning and tomography. *Geophysics: the Leading Edge of Exploration*, **6**, 36–41.

Woollard, G.P. (1969) Regional variations in gravity. *In*: Hart, P.J. (ed.), *The Earth's Crust and Upper Mantle*. Amer. Geophys. Un. Monograph 13.

Wright, C., Barton, T., Goleby, B.R., Spence, A.G. & Pfister, D. (1990) The interpretation of expanding spread profiles: examples from central and eastern Australia. *Tectonophysics*, **173**, 73–82.

Yüngül, S. (1954) Spontaneous polarisation survey of a copper deposit at Sariyer, Turkey. *Geophysics*, **19**, 455–58.

Zalasiewicz, J.A., Mathers, S.J. & Cornwell, J.D. (1985) The application of ground conductivity measurements to geological mapping. *Q. J. Eng. Geol. Lond.*, **18**, 139–48.

Ziolkowski, A. (1983) *Deconvolution*. IHRDC, Boston.

Zohdy, A.A.R. (1989) A new method for the automatic interpretation of Schlumberger and Wenner sounding curves. *Geophysics*, **54**, 245–53.

Index

ARY POWDER

Also by Jo Nesbø

Doctor Proctor's Fart Powder:
Time-Travel Bath Bomb

Doctor Proctor's Fart Powder:
The End of the World. Maybe

Doctor Proctor's Fart Powder:
The Great Gold Robbery

JO NESBØ

SIMON AND SCHUSTER

First published in Great Britain in 2010 by Simon and Schuster UK Ltd,
A CBS COMPANY

First published in the USA in 2010 by Aladdin, an imprint of
Simon & Schuster Children's Publishing Division
Originally published in Norway in 2007 as *Doktor Proktor's Prompepulvet*
by H. Aschehoug & Co.

This edition published in Great Britain in 2014 by Simon and Schuster UK Ltd

3 5 7 9 10 8 6 4 2

Simon & Schuster UK Ltd
1st Floor, 222 Gray's Inn Road
London
WC1X 8HB

Simon & Schuster Australia, Sydney
Simon & Schuster India, New Delhi

A CIP catalogue record for this book
is available from the British Library.

PB ISBN: 978-1-47112-535-5
eBook ISBN: 978-0-85707-712-7

Printed and bound by CPI Group (UK) Ltd, Croydon, CR0 4YY

www.simonandschuster.co.uk
www.simonandschuster.com.au

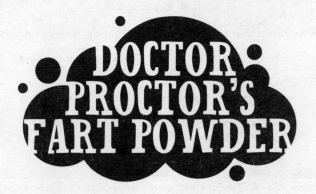

The New Neighbour

IT WAS MAY, and once the sun had shone for a while on Japan, Russia and Sweden, it came up over Oslo – the very small capital city, of a very small country called Norway. The sun got right to work shining on the yellow, and fairly small, palace that was home to the king of Norway, who didn't rule over enough for it to amount to anything, and on Akershus Fortress. There it shone on the old cannons that were aimed out over the Oslo Fjord, through the window of the Commandant's office and then onto the most remote of doors. The door that ultimately led to the city's most feared jail cell, the Dungeon of the Dead, where only the most dangerous and worst criminals were kept. The cell was empty, apart from a *Rattus norvegicus*, a little Norwegian rat that was taking its morning bath in the toilet.

The sun rose a tiny little bit higher and shone on some children in a school marching band who had practised waking up very early and putting on uniforms that itched, and who were now practising marching and playing almost in time. Because soon it would be the seventeenth of May, Norwegian Independence Day, and that was the day when all the school marching bands in the whole small country would get up very early, put on uniforms that itched and play almost in time.

And as the sun climbed a tiny little bit higher it shone on the wooden wharves on the Oslo Fjord, where a ship from Shanghai, China, had just docked. The wharf planks swayed and creaked from all the busy feet running back and forth unloading goods from the ship. Some of the sun's rays even made their way between the planks and down under the wharf to a sewer pipe that stuck out into the water.

And one single ray of sunlight made its way into the darkness of the sewer pipe making something in there gleam. Something that was white, wet and *very* sharp. Something that bore a nasty resemblance to a row of teeth. And if you knew something about reptiles, but were otherwise very stupid, you might have thought that what you were seeing were the eighteen fangs found in the jaws of the world's biggest and most feared constrictor. The anaconda. But nobody's that stupid. Because anacondas live in the jungle, in rivers like the Amazon in Brazil, not in the sewer pipes running every which way beneath the small, peaceful, northerly city called Oslo. An anaconda in the sewer? Sixteen metres of constricting muscles, a jaw the size of an inflatable swim ring and teeth like upside-down ice-cream cones? Ha, ha! Yeah, right, that would've been a sight!

And now the sun was starting to shine on a quiet street called Cannon Avenue. Some of the sun's rays shone on a red house there, where the Commandant of Akershus Fortress was eating breakfast with his wife and their daughter, Lisa. Some of the rays shone on the yellow house on the other side of the street, where Lisa's best friend used to live. But as her best friend had just moved to a town called Sarpsborg, seeing the yellow house empty made Lisa feel very lonely. Because now there wasn't anyone for Lisa to play with on Cannon Avenue. The only other kids in the neighbourhood were Truls and Trym Trane. They were the twins who lived in the big house with the three garages at the bottom of the hill, and they were two years older than Lisa. In the winter they threw rock-hard snowballs at her little red-haired head. And when she asked if they wanted to play, they pushed her down into the snow. And with icy mittens they rubbed snow into her face while christening her Greasy Lisa, Flatu-Lisa or Commandant's Debutante.

Maybe you're thinking that Lisa should've mentioned this to Truls and Trym's parents so they would rein the boys in. But that's because you don't know Truls and Trym's father, Mr Trane. Mr Trane was a fat and angry man, even fatter than Lisa's father and way, way angrier. And at least ten times as rich. And because he was so rich, Mr Trane didn't think anyone had any business coming and telling him *anything* whatsoever, and especially not how he ought to be raising his boys! The reason Mr Trane was so rich was that he had once stolen an invention from a poor inventor. The invention was a very hard, very mysterious and very secret material that was used, among other things, on prison doors to make prisons absolutely escape-proof. Mr Trane had used the money he'd made from the invention to build the big house with the three garages and to buy a Hummer. A Hummer, in case you don't know, is a big, angry car that was made to use in wars and that took up almost the whole road when Mr Trane drove up Cannon Avenue. Hummers are also awful polluters.

GRRR

But Mr Trane didn't care, because he liked big, angry cars. And besides, he knew that if he crashed into someone his car was a lot bigger than theirs, so it would be too bad for them.

Luckily it would be a while until Truls and Trym could christen Lisa with snow again, because the sun had long since melted it on Cannon Avenue, and now the sun was shining on the gardens, which were green and well groomed. All, that is, except for one. The garden was scraggly, drab and unkempt, but was pleasant anyway because it had two pear trees and belonged to a small, crooked house that might possibly have been blue at one time and that was missing a fair number of roof tiles. The neighbours on Cannon Avenue rarely saw the man who lived there. Lisa had only met him a couple of times, he'd smiled and otherwise looked sort of like his garden – scraggly, drab and unkempt.

"What's that?" grumbled the Commandant as the

roar of a large engine disturbed the morning quiet. "Is that that awful Hummer of Mr Trane's?"

His wife craned her neck and peered out the kitchen window. "No. It looks like a moving van."

Lisa, who was generally a very well-behaved girl, got up from the table without having finished what was on her plate or having been excused. She ran out onto the front steps. And sure enough a moving van with the name CRAZY-QUICK written on its side was parked in front of the empty, yellow house that used to be her best friend's house. Movers were unloading cardboard boxes from the back. Lisa went down the stairs and over to the so-called apple tree in her garden by the fence to get a closer look. The men in overalls were carrying furniture, lamps and big, ugly pictures. Lisa noticed one of the movers showing the other a dented trumpet that was sitting on top of one of the cardboard boxes and then they both laughed. But she couldn't see any sign of what she'd been hoping to see — dolls, small bicycles, a pair of short skis. And that could only mean that whoever was moving in didn't have kids, at least no girls her age. Lisa sighed.

Just then she heard a voice.

"Hi!"

She looked around in surprise, but didn't see anyone.

"Hi there!"

She looked up at the tree her father said was an apple tree, but that no one had ever seen any apples on. And that now appeared to be talking.

"Not there," the voice said. "Over here."

Lisa stretched up on her tiptoes and peered down on the other side of the fence. And there was a little boy with red hair standing there. Well, not just red, actually, but bright red. And he wasn't just small, he was tiny. He had a tiny face with two tiny blue eyes and a tiny turned-up nose in between.

The only things on his face that were big, were the freckles.

"I'm Nilly," he said. "What do you have to say about that?"

He was supposed to be named William, but the priest refused to give such a tiny boy such a long name. So Billy would have to do. But the ringer of the church bells came up with a brilliant idea: a boy who was so tiny that he was *nearly* invisible should be called Nilly! His parents had just sighed and said okay, and thus the bell ringer got his way.

Lisa asked, "What do I have to say about what?"

"About my being called Nilly. It's not exactly a common name."

Lisa thought about it. "I don't know," she said.

"Good." The boy smiled. "It rhymes with 'silly', but let's just leave it at that. Deal?"

Lisa nodded.

The boy stuck his right index finger in his left ear. "And what's your name?"

"Lisa," she said.

Nilly's index finger twisted back and forth as he watched her. Finally he pulled his finger out, looked at it, gave a satisfied nod and rubbed it on his trouser leg.

"Jeez, I can't think of anything interesting that rhymes with Lisa," he said. "You're lucky."

"Are you moving into Anna's house?"

"I don't know who Anna is, but we're moving into that yellow shack over there," Nilly said, pointing over his shoulder with his thumb.

"Anna's my best friend," Lisa said. "She moved to Sarpsborg."

"Whoa, that's far," Nilly said. "Especially since she's your best friend."

"It is?" Lisa said. "Anna didn't think it was that far. She said I should just go south on the highway when I visit her."

Nilly shook his head, looking gloomy. "South is right, but the question is if the highway even goes that far. Sarpsborg is actually in the Southern Hemisphere."

"The Southern what-i-sphere?" Lisa said, shocked.

"Hemisphere," Nilly said. "That means it's on the other side of the world."

"Whoa," Lisa said, taken aback. After she thought about it a minute, she said, "Dad says that it's super warm in the south all year round, so I bet Anna can go swimming all the time now, whether it's summer or winter."

"No way," Nilly said. "Sarpsborg is so far south that it's practically at the South Pole. It's freezing. Penguins live on people's roofs down there."

"You mean it snows all year round in Sarpsborg?" Lisa asked.

Nilly nodded and Lisa shivered. Nilly pursed his lips together while at the same time pressing air out between them. It sounded like a fart. Lisa furrowed her brow, remembering how the twins had called her Flatu-Lisa. "Are you trying to tease me?" she asked. "About my nickname?"

Nilly shook his head. "Nope, I'm practising," he said. "I play the trumpet. That means I have to practise all the time. Even when I don't have my trumpet."

Lisa cocked her head to the side and looked at him. She wasn't really sure anymore if he was telling the truth.

"Lisa, you have to brush your teeth before you go to school," she heard a voice rumble. It was her dad, who'd put on his blue Commandant's uniform and was waddling towards the door with his big belly. "The ship with the gunpowder for our cannons arrived from Shanghai this morning, so I'll be home late. You be a good girl today."

"Yes, Dad," said Lisa, who was always good.

She knew it was a special day when the gunpowder arrived. It had sailed halfway around the world and had to be handled very carefully and respectfully, since it was used to fire off Akershus Fortress's Big and Almost World-Famous Royal Salute on May seventeenth, Norway's Independence Day.

"Dad," Lisa called to him. "Did you know that Sarpsborg is in the Southern . . . uh, Hemisphere?"

The Commandant stopped, looking puzzled. "Says who?"

"Nilly."

"Who's that?"

She pointed. "Nill . . ." she started, but stopped suddenly when she discovered that she was pointing at a stretch of Cannon Avenue where there was only Cannon Avenue and absolutely no sign of Nilly.

Seasick Goats

WHEN NILLY HEARD Lisa's dad tell her she had to go to school, he remembered he was supposed to go to school himself. Wherever it might be. And if he was fast maybe he would have time to eat breakfast, find his backpack and, if absolutely necessary, brush his teeth and still tag along with someone who knew the way to his new school.

He squeaked between the moving guys' legs and into the house. And there, in a cardboard box in the hallway, he saw his trumpet. He exhaled in relief, snatched it up and clutched it to his chest. Nilly and his sister and mum had arrived with the first load of stuff the night before, and the only thing he'd been worried about was whether the movers would forget the box with his trumpet in it.

He cautiously placed his lips against the mouth-piece.

"A trumpet should be kissed. Like a woman," his grandfather had always said. Nilly had never kissed a woman in his whole life, at least not like that, not right on the mouth. And truth be told he hoped he wouldn't have to either. He pressed the air into the trumpet. It bleated like a seasick goat. There aren't many people who've heard a seasick goat bleat, but that's exactly what it sounded like.

Nilly heard someone banging on the wall and knew

it was his mum, who hadn't got up yet. "Not now, Nilly!" she yelled. "It's eight a.m. We're sleeping."

She pretty much always said "we", even if she was alone in her bedroom: "We're going to bed now" and "We're going to make ourselves a cup of coffee." As if Nilly's dad weren't gone at all, as if she still had him in there — stored in a little box, and every once in a while when Nilly wasn't there, she would take him out. A tiny miniature dad who looked like the dad Nilly had seen in pictures. Miniature meant that something was really small and it made sense that of all people, Nilly would have a miniature dad, since Nilly was the smallest boy Nilly had ever seen.

He went down to the kitchen and made himself some breakfast. Even though they'd just moved in the day before, he found everything he needed, because they'd moved so many times, he pretty much knew where his mum would put stuff. The plates in the cupboard on the left, the silverware in the top drawer and the bread in the drawer below that. He was about to sink his teeth into a thick slice of bread with salami on top when it was snatched out of his hands.

"How you doing, dwarf?" Eva asked, sinking her teeth exactly where Nilly had been planning to sink his teeth. Eva was Nilly's sister. She was fifteen and when she wasn't bored, she was mad. "Did you know that the pit bull is the world's dumbest dog?" Nilly asked. "It's so dumb that when it takes food from the dwarf poodle, which happens to be the world's smartest dog, it doesn't get that it's been tricked."

"Shut up," Eva said.

But Nilly didn't shut up. "When the dwarf poodle knows the pit bull smells bread and salami and she's coming to take it away from him, he usually smears slime from elephant snails on the bottom of the slice of bread."

"Elephant snails?" Eva scoffed, eyeing him with suspicion. Unfortunately for her, Nilly read books and thus knew quite a few things she didn't know, so Eva could never be totally sure if what he was saying was a Nilly invention or something from one of those old books of their grandfather's. For example, this might be something from the book Nilly read the most, a thick, dusty one called *Animals You Wish Didn't Exist*.

"Haven't you ever seen an elephant snail?" Nilly yelled. "All you have to do is look out the window — there's a ton of them in the lawn. Big, ooky ones. When you squish them between two books, something oozes out of them that looks like the yellowish-green snot that runs out of the noses of people who have grade-three Beijing influenza. There's no snot worse than authentic third-degree Beijing snot. Well, apart from elephant-snail slime, that is."

"If you lie anymore, you're going to go to hell," Eva said, sneaking a quick peek at the bottom of the slice of bread.

Nilly hopped down from the chair. "Fine with me,

as long as they have a band there," he said, "and I get to play the trumpet."

"You're never going to get to play in any band!" Eva yelled after him. "No one wants a trumpet player who's so small, he doesn't even come up to the top of the bass drum. No band has uniforms that small!"

Nilly put on the itty-bitty shoes that were sitting in the hallway and went out onto the front steps, stood to attention, pressed his lips together, placed them against the trumpet and blew a tune his grandfather had taught him. It was called "Morning Reveille" and was designed to wake up sleepyheads.

"Attention!" Nilly yelled when he was done, because his grandfather had taught him that too. "I want to see both feet on deck and eyes front! Everyone ready for morning inspection, fall in and prepare for the playing of the royal anthem. *Attention!*"

The movers obeyed, snapping to attention on

the gravel walkway and standing stiffly with Nilly's mum's five-seater oak sofa between them. For a few seconds it was so quiet that all you could hear were cautious birds singing and a garbage truck that was making its way up Cannon Avenue.

"Interesting," Nilly heard a jovial, accented voice say. "There's a new Commandant on the street."

Nilly turned around. A tall, thin man was leaning against the wooden fence of the house next door. His white hair was just as long and unkempt as the grass in his garden. He was wearing a blue coat like the one the woodwork teacher at Nilly's last school had worn and he was also wearing something that looked liked swimming goggles. Nilly thought he was either a Santa Claus who'd lost weight or a crazy professor.

"Was I bothering you?" Nilly asked.

"Quite the contrary," the bushy-haired man said. "I came to see who was playing so well. The sound brought back wonderful memories of a boat trip on a river in France many, many years ago."

"A boat?" Nilly asked.

"Precisely." The man closed his eyes dreamily, facing the sun. "A riverboat that was carrying me, my beloved, my motorcycle and a bunch of goats. The sun was just starting to set, the wind was picking up, the water was a little choppy, and then the goats started bleating so vigorously. I'll never forget the sound."

"Hi," Nilly said. "I'm Nilly. I'm not sure what to say to that."

"No need to say anything," the man replied. "Unless you want to say something, of course."

And that's how Nilly met Doctor Proctor. Doctor Proctor wasn't Santa Claus. But he *was* kind of crazy.

The First Powder Test

"I'M DOCTOR PROCTOR," the professor said at last. His accent was guttural, making his voice sound like a badly oiled lawnmower. "I'm a crazy professor. Well, almost, anyway." He laughed a hearty, snorting sort of laugh and started watering his unmowed lawn with a green watering can.

Nilly, who was never one to say no to an interesting conversation, set down his trumpet, ran down his front steps and over to the fence, and asked, "And just what makes you so sure that you're almost crazy, Mr Proctor?"

"*Doctor* Proctor. Did you ever hear of a professor trying to invent a powder to prevent hay fever, but ending up inventing a farting powder instead? No, I didn't think so. Quite a failure . . . and pretty outrageous, isn't it?"

"Well, it depends," Nilly said, hopping up to sit on the fence. "What does your farting powder do? Does it keep people from farting?"

The professor laughed even louder. "Ah, if only it did. I could probably have found someone to buy my powder, then," he said. Suddenly he stopped watering the grass and stroked his chin, lost in thought. "You're on to something there, Nilly. If I'd made the powder so it kept people from farting, then people could take it before going to parties or funerals. After all there are lots of occasions when farting is inappropriate. I hadn't thought of that." He dropped the watering can in the grass and hurried off towards his little blue house. "Interesting," he mumbled. "Maybe I can just reverse the formula and create a non-fart powder."

"Wait!" Nilly yelled. "Wait, Doctor Proctor."

Nilly jumped down from the fence, tumbling into the tall grass and when he got up again, he couldn't see the professor – just his blue house and a side staircase that led down to an open cellar door. Nilly ran to the door as fast as his short legs could carry him. It was dark inside, but he could hear clattering and banging. Nilly knocked hard on the door frame.

"Come in!" the professor yelled from inside.

Nilly walked into the dimly lit cellar. He could vaguely make out an old, dismantled motorcycle with a sidecar by one wall. And a shelf with various Mickey Mouse figurines and a jar full of

a light-green powder, with a label in big letters that read DR PROCTOR'S LIGHT-GREEN POWDER! and underneath, in slightly smaller letters: "A bright idea that may make the world a little more fun."

"Is this the fart powder?" Nilly asked.

"No, it's just a phosphorescent powder that makes you glow," said Doctor Proctor from somewhere in the darkness. "A rather unsuccessful invention."

Then the professor emerged from the darkness with a lit torch in one hand and a snorkel mask in the other. "Wear this for safety during the experiment. I've reversed the process so that everything goes backwards. Shut the door and watch out. Everything is connected to the light switch."

Nilly put on the face mask and pulled the door shut.

"Thanks," the professor said, flipping the light switch. The light came on and a bunch of iron pipes that ran back and forth between a bunch of barrels, tanks, tubing, funnels, test tubes and glass containers started trembling and groaning and rumbling and sputtering.

"Remember to duck if you hear a bang!" Doctor Proctor shouted over the noise. The glass containers had started simmering and boiling and smoking.

"Okay!" Nilly yelled and right then and there was a bang.

The bang was so loud that Nilly felt like earwax was being pressed into his head while at the same time his eyes were being pressed out. The light went off and it was pitch-black. And totally silent. Nilly found the torch on the floor and shone it on the professor, who was lying on his stomach with his hands over his head. Nilly tried to say something, but when he couldn't hear his own voice, he realised he had gone deaf. He stuck his right index finger into his left ear and twisted it around. Then he tried talking again. Now he could just barely hear something far away, as if there were a layer of elephant-snail slime covering his eardrum.

"That was the loudest thing I've ever heard!" he screamed.

"Eureka!" Doctor Proctor yelled, leaping up, brushing off his coat and pulling off the glasses that Nilly now realised weren't swimming goggles, but motorcycle goggles. The professor's whole face was coated in blackish-grey powder, except for two white rings where his goggles had been. Then he dashed over to one of the test tubes and poured the contents into a glass container with a strainer on top.

"Look!" Doctor Proctor exclaimed.

Nilly saw that there was a fine, light-blue powder left in the strainer. The professor stuck a teaspoon into the powder and then into his mouth. "Mmm," he said. "No change in the flavour." Then he gritted his teeth and closed his eyes. Nilly could see the professor's face slowly turning red underneath the black soot.

"What are you doing?" Nilly asked.

"I'm trying to fart," the professor hissed through his clenched teeth. "And it's not working. Isn't it great?"

He smiled as he tried one more time. But as we all know, it's very hard to smile and fart at the same time, so Doctor Proctor gave up.

"Finally I've invented something that can be used for something," he said, smiling. "An anti-fart powder."

"Can I try?" Nilly asked, nodding towards the strainer.

"You?" the professor asked, looking at Nilly. The professor raised one bushy eyebrow and lowered the other bushy eyebrow so that Nilly could tell he didn't like the idea.

"I've tested anti-fart powder before," Nilly quickly added.

"Oh, really?" the professor asked. "Where?"

"In Prague," Nilly said.

"Really? How did it go?" the professor asked.

"Fine," Nilly replied, "but I farted."

"Good," the professor said.

"What's good?" Nilly asked.

"That you farted. That means there isn't anything that *prevents* farting yet." He passed the spoon to Nilly. "Go ahead. Take it."

Nilly filled the spoon and swallowed a mouthful.

"Well?" the professor asked.

"Just a minute," Nilly mumbled with his mouth full of powder. "It sure is dry."

"Try this," the professor said, holding out a bottle.

Nilly put the bottle to his lips and washed the powder down.

"Whoa, that's good," Nilly said, looking in vain for a label on the bottle. "What is this?"

"Doctor Proctor's pear soda," the professor said. "Mostly water and sugar with a little dash of wormwood, elephant-snail slime and carbonation. . . Is something wrong?"

The professor looked worriedly at Nilly, who had suddenly started coughing violently.

"No, no," said Nilly, his eyes tearing up. "It's just that I didn't think elephant snails really existed . . ."

Bang!

Nilly looked up, frightened. The bang wasn't as loud as the first one that had made him deaf for a minute, but this time Nilly had felt a strong tug on the seat of his trousers and the cellar door had blown open.

"Oh, no!" Doctor Proctor said, hiding his face in his hands.

"What was that?" Nilly asked.

"You farrrrrrted!" the professor yelled.

"That was a fart?" Nilly whispered. "If it was, that's the loudest fart I've ever heard."

"It must be the pear soda," the professor said. "I should have known the mixture could be explosive."

Nilly started filling the spoon with more powder, but Doctor Proctor stopped him.

"I'm sorry, this isn't appropriate for children," he said.

"Sure it is," Nilly said. "All kids like to fart."

"That's absurd," Doctor Proctor said. "Farts smell bad."

"But these farts don't smell," Nilly said.

The professor sniffed loudly. "Mmm," he said. "Interesting, they don't smell."

"Do you know what this invention could be used for?" Nilly asked.

"No," Doctor Proctor said, which was the truth. "Do you?"

"Yes," Nilly said triumphantly. He crossed his arms and looked up at Doctor Proctor. "I do."

And that was the beginning of what would become Doctor Proctor's Fart Powder.

But now Nilly's mother was standing on the steps, yelling that he had to hurry because this was his first day at his new school. And that's what the next chapter is about.

The New Boy in Mrs Strobe's Class

THE BIRDS WERE chirping and the sun was shining outside the classroom, but inside it was very quiet. Mrs Strobe nudged her glasses down her unbelievably long nose and peered at the new boy.

"So, you're Nilly then?" she said in a slow, raspy voice.

"Yup, what of it?" Nilly responded.

A few people laughed, but when Mrs Strobe did her signature move, slapping her hand against her desk, it got very quiet again in an instant.

"Could you please stand up straight, Nilly," her voice rasped. "I can hardly see you sitting there behind your desk."

"I'm sorry, Mrs Strobe," Nilly said. "But I *am* sitting up straight. The problem, as you can see, is that I'm tiny."

Now the other students were laughing even louder.

"Silence!" Mrs Strobe fumed. She nudged her glasses even further down her nose, which she could safely do because there was still plenty of nose left to go. "Since you're new, why don't you please tell us all a little bit about yourself, Mr Nilly?"

Nilly looked around. "New?" he said. "I'm not new. If you ask me, you guys are the ones who are new. Apart from Lisa, that is. I've met her already."

Everyone turned around to look at Lisa, who mostly wanted to sink down on to the floor.

"Besides, I'm ten years old," Nilly said. "So, for example, if I were a pair of shoes, I wouldn't be new at all. I'd be extremely old. My grandfather had a dog who got sent to the old age home when she was ten."

Mrs Strobe didn't make any attempt to stop the snide laughter that followed, but just looked at Nilly thoughtfully until the laughter had subsided.

"Enough clowning around, Mr Nilly," she said, a thin smile spreading over her thin lips. "Considering your modest size, I suggest that you stand on your desk while you address the class."

To Mrs Strobe's surprise, Nilly didn't wait to be asked twice, but leaped up on to his desk and hoisted his trouser up by his braces.

"I live on Cannon Avenue with my sister and mother. We've lived in every county in Norway, plus a couple that aren't in Norway anymore. By which I

mean, they were in Norway during the Ice Age, but once the ice started melting, big pieces broke off and drifted away in the ocean. One of the biggest pieces is called America now and over there they have no idea that they're living on a chunk of ice that used to be part of Norway."

"Mr Nilly," Mrs Strobe interrupted. "Stick to the most important details, please."

"The most important," Nilly said, "is playing the trumpet in the Norwegian Independence Day parade on the seventeenth of May. Because playing the trumpet is like kissing a woman. Can anyone tell me where I can find the nearest marching band?"

But everyone in the classroom just stared at him with their mouths hanging open.

"Oh, yeah. I almost forgot," Nilly said. "I was there this morning when one of the world's greatest inventions was invented. The inventor's name is Doctor Proctor and I was selected to be his assistant.

We're calling the invention Doctor Proctor's Far—"

"Enough!" Mrs Strobe yelled. "You can take your seat, Mr Nilly."

Mrs Strobe spent the rest of the class explaining the history of Norwegian Independence Day, but none of the children in the classroom were listening. They were just staring at the little bit of Nilly they could see sticking up over his desk. Then the bell rang.

DURING BREAK-TIME NILLY stood by himself watching the other children play tag and hopscotch. He noticed Lisa, who was also just standing there watching. Nilly was just about to go over to her when two large boys with crew cuts and barrel-shaped heads suddenly stepped in front of him, blocking his way. Nilly already had an idea of what was coming next.

"Hello, pip-squeak," one of them said.

"Hello, O giants who wander the earth with heavy

footsteps, blocking out the sun," Nilly said without looking up.

"Huh?" the boy said.

"Nothing, pit bulls," Nilly said.

"You're new," the other boy said.

"So what?" Nilly asked quietly. Even though he already knew more or less what the answer to the so-what question would be.

"New means we dunk you in the drinking fountain," the other boy said.

"Why?" Nilly asked, even more quietly. He knew the answer to that too.

The first boy shrugged. "Because . . . because . . ." he started, trying to think of the reason. And then all three of them – the two boys and Nilly, that is – all exclaimed in unison: "*Because that's just the way it is.*"

The two boys looked around, obviously checking if any of the teachers were nearby. Then the bigger of the two boys grabbed Nilly's collar and lifted him up. The other one took hold of Nilly's legs, and then they carried him off towards the drinking fountain in the middle of the playground. Nilly hung there like a limp sack of flour between those two, studying a little white cloud that looked like an overfed rhinoceros up in the breathtakingly blue sky. He could hear children joining the procession, mumbling quietly in anticipation. He watched them fight for a chance to plug the openings of the other fountains with their fingers so that only one, powerful stream of water was left, shooting almost three metres into the air. Nilly felt himself being lifted up and could feel the cold gust of air next to the stream of water. People started cheering.

"We christen you . . ." said the guy holding Nilly's legs.

"Flame Head the Pygmy," the other said.

"Nice one, dude!" the first one yelled. "Guess we'd better put out his flame!"

The boys laughed so hard, it made Nilly shake up and down. Then they held him over the fountain of water, which shot Nilly right in the face, hitting his nose and mouth. He couldn't breathe and for a second he thought he was going to drown, but then the hands lifted him up out of the stream. Nilly looked around at all the children near the drinking fountains and at Lisa, who was still standing by herself at the edge of the playground.

"More, more!" the kids yelled. Nilly sighed and took a deep breath. Then they dunked him down into the water again.

Nilly didn't put up any opposition and didn't say a word. He just closed his eyes and mouth. He tried to imagine he was sitting at the front of his grandfather's motorboat with his head hanging out over the side, so the sea spray hit him in the face.

When the boys were done, they set Nilly down again and went on their way. Nilly's wet red hair stuck to his

head and his shoes squished. The other kids crowded around and watched, laughing at him, while Nilly pulled his T-shirt up from between the braces.

"Weak drinking fountains you guys have here," he said loudly.

It got awfully quiet around him. Nilly dried his face. "At Trafalgar Square in London they have a drinking fountain that shoots ten metres straight up," he said. "A friend of mine tried to drink from it. The water knocked out two teeth and he swallowed his own retainer. We saw an Italian guy get his wig knocked off when he went to take a drink."

Nilly paused dramatically as he wrung out his wet T-shirt. "True, some people said it wasn't a wig, that it was the Italian guy's own hair that had been pulled right out. I decided to try sitting on the fountain." Nilly leaned to the side to get the water out of his ear.

Finally one of the kids asked, "What happened?"

"Well," Nilly said, holding his nose and blowing

hard, first through one nostril and then the other.

"What did he say?" one of the kids who was standing farther back asked. The ones who were standing in front said, "Shhh!"

"From up where I was sitting, I could see all the way to France, which was more than five hundred miles away," Nilly said, shaking his bangs and sending out a spray of water. "That may sound like an exaggeration," he said, pulling a comb out of his back pocket and running it through his hair. "But you have to remember that it was an unusually clear day and that that part of Europe is extremely flat."

Then Nilly plowed his way through the crowd of kids and walked over to Lisa at the edge of the playground.

"Well," she said with a little smile. "What do you think of our school so far?"

"Not so bad," he said. "No one's called me Silly Nilly yet."

"Those two were Truls and Trym," Lisa said. "They're twins and, unfortunately, they live on Cannon Avenue."

Nilly shrugged. "Truls and Trym live everywhere."

"What do you mean?" Lisa asked.

"Every street has Trulses and Tryms. You can't get away from them, no matter where you move," Nilly explained.

Lisa thought about it. Could there be Trulses and Tryms in Sarpsborg too?

"Did you find a new best friend yet?" Nilly asked.

Lisa shook her head. They stood there next to each other in silence, watching the other kids play, until Lisa asked, "Was that really true, what you said about Doctor Proctor and the invention?"

"Of course," Nilly said with a wry smile. "Almost everything I say is true."

Right then the bell rang.

Nilly Has an Idea

THAT AFTERNOON NILLY knocked hard on the cellar door at the blue house. Three firm knocks. That was the signal they'd agreed on.

Doctor Proctor flung open the door and when he saw Nilly, he exclaimed, "Wonderful!" Then he raised one bushy eyebrow and lowered another bushy eyebrow, pointed, and asked, "Who is that?"

"Lisa," Nilly said.

"I can see that," the professor said. "She lives across the street if I'm not mistaken. What I mean is: what's she doing here? Didn't we agree that this project was top secret?"

"Obviously it's not that secret," Lisa said. "Nilly told the whole class about it today."

"What?" the professor exclaimed, frightened. "Nilly, is that true?"

"Uh," Nilly said. "A little, maybe."

"You told . . . you told . . ." the professor sputtered, waving his arms around in the air, while Nilly stuck out his lower lip and made his eyes look big, as if he were on the verge of tears. This facial expression, which Nilly had practised especially for situations like this, made him look like a tiny, little, very depressed camel. Because everyone knows that it's absolutely impossible to be mad at a very depressed camel. The professor groaned, giving up, and lowered his arms again. "Well, well, maybe it's not so terrible. And you are my assistant after all, so I suppose it's all right."

"Thanks," Nilly said quietly.

"Sure, sure," said the professor, waving his hands at Nilly. "You can stop trying to look like a camel now. Come in and close the door behind you."

They did as he said, while Doctor Proctor hurried over to the test tubes and glass containers that were bubbling and smoking with something that smelled like cooked pears.

Lisa stopped just inside the door and looked around. There was a potted plant with white petals on the windowsill. And on the wall next to it hung a picture of a motorcycle with a sidecar in front of what she assumed must be the Eiffel Tower in Paris. A smiling young man who looked like the professor was sitting on the motorcycle seat and there was a sweet, smiling girl with dark hair in the sidecar.

"What are you doing?" Nilly asked Doctor Proctor.

"I'm perfecting the product," he said, stirring some mixture in a big barrel. "Something that ought to give it even more pep. A concoction of the more explosive type, you might say."

The professor dipped a finger in and then brought it to his mouth. "Hmm. A little more wormwood."

"Can I taste?" Lisa asked, peering over the edge of the barrel.

"Sorry," the professor said.

"Sorry," Nilly said.

"Why not?" Lisa asked.

"Are you a certified fart powder tester, perhaps?" Nilly asked.

Lisa thought for a second and said, "Not as far as I know."

"Then I recommend that you leave the testing to me for the time being," Nilly said, pulling on his braces. Then he took a spoon and stuck it down into the barrel.

"Careful," the professor said. "Start with a quarter of a spoonful."

"Sure," Nilly said, putting a quarter spoonful of powder in his mouth.

"Then we'll start the countdown," Doctor Proctor said and looked at the clock. "Seven — six — five — four — three . . . hey, don't stand right behind him, Lisa!"

Right then there was a bang. And Lisa felt a blast of air hitting her before she lost her balance and sat down hard on her bottom on the cold floor.

"Oh," Nilly said. "Lisa, are you okay?"

"Yeah," she said, a little dazed as the professor helped her back on to her feet. "Well, I'd call that some pep!"

Nilly laughed out loud. "Well done, Doctor!"

"Thank you, thank you," the professor said. "I think I'll conduct a little test myself . . ."

The professor took half a teaspoon and counted down. At zero there was another bang, but this time Lisa was careful to stand by the door.

"Wow," the professor said, picking up the plant, which didn't have leaves on it anymore. "I think we'll do the next test outside."

They poured the powder into a biscuit tin and brought it outside.

"Give me the teaspoon," Nilly said.

"Careful with the dose . . ." Doctor Proctor started to say, but Nilly had already gobbled up a full teaspoon.

"I feel a tingling in my stomach," said Nilly, who

was so excited that he was whining and jumping up and down.

"Seven — six — five," the professor counted.

When the bang came, all the songbirds in the professor's pear tree took off and flew away in alarm. And this time it wasn't Lisa but Nilly who got knocked over and disappeared in the tall grass.

"Where are you?" Doctor Proctor yelled, searching in the grass. "How did it go?"

They heard a gurgling noise and then Nilly popped up, totally red in the face from laughing.

"More!" he yelled. "More!"

"Look, Professor!" Lisa pointed. "It ripped the seat of Nilly's trousers!"

And indeed it had. Nilly's trousers were practically torn apart. The professor looked at the results with concern and decided that they should stop the testing for today. He asked them to search for his lawn furniture, which was in the grass somewhere, and then went inside. When he came back out he brought bread, butter, liverwurst and juice. Lisa had found the lawn furniture and while they sat in the crooked white-painted chairs and ate, they contemplated what the invention could be used for. The professor had the idea of trying to sell the powder to farmers. "They could eat a half teaspoon of fart powder," he explained, "and hold the sack of seed grain in front of the . . . uh, launch site. Then the air pressure would spread the seeds over the whole field. It'll save a ton of time. What do you guys think?"

"Excellent!" Nilly said.

"To be completely honest," Lisa said, "I don't think people are really going to want to eat food that comes from seeds that have been farted on."

"Hmm," the professor said, scratching his mop of white hair. "You're probably right about that."

"What about making the world's fastest bicycle pump?" Nilly yelled. "All you have to do is take a hose, fasten one end to your bum and the other to the valve on the bike tyre and then . . . *kaboom!* The tyre is filled in a fraction of a second!"

"Interesting," said the professor, stroking his goatee. "But I'm afraid it's the kaboom that's the problem. The tyre's going to explode too."

"What if we use the fart powder to dry hair?" Lisa suggested.

Nilly and the professor looked at Lisa while she explained that the whole family could draw straws, everyone from the littlest to Grandma, to see who would eat the fart powder after everyone had showered in the morning. And then everyone else could just stand behind that person.

"Good idea," said the professor. "But who's going to dry the farter's hair?"

"And what if the blast knocks Grandma over and she breaks her hip?" Nilly said.

They kept tossing out one suggestion after

another, but all of the suggestions had some kind of annoying drawback or other. In the end they were all sitting there quietly chewing their sandwiches when Nilly suddenly exclaimed, "I have it!"

Lisa and Doctor Proctor looked at him without much enthusiasm, since this was the fourth time in only a couple of minutes that Nilly had said he had it and so far he definitely hadn't had it. Nilly leaped up on to the table. "We could just use the powder for the same thing we've been using it for so far!" he said.

"But we're not using it for anything," the professor said.

"We're just making meaningless bangs," Lisa said.

"Exactly!" Nilly said. "And who likes meaningless bangs better than anything?"

"Well," the professor said. "Kids, I guess. And adults who are a little childish."

"Exactly! And when do they want things that bang?"

"New Year's Eve?"

"Yes!" Nilly shouted, excited. "And . . . and . . . and?"

"Norwegian Independence Day!" Lisa blurted out, jumping up on to the table next to Nilly. "That's only a few days away! Don't you see, Professor? We don't need to come up with anything at all, we can just sell the powder the way it is!"

The professor's eyes widened and he stretched his thin, wrinkled neck so that he looked like some kind of shorebird. "Interesting," he mumbled. "Very interesting. Independence Day . . . children . . . things that go boom . . . it's . . . it's . . ." With a bounce he leaped up on to the table too. "Eureka!"

And as if on cue the three of them started dancing a victory dance around the table.

Conductor Madsen and the Dølgen School Marching Band

MR MADSEN WAS standing in the gym with both arms out in front of him. Facing him sat the twenty students who made up the Dølgen School Marching Band. Mr Madsen squeezed a baton between his right thumb and index finger, his other eight fingers splayed in all directions. He had closed his eyes and for a second he imagined he was far away from the bleachers, worn-wood floor and stinky gym mats, standing before a sold-out audience in a concert hall in Venice, with chandeliers hanging from the ceiling and cheering people in formal clothes in the balcony seats. Then Mr Madsen opened his eyes again.

"Ready?" he yelled, wrinkling his nose so his dark sunglasses wouldn't slide down. Because unlike Mrs Strobe, Mr Madsen had a short, fat nose with black pores.

None of the twenty faces in the chairs in front of him looked like they were ready. But they didn't protest either so Mr Madsen counted down as if for a rocket launch.

"Four – three – two – one!"

Then Mr Madsen swung his baton as if it were a magic wand and the Dølgen School Marching Band began to play. Not like a rocket, exactly. More like a train that, snorting and puffing, started to move. As usual, the drums had started playing long before Mr Madsen got to "one." Now he was just waiting for the rest of the band. First came a screech of a trombone, then a French horn bleated in the wrong key, before two clarinets played almost the same note. The two trumpet players, the twins Truls and Trym Trane, were picking their noses. Finally Petra managed to get her tuba to make a sound and Per made a hesitant tap on the base drum.

"No, no, no!" Mr Madsen called, losing hope and waving his baton defensively. But just like a train, the Dølgen School Marching Band was hard to stop once it got going. And when they tried to stop, it sounded like a ton of kitchen implements falling on the floor. *Crash! Bang! Toooot!* When it was finally quiet and the windows at Dølgen School had stopped vibrating, Mr Madsen took off his sunglasses.

"My dear ladies and gentlemen," he said. "Do

you know how many days there are left until Independence Day?"

No one said anything.

Mr Madsen groaned. "Well, I wouldn't expect you to either, since you don't even seem to know what song we're playing. What song is this, Trym?"

Trym stopped picking his nose and glanced over at his brother questioningly.

"Well, Truls," Mr Madsen said. "Can you help Trym out?"

Truls scratched his back with his trumpet and squinted at the music stand. "I've got some rain on my music, Mr Madsen. I can't see nothin'," he said.

"Right," Mr Madsen said. "For crying out loud, this is the national anthem. Is there really no one here besides Lisa who can read music? Or at least play in key?"

Lisa cowered behind her clarinet as she felt everyone else looking at her. She knew what those looks were saying. They were saying that even if Mr Madsen said she was good, she shouldn't think that any of them wanted to be friends with her. In fact, the opposite was true.

"If we don't improve by Independence Day, we're going to have to give up the idea of a band camp this summer," Mr Madsen said. "I don't want to be made into a laughing stock in front of dozens of other band conductors. Understood?"

Mr Madsen saw the faces in front of him start gaping. This was a shock to them, that much was clear. After all he had talked so much and so positively about the big band competition in Eidsvoll and they were all really looking forward to it. But he had made it clear to them from the very beginning. Nikolai Amadeus Madsen was not playing around, conducting a rattling, old military band. So unless a miracle occurred, no one at Eidsvoll was going to hear so much as a triangle clang from the Dølgen School Marching Band. And unfortunately, since Mr Madsen's baton wasn't a magic wand, there wasn't going to be any miracle.

"Let's take it again from the top," Mr Madsen said with a sigh, raising his baton. "Ready?"

But they simply were not ready. In fact, they were all staring at the door to the locker room that was right behind Mr Madsen's back. Irritated he turned around but he couldn't see anyone. He turned back towards the band and was just about to count off when his brain realised that it had seen something in the doorway after all. Something down by the floor. He turned around, took off his sunglasses and looked at the tiny little boy with the red hair.

"What are you doing here?" Mr Madsen asked curtly.

"Shouldn't you ask *who* I am first?" Nilly said, holding out an old, beaten-up trumpet. "I'm Nilly. I can play the trumpet. You want to hear me play a little?"

"No!" Mr Madsen said.

"Just a little . . ." Nilly said, raising his trumpet and forming his lips as if for a kiss.

"No! No! No!" growled Mr Madsen, who was bright red in the face and slapping his thigh with the baton. "I am an artist!" he yelled. "I have arranged marches for the big marching-band festival in Venice. And now I'm conducting a school band for tone-deaf brats and I don't need to hear one more tone-deaf brat. Understood? Now get out!"

"Hmm," Nilly said. "That sounded like an A. I have perfect pitch. Just check with your tuning fork."

"You're not only tone-deaf, you're deaf!" Mr Madsen sputtered, shaking and spitting in agitation. "Shut that door again and don't ever come back here! Surely you don't think any band would take someone so small that . . . that . . ."

"That there isn't even room for the stripe on the side of his uniform trousers," Nilly said. "So short that his band medals would drag on the ground. So teensy-weensy that he couldn't see what was on the music stand. Whose uniform hat would fall down over his eyes."

Nilly smiled innocently at Mr Madsen, who was now rushing straight towards him in long strides.

"So he can't see where he's going," Nilly continued. "And suddenly he finds himself on Aker Street while the rest of the band is marching down Karl Johan Street."

"Exactly!" Mr Madsen said, grabbing hold of the door and flinging it shut right in Nilly's face. Then he stomped back over to his music stand. He noted the big grins on Truls and Trym's faces before he raised his baton.

"So," Mr Madsen said. "Back to the national anthem."

That Night, in a Sewer Beneath Oslo

THERE ARE BIG animals in the sewers that run every which way beneath Oslo. So big that you probably wouldn't want to bump into them. But if you pick up a manhole cover on one of Oslo's streets and shine a torch down into the sewer world, it just might happen that you'll see the light catch the teeth in the jaws of one of the huge, slimy beasts before it scurries away. Or before it sinks its teeth into your throat. Because they are quite speedy beasts. And we're not talking now about the regular, innocent *Rattus norvegicus* i.e. little Norwegian rats, but about properly beastly beasts. Like Attila. Attila was an old Mongolian water vole who'd lived for thirty-five years and weighed more than fourteen kilograms. If you want to read more about water voles, turn to page 678 of *Animals You Wish Didn't Exist*.

As it so happened, Attila liked to eat a little *Rattus norvegicus* for breakfast and was the king of the Oslo Municipal Sewer and Drainage System. That is to say, Attila thought it was, until now. Attila's reign had started many years ago, but this water vole hadn't always been king. When Attila was a few months old and was a cute, tiny fur ball weighing only a few grams, it had been bought in a pet store by a family in Hovseter, Norway. They bought the Mongolian water vole because the fat little boy in the family had pointed at Attila and yelled that he wanted a rat like that. And his parents had done what the little boy ordered. They had fed Attila fish balls, the worst thing Attila knew of, and put a metal collar on the rat with the name ATTILA engraved on it, and the fat little boy had tormented the poor water vole every single day by poking sticks into the cage. Every single day, until the day Attila had got so big from eating fish balls that it needed a new cage while it could still fit through the opening of the old one. Attila had been looking forward to this day. And when the boy stuck his hand into the cage to drag Attila out, Attila had opened its mouth as wide as it could and sunk its teeth into the delicious, soft, white human meat. This was a totally different kind of meatball! And while the little boy screamed and his blood gushed, Attila was out of the cage, as fast as a Mongolian water vole could go, out of the house, away from the above-ground part of Hovseter and down into the sewer. And from there the water vole had found his way to downtown Oslo, where its beastly behaviour had quickly earned the water vole respect. Attila was feared by Norwegian rats citywide, from the manhole covers at Majorstua subway station to the sewage treatment plant at Aker.

But on this night, deep beneath Oslo, while Lisa and Nilly were sleeping like babies, it was Attila who was gripped by fear. The vole was sitting in the corner of a sewer pipe, shaking. Because in a flash of light it had seen something right in front of him. A glimpse of teeth. Teeth even bigger than its own. Could the legend he'd been hearing for all these years in the Oslo Municipal Sewer and Drainage System be true after all? Attila felt its Mongolian water vole heart pounding in fear and it was so dark, so dark. And for the first time Attila realised that it actually smelled pretty bad down here in the sewer and that it really would prefer to be anywhere else besides this sewer pipe right now. Even its old cage in Hovseter. So Attila tried to comfort itself. Obviously the legend must be made up. An anaconda? What rubbish. An anaconda is a boa constrictor that is found in the Amazon, where it lives off of huge water voles and such, not here deep beneath Oslo, where there aren't any water voles at all. Apart from the one, that is. Attila contemplated this briefly.

And while Attila was thinking, something moved towards the water vole. It was huge, like an inflatable swimming ring, surrounded by jagged teeth the size of ice-cream cones, and it was hissing and had such bad breath that the rest of the sewer smelled like a flower bed in comparison.

It was so frightening that Attila quite simply squeezed its eyes shut.

When the vole opened them again something was dripping and dripping all around. And it was excessively dark. Just as if Attila weren't sitting in a sewer pipe, but inside something that was even darker. And it was as if the walls were moving, pulling in closer and slithering. As if the water vole were already inside the stomach of . . . of . . .

Attila screamed at least as loud as that fat little boy he had bitten so long ago.

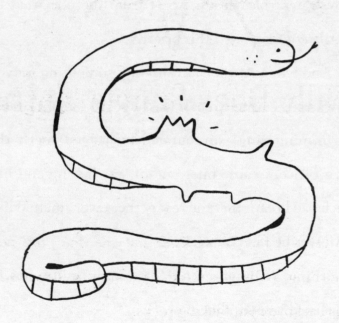

Nilly Does Simple Maths

WHEN LISA WALKED out the door the next morning, Nilly was standing across the street with his backpack on, kicking rocks.

"What are you waiting for?" Lisa asked.

Nilly shrugged and said, "To see if anyone walks by who's going the same way as me."

"No one's going to come by," Lisa said. "This is a dead-end street and we live at the end of it."

"Well then," Nilly said, and they started walking down Cannon Avenue together.

"Proctor invited us to come over after school for the Last Big Powder Test," Nilly said. "Are you coming?"

"Of course," Lisa said. "Are you excited?"

"As excited as a little kid," said Nilly.

When they'd made it almost all the way down to the main road Lisa stopped and pointed at the house at the bottom of Cannon Avenue.

"That's where Truls and Trym live," she said. "If I see them come out, I usually wait here until they're gone. If I don't see them, I run quickly past. Come on . . ."

Lisa took Nilly's hand and was about to run, but Nilly held her back.

"I don't want to run," he said. "And I don't want to wait either."

"But . . ." Lisa began.

"Remember, there's two of us," Nilly said. "There's just as many of us as Truls and Trym. At least. It's simple maths."

So they walked past Truls and Trym's house. Nilly was walking really, really slowly, Lisa thought. She could still tell that he was a little scared though, because he was constantly looking over at the house. But luckily neither Truls nor Trym came out and Lisa looked at her watch and realised they must have gone to school already.

"Do you know what time it is?" she exclaimed in alarm, because she was a good girl and wasn't used to being late.

"I don't have a watch," Nilly said.

"Mrs Strobe is going to be super angry. Hurry!"

"Aye, aye, boss," Nilly said.

And they ran so fast that they got there in the time it took you to read from the beginning of this chapter to here.

UNFORTUNATELY TIME DIDN'T pass as quickly the rest of the day. Nilly was so impatient to get home for the Last Big Powder Test that he sat there in the classroom counting the seconds as he watched Mrs Strobe's mouth moving. He wasn't paying attention, so when he suddenly realised that Mrs Strobe was pointing at him and that everyone else in class was looking at him, Nilly figured that Mrs Strobe had probably asked him a question.

"Two thousand six hundred and eighty-one," Nilly said.

Mrs Strobe wrinkled her brow and asked, "Is that supposed to be the answer to my question?"

"Not necessarily," Nilly said. "But that's how

many seconds have passed during this class. Well, now four more have gone by, so now two thousand six hundred and eighty-five seconds have passed. It's simple maths."

"I understand that," Mrs Strobe started. "But Nilly . . ."

"Excuse me. That isn't the right answer anymore," Nilly said. "The right answer is now two thousand six hundred and eighty-nine."

"To me it sounds like you're trying to talk your way out of what I asked you about," Mrs Strobe said. "Because you heard what I asked you, right, Nilly?"

"Of course," Nilly said. "Two thousand six hundred and ninety-two."

"Get to the point," Mrs Strobe said, sounding a little irritated now.

"The point," Nilly said, "is that since there are sixty seconds in a minute and forty-five minutes in each class, I won't have time to answer your question, since sixty seconds times forty-five is two thousand seven hundred seconds, and that means the bell is going to ring right . . ."

No one heard the rest of what Nilly said, because the bell started ringing right then, loud and shrill. Mrs Strobe tried looking sternly at Nilly, but when she yelled, "All right, everyone out!" he could see that she couldn't quite help but smile anyway.

AFTER LISA AND Nilly had spent sixteen thousand and two hundred seconds together in the classroom and two thousand seven hundred seconds on the playground, they ran away from the school as quickly as they had run towards it. They parted on Cannon Avenue, each opening their own gate, each running up their own front steps and each flinging their backpack in their own hallway. Then they met again in front of Doctor Proctor's gate.

"I'm almost dreading it a little," Lisa said.

"I'm almost looking forward to it a little," Nilly said.

Then they stormed into the garden and through the tall grass.

"There you guys are!" called the doctor joyfully in his remarkable accent. He was sitting at the picnic table under the pear tree. In front of him lay three tablespoons and a teaspoon, an ice hockey helmet, two knee pads, a jar full of powder, a pair of motorcycle pants and a metre long, rectangular, homemade jelly bathed in caramel sauce. "Are you guys ready for the Last Big Powder Test?" he asked.

"Yes!" Lisa and Nilly shouted in unison.

"But first, jelly," said the doctor.

They sat down around the table and each grabbed a spoon.

"On your marks, get set . . ." Doctor Proctor said.

"Go!" Nilly yelled, and they flung themselves at the jelly. If Nilly had been counting, he wouldn't have got any further than thirty seconds before the metre long jelly had vanished completely.

"Good," Nilly said, patting his stomach.

"Good," Lisa said, patting her stomach.

"I've made a few tiny adjustments to the powder mixture," Doctor Proctor said.

"I'm ready," Nilly said, taking the lid off the jar.

"Hold on!" the professor said. "I don't want you to ruin your trousers again, so I made these."

He held up the motorcycle trousers. They were very normal, aside from the fact that the seat of the trousers had a Velcro flap.

"So the air can pass through unobstructed," the doctor explained. "I remodelled my old motorcycle gear."

"Niiice," Nilly said once he'd put on the trousers,

which were way too big for him. Lisa just shook her head.

"These too," the doctor said, and passed Nilly the hockey helmet and the knee pads. "In case you get knocked over again."

Nilly put everything on, then crawled up onto the table and over to the jar.

"Only one teaspoon!" Doctor Proctor yelled.

"Yeah, yeah!" Nilly said, filling the spoon he was holding in his hand and sticking it into his mouth.

"Okay," the doctor said, looking at his watch. "We'll start the countdown then. Seven. Six."

"Doctor Proctor . . ." Lisa said warily.

"Not now, Lisa. Nilly, hop down from the table and stand over there so you don't ruin anything. Four. Three," the doctor continued.

"He didn't use the teaspoon," Lisa practically whispered.

"Two," the doctor said. "What did you say, Lisa?"

"Nilly used that big tablespoon he ate his jelly with," Lisa said.

The doctor stared at Lisa with big, horrified eyes. "One," he said. "Tablespoon?"

Lisa nodded.

"Oh, no," Doctor Proctor said, running towards Nilly.

"What now?" Lisa whispered.

"Simple maths," Nilly yelled happily. "Zero."

And then came the bang. And if the earlier bangs had been loud, they were nothing compared to this. This was as if the whole world had exploded. And the air pressure! Lisa felt her eyelids and lips distort as she was peppered with dirt and pebbles.

When her eyes settled back into place, the first thing Lisa noticed was that the birds had stopped singing. Then she noticed Doctor Proctor, who was sitting in the grass with a confused look on his face. The leaves from the big pear tree wafted down around him as if it were suddenly Autumn. But she didn't see Nilly. She looked to the right, to the left and behind her. And finally she looked up. But Nilly was nowhere to be seen. Then the first bird cautiously started singing again. And that's when it occurred to Lisa that she might never, ever see Nilly again and that that would actually be almost as sad as Anna having moved to Sarpsborg.

The Fartonaut

WHEN NILLY SAID "zero" he felt an absolutely wonderful tickle in his stomach. It was as if the fart was a giant, burbling laugh that just had to get out. Sure he had seen Lisa's worried expression and Doctor Proctor coming running towards him, but he was so excited that it hadn't occurred to him that something might be wrong. And when the bang came, it was so delightfully liberating that Nilly automatically shut his eyes. The previous farts had been short explosions, but this one was more drawn out, like when you let the air out of a balloon. Nilly laughed out loud because it felt just like he'd blasted off from the ground, like he was an astronaut who'd been shot up and propelled into space. He could feel the air rushing past his face and hair and it was as if his arms were being pressed in against his body. It felt totally real. And when Nilly finally opened his eyes he discovered that it was very real in reality too. He blinked twice and then he understood that not only was it very real, it was utterly, incredibly real. It was as if he were sitting on a chair of air that was shooting upwards. The blue sky arched above him and below him he saw a big cloud of dust in what looked like a tiny copy of Doctor Proctor's garden. The fart howled like a whole pack of wolves and Nilly

realised he was still going up, because the landscape down below was starting to look like a smaller and smaller version of Legoland.

Then the fart turned into a low rumbling, the chair of air disappeared from underneath him and for just a second, Nilly felt like he was totally weightless. A crow turned its head as it flew by, staring at him with astonished crow eyes.

Nilly tipped forwards and then felt the descent begin. Headfirst. Slowly at first, then faster.

Uh-oh, Nilly thought, no longer finding any reason to smile. *Hockey helmet or not, I'm never going to survive this.*

Legoland got bigger and bigger, and with perilous speed it started to resemble the Cannon Avenue that Nilly had just left. And things very surely would have gone really badly for our friend Nilly if he hadn't been such a quick-witted little guy and remembered what it was that had sent him up in the first place. Because although the fart was no longer howling like a pack of wolves and was now just a tame sputtering, it was still going. And remember that when I say sputtering, that's compared to an enormous bang and not compared to one of your farts. Because even if you've been eating un-ripe apples and think you just farted the loudest fart anyone has ever farted, it would be considered a gentle breeze compared to the tamest sputtering caused by Doctor Proctor's Fart Powder. Once Nilly had thought about all this, he swung himself quickly back into the sitting position he'd been in when he'd flown up. And once the seat of his trousers, with the open Velcro flap, was pointed straight at the ground, to his relief he immediately started slowing down, thanks to the air pressure of the fart. But he also knew that the fart was going to be over soon and there was still a way to go until he was back on the ground. Nilly tried as hard as he could to keep it going, because even an eight-metre fall is very high for such a small boy. And that's exactly how high he was above the ground when the fart finally came to an end.

"NILLY!" LISA YELLED.

"Nilly!" Doctor Proctor yelled. They were still looking around for him like crazy.

"Do you think the powder exploded him into smithereens?" Lisa asked.

"If so the pieces must be so small that we can't see them," Doctor Proctor said, adjusting his motorcycle glasses and studying the ground where Nilly had been standing when the fart happened. All of the grass was torn up and there was a little pit there.

"We're never going to see him again," Lisa said. "And it's my fault. I should've noticed that he was holding the tablespoon."

"No, no. It's my fault," Doctor Proctor said, getting up again. "I should never have tinkered with the formula."

"Nilly!" Lisa yelled.

"Nilly!" Doctor Proctor yelled.

"What's all the commotion?" someone complained from over by the fence along the road. "And what are you doing here, Lisa? Dinner's on the table."

It was Lisa's father, the Commandant. He looked gruff.

Doctor Proctor stood up. "My good sir, the whole situation is hopeless—" he started, but was interrupted by a voice barking from behind the fence at Nilly's house.

"What's all the commotion?" It was Nilly's mother. She looked mad. "Dinner's on the table. Has anyone seen Nilly?"

Doctor Proctor turned to face her. "My good ma'am, the whole situation is hopeless. You see, your son, Nilly, he . . . he . . ."

Then Doctor Proctor was interrupted for the third time and this time by a high-pitched boy's voice that came from above: "He's sitting up here wondering what's for dinner."

All four of them looked up. And there, on top of Doctor Proctor's roof, stood Nilly with his arms crossed, wearing a hockey helmet, knee pads and leather trousers with the bottom flapping around.

"Don't move," called Doctor Proctor, running into the cellar.

"What in the world are you doing up there, Nilly?" his mother squealed.

"Playing hide-and-seek obviously," Nilly said. "What's for dinner?"

"Meatballs," Nilly's mother said to Nilly.

"Fish au gratin," Lisa's father told Lisa.

"Yippee!" said Nilly.

"Yippee!" said Lisa.

"You guys can go back to playing after dinner," Lisa's father growled.

"But not up there," Nilly's mother said. "Get down here right now."

"Yes, Mom," Nilly said.

The doctor came running back out of the cellar with a ladder that he immediately leaned up against the wall of his house so that it was resting against the gutter. Nilly crawled to the ladder and then down the rungs, smiling and as proud as an astronaut climbing down from his spaceship after a successful landing following an expedition to somewhere in space where no one — or at least very few people — had ever been before him.

And three minutes later, which a little simple maths can tell you is the same as a hundred and eighty seconds, Lisa was sitting with freshly washed, completely clean hands, eating fish au gratin, and Nilly, with pretty clean hands, was eating meatballs. Neither of them had ever eaten so fast before.

WHEN THEY GOT back to Doctor Proctor's yard, the professor was sitting on the bench, reviewing everything as he jotted some things down and did some calculations on a piece of paper. Nilly looked at all the numbers and squiggles. This maths wasn't quite so simple.

"With the new formula the effect of the powder is seven times stronger," Proctor said in his heavy accent. "That's why I said you should use the teaspoon, not the tablespoon."

Nilly shrugged. "It worked out fine. The fart ended when I was on my way down, just as I reached the roof of your house."

"Hmm," the professor said, looking at the numbers. "But I'm puzzled about why you took off like a rocket."

"It was a looong fart," Nilly said. "It was like sitting on a column of air that was pushing me up. And it was the same column that slowed me down on the way back down too."

"Hmm," the doctor said, scratching his chin. "Because of the new formula, the powder seems to have a much longer reaction time. Interesting."

"Maybe we should go back to the original formula," Lisa suggested hesitantly.

"I suppose you're right, Lisa," the doctor said. "It would be dangerous to sell this powder mixture to children. Or adults, for that matter."

"I've got it," Nilly said. "We make two kinds of powder. A Doctor Proctor's Fart Powder that we sell to all the kids for Independence Day. And a Doctor Proctor's Rather Special Rocket Mixture that we don't sell to anyone. That we just do some tiny little tests with here in the garden."

Doctor Proctor didn't look like he liked the last part of the idea that much.

"Just every once in a while, I mean," Nilly said. "When we're really super bored."

Doctor Proctor still looked like he didn't like the idea.

"Or," Lisa said, "we could sell it to NASA."

"NASA?" Nilly and Doctor Proctor asked in unison.

"The U.S. National Aeronautics and Space Administration," Lisa said without tripping over a single syllable. "They're the ones who send astronauts into space. My dad said it costs more to build one small spaceship than all of Akershus Fortress put together. Just think how happy they'll be when they find out you can send astronauts up without a spaceship."

"Hmm," Doctor Proctor said. "Interesting."

"And maybe we could do something about the name of the rocket powder too," Lisa said. "What about Doctor Proctor's Fartonaut Powder?"

"That's it, Lisa!" Nilly yelled. "You're a genius!"

"Excellent," Doctor Proctor said. "This calls for a celebration. . . ."

And while Doctor Proctor shuffled into the house to get the other metre long portion of the jelly he'd made, Lisa beamed. Because it's always nice to be praised when you've been extra clever.

Nilly Gets Tricked and Juliette Margarine

THE NEXT DAY rumours started flying on the playground, about a powder that makes you fart louder than you ever have before. And you didn't even have to try hard. And best of all: There was absolutely no smell. Supposedly the bang was louder than thirteen firecrackers, three bangers, and a half stick of dynamite put together *and* the powder cost less than a bottle of fizzy pop. Plus it was totally harmless and was totally legal. In short, the kids at school thought it was too good to be true.

But none of them knew where they could get hold of this powder. They only knew that Lisa and Nilly, that new little kid with the red hair, knew everything they didn't know.

And Lisa and Nilly wouldn't say anything.

The other kids nagged them between each class, but Lisa just smiled slyly while Nilly said things like: "I wonder what the weather's going to be like tomorrow." Or, "I hear it's going to be spaghetti and meatballs for lunch in the cafeteria today."

During break Truls and Trym came over to Nilly and Lisa, who were standing by the drinking fountain.

"Well, pip-squeaks," Truls said, towering over them. "What's all this we're hearing about some new powder? Spit it out."

Nilly raised his head and peered up at them, shielding his eyes: "I do believe I can just make out two specimens of Idiotus Colossus. Interesting."

"What did you call us?" Truls asked, moving a step closer. Lisa automatically stepped back, but Nilly didn't budge.

"Idiotus Colossus," he said, smiling. "A dinosaur that lived in the seventeenth century. Very strong and very big. I wouldn't be insulted if I were you."

"Oh?" Truls said, squeezing one eye shut so that he looked like a one-eyed troll. "How strong, huh?"

"Unbelievably strong," Nilly said. "Idiotus Colossus had so many tons of muscles that it was known to have the smallest brain in history in proportion to its body weight."

"Hey!" Trym yelled at Truls. "That dwarf just said 'small brain'!"

"Hey!" Truls yelled at Nilly, grabbing hold of his shirt collar. "You said 'small brain.'"

Nilly sighed. "You guys need to listen more carefully. Idiotus Colossus actually had a brain that's three times the size of your two brains combined. But that's still a small brain in proportion to eighty tons of muscles. Get it? It's simple maths."

Truls and Trym looked at each other uncertainly.

"Enough brain talk," Truls said, letting go of Nilly's collar. "Where's the powder, Mr Tinypants."

Nilly looked around cautiously. "Okay," he whispered. "Since we're practically neighbours, you guys can find out what no one else knows."

Truls and Trym moved in closer to hear what Nilly said.

"Tomorrow, here by the drinking fountain," Nilly

whispered. "Lisa and I are going to tell all the kids at school everything you guys need to know. But only you know this. Okay? Don't tell anyone."

"Cross my heart," Trym said.

Truls looked at Nilly as if there was something he didn't really like, but couldn't quite put his finger on it. And luckily, before he managed to, the bell rang.

That afternoon Lisa, Nilly and Doctor Proctor planned and prepared until sundown. They made a sign to put on the gate so that everyone could find the sale, set up the table with a cash box and change and got the fart powder ready. They filled little plastic bags with one tablespoon of powder from the jar of Doctor Proctor's Totally Normal Fart Powder and decided to sell them for five kroner each. A krone, in case you don't know, is the money used in the very small country of Norway. Although Lisa and Nilly had said that Doctor Proctor should keep the money, the doctor had insisted that they should split what they earned three ways.

"Make sure you don't take it from the wrong jar and put fartonaut powder into the bags instead," Doctor Proctor chuckled.

"No way," said Nilly, who was responsible for putting one teaspoon of the special-formula fartonaut powder into three different envelopes that just needed stamps, since Lisa had already written on them: *To NASA. United States of America. Keep out of reach of children.*

"What are you guys going to do with your share of the money?" Lisa asked.

"I'm going to buy myself a uniform so I can play in the school band," Nilly said.

"I'm going to drive my motorcycle to Paris with the sidecar," Doctor Proctor said. "What about you, Lisa?"

"I'm going to buy an airplane ticket to Sarpsborg and visit Anna," she said. "If we get that much, I mean."

Doctor Proctor laughed. "If not, you can have my third. There's no hurry for my trip to Paris."

"My third too," Nilly said. "I'm sure my mum can sew a band uniform for me."

"Thanks," Lisa said, feeling so happy that her cheeks turned red. Not just because she realised that now she was sure to get enough money to visit Anna, but because she realised that Doctor Proctor and Nilly were so nice to her because they liked her. Lisa liked being liked. Most people do. But she noticed that she especially liked being liked by Nilly and Doctor Proctor.

"What are you going to do in Paris, Doctor?" Nilly asked as he carefully poured powder into one of the bags and then taped it shut.

"Oh, it's a long story," the doctor said, a distant look coming over his eyes. "A long, long story."

"Does it have anything to do with that picture that's hanging in the cellar?" Lisa asked. "The one with you and the girl on the motorcycle in front of the Eiffel Tower?"

"That's right, Lisa," Proctor said.

"Well, let's hear it," Lisa encouraged him.

"Oh, there's not that much to tell," Proctor said. "I had a girlfriend there. Her name was Juliette. We were going to get married."

"Tell us," Lisa whispered eagerly. "Tell us, Doctor Proctor."

"It's just a boring old story, I'm afraid," Doctor Proctor said.

But Lisa didn't back down, and in the end Doctor Proctor gave in. And this is how he told it.

"When I was studying chemistry in Paris many, many years ago, I met Juliette Margarine. She was studying chemistry too, and when we saw each other the first time, there was a . . . uh, 'bang'! She was a brown-eyed beauty and I was . . . well, I was younger than I am now, anyway. And I must have had a certain charm, I guess, because Juliette and I started dating after just a short time. We were inseparable, like two oppositely charged particles in an atom."

"Huh?" Lisa asked.

"Sorry. Like a magnet and a refrigerator door," the professor explained.

"Oh, right," Lisa said.

"Juliette and I were determined to get married when we finished school. But there was one problem. Juliette's father, the Duke of Margarine, was a rich and powerful man who was on the board of the university, and he had totally different plans for Juliette than her marrying a penniless Norwegian without a drop of blue blood in his veins. The day Juliette went to tell her dad that he couldn't stop her from marrying me, she never came back. When I called, they told me that Juliette was sick and couldn't talk to anyone. And especially not me. The next day I got a letter from the board of the university saying that I'd been expelled from the university, because of an experiment that went just a little bit wrong. Well it's not like it was any big deal or anything, just a nitroglycerine mixture that I happened to shake a little too hard, so it exploded and . . . well, caused a bit of damage. But that kind of thing happened all the time and it had been months since it had happened, so I was very surprised. That night a phone call woke me up. It was Juliette. She whispered that she loved me and that she would wait for me. Then she hung up in a hurry. It wasn't until a few days later when the police came to get me that I understood who was behind the whole thing. They gave me a letter that said that I couldn't stay in France anymore, since I wasn't going to school and didn't have a job. Then they drove me to the airport, put me on the first flight back to Norway and said I couldn't come back until I was rich, noble or famous. And since I'm not especially good with money and don't have any aristocratic blood in my veins, I decided to become a famous inventor. Which isn't that easy because so many things have been invented already, but I've been working day and night trying to invent something that is totally and completely new. So that I can go back and find my Juliette."

"Oh," Lisa said when Doctor Proctor was done telling the story. "How romantic."

"You know what?" Nilly asked. "Doctor Proctor's Fartonaut Powder will make you world famous. That's for sure."

"Well we'll see about that," the doctor said.

They heard a grasshopper rubbing its legs together. It was the first one they'd heard that year and it made them realise that summer wasn't far off. Then they glanced up at the moon, which hung pale and almost transparent over the pear tree.

The Big Fart Powder Sale

NILLY STOOD UP on the drinking fountain so that all the kids could see and hear him.

"Doctor Proctor's Fart Powder will be for sale up at the top of Cannon Avenue. There'll be a sign on the gate!" Nilly yelled, even though it was so quiet that he could have spoken in a totally normal voice.

"We'll start at six p.m. and keep going until seven! No pushing, let the little kids go first and no farting until you've left. Understood?"

"Understood!" they all yelled.

"Any questions?" Nilly asked. He glanced out over the crowd and saw a hand sticking up in the air way in the back. "Yes?"

"Is it dangerous?" a small voice asked.

"Yes," Nilly said seriously. "Unfortunately there is one thing that is dangerous about using this powder."

The faces before him got long, their mouths hanging open.

"You might laugh yourself to death," Nilly said.

A sigh of relief ran through the crowd. The bell rang.

"See you this evening!" Nilly yelled, hopping down from the water fountain. Several people clapped and shouted "Hurray" and a murmur of anticipation rose from the crowd, which slowly dissipated, heading towards the different doors back into the school.

"Do you think anyone's going to come?" Lisa asked Nilly, who was whistling the national anthem to himself in satisfaction.

"You should be asking if there's anyone who *won't* come," Nilly said. "Didn't you see the gleam in their eyes? You might as well go ahead and book that plane ticket to Sarpsborg, Lisa."

"Well all right then," said Lisa, even though deep down she wasn't so sure. But then Lisa was almost never totally sure about anything. That's just the way she was.

"Absolutely positive," Nilly said, raising his hands as if he were playing the trumpet. That's just the way he was.

AFTER SCHOOL LISA and Nilly ran home to complete the final preparations. After dinner they ran back to the doctor's garden where they found

Proctor asleep on the bench. They let him sleep while they attached a sign to the gate. It said:

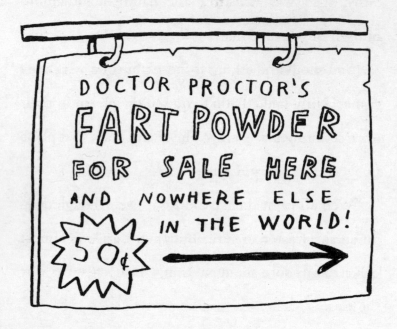

DOCTOR PROCTOR'S
FART POWDER
FOR SALE HERE
AND NOWHERE ELSE
IN THE WORLD!
50¢ →

They took the lids off the shoeboxes and cartons in which the bags of powder were neatly stacked and set them on the picnic table. Then they each sat down in a chair behind the table and started waiting.

"It's ten to six," Lisa stated.

"Excited?" Nilly asked, smiling.

Lisa nodded.

When it was five minutes to six, Lisa told Nilly that it was five minutes to six. The birds were singing in the pear tree. When it was six o'clock, Lisa told Nilly it was six o'clock. And when it was two minutes past six, Lisa looked at her watch for the ninth time since six o'clock.

"Where is everyone?" she asked, worriedly.

"Relax," Nilly said. "We have to give them time to get here." He'd crossed his arms and was dangling his legs contentedly.

"It's five past," Lisa said.

Nilly didn't respond.

At ten past six, they heard Doctor Proctor grunt from the bench. And saw him blinking his eyes. And then suddenly he leaped up, exclaiming, "Good heavens! Did I oversleep?"

"Actually, no," Lisa said. "No one came."

"Yet," Nilly said. "No one has come *yet*. Just wait."

At quarter past six Doctor Proctor sighed almost inaudibly.

At twenty past six, Nilly scratched the back of his head and mumbled something about how kids these days weren't very punctual.

At twenty-five past six, Lisa put her forehead down on the tabletop. "I knew it," she whined.

At six thirty they agreed to pack up.

"Well," Doctor Proctor said, smiling sadly as they put the lid on the last box. "We'll try again another day."

"They're never going to come," Lisa said, sounding choked up. She was on the verge of tears.

"I don't get it," Nilly said, shaking his head.

"Chin up," Proctor said. "I've been inventing things no one wants for years. It's not the end of the world. The main thing is not to give up. Tomorrow I'll invent something that's even more fantastic than Doctor Proctor's Fart Powder."

"But there can't be anything more fantastic than Doctor Proctor's Fart Powder," Nilly said.

"I'm going to go home and go to bed," Lisa whispered, and started walking towards the gate in the front garden with her head down and her arms hanging at her sides.

"Good night," Nilly and Doctor Proctor said.

They sat down on the bench.

"Well," the doctor said.

"Well," Nilly said.

"Maybe I should do a little more work on that time machine I started last year," Proctor said and looked up at the swallows.

"How hard do you think it would be to invent a machine that makes jelly out of air?" Nilly asked and looked up at the swallows.

And that's what they were doing when they heard Lisa's voice from over by the gate.

"You guys . . ." she said.

"Yeah?" the doctor and Nilly said in unison.

"Someone did come," Lisa said.

"Who?"

"You kind of have to come see for yourselves," Lisa said.

Nilly and the doctor got up and went over to the gate.

"Good heavens," Doctor Proctor said, dumbfounded. "What do you say, Nilly?"

But Nilly didn't say anything, because something

extremely rare had happened to Nilly. He was speechless. He couldn't utter a single word. Outside the gate there was a line of children that reached as far as the eye could see. At any rate, as far as you could see on Cannon Avenue.

"Why are you guys so late?" asked the kid at the front of the line, a boy in a cap with the Tottenham football team's logo. "We've been standing here for over half an hour."

Then Nilly finally found his voice again.

"But . . . but why didn't you guys come in?"

"Because it says *here* on the sign, doesn't it?" the boy in the Tottenham hat said. "It says that Doctor Proctor's Fart Powder is for sale *here* and *nowhere else in the world*."

"Yeah, so?" said Nilly, confused.

"And *here* is *here*, right?" the boy said. "And not in *there*." The other kids in line behind him nodded. Then Lisa pulled a marker out of her bag, went over to the sign, drew a line through HERE and wrote THERE in capital letters.

"Then let's get to it!" she yelled so they heard her almost to the end of the line. "No pushing, let the little ones in first and have your money ready!"

THERE WAS STILL a line out there at seven o'clock when Nilly shut the gate, but they were totally out of powder.

"Sold out!" Lisa shouted and said that anyone

who hadn't been able to buy fart powder could come back tomorrow, once Doctor Proctor had made some more. And even though naturally a few people were a little disappointed, they quickly started looking forward to the next day. Because all the way down Cannon Avenue you could already hear the farts banging and the laughter from the kids who had bought the powder.

"Phew," Lisa said, flopping down into a garden chair once everyone was gone.

"Phew," Nilly said.

"You know what?" Doctor Proctor said. "We have to celebrate this. What would you guys say to a little . . ."

"Jelly!" Lisa yelled in delight.

"A three-metre-long jelly!" Nilly yelled, jumping up and down in his chair.

The doctor disappeared, but returned quickly with the longest jelly Nilly and Lisa had ever seen.

"I made this just in case," Proctor said, smiling slyly.

And as the swallows drew strange letters in the evening sky over the pear tree, silence settled over Doctor Proctor's garden. In the end all you could hear was the smacking noise of three mouths devouring a two-and-a-half-metre-long jelly.

Truls and Trym
Blast Off

WHEN LISA WALKED out her front gate the next morning Nilly was standing there with his backpack on.

"Waiting for someone who's going the same way?" Lisa asked.

"Yup," Nilly said.

Then they started walking.

"My mum and dad asked me what was going on at Doctor Proctor's yesterday," Lisa said.

"Did you tell them?" Nilly asked.

"Yeah, of course," Lisa said. "I mean, it's not a secret is it?"

"Nooooo," Nilly said hesitantly. "I just don't usually risk telling my mum about things I think are really fun. Because she almost always decides they're dangerous or naughty or something."

"She may almost always be right, you know," Lisa said.

"Yeah, that's what's so irritating," Nilly said, kicking a rock. "What did your parents say?"

"Dad said it was just fine if I earned some money of my own, then he wouldn't have to earn it for me."

"Oh? So he didn't think it was dangerous?" Nilly asked, a bit skeptical.

"A little farting? Not at all," Lisa said. They walked for a while before Lisa added, "Of course, I didn't tell him about the fartonaut powder."

Nilly nodded. "Probably just as well."

"Anyway I have an idea," Lisa said.

"Well that's definitely good," Nilly said.

"Why?"

"Because you pretty much only ever have good ideas," Nilly said.

"I was thinking that the fart powder doesn't really taste like anything," Lisa said.

"It has absolutely no taste," Nilly said.

"That's what I'm saying. I mean, the farting is fun," Lisa said. "But what if we added a flavour to it, so it tastes good when you eat it too?"

"Like I said," Nilly replied. "Only good ideas. But what kind of flavour?"

"Simple," Lisa said. "What's the best thing you've tasted recently?"

"Simple," Nilly answered. "Doctor Proctor's jelly."

"Exactly! So what we do is add five per cent essence of jelly to the fart powder."

"Brilliant!" Nilly exclaimed.

"Brillll-yant?" they heard a voice say right behind them. "Don't you think that sounds brillll-yant, Trym?"

"It sounds like gobbledygook," said another voice, which may possibly have been even closer.

Nilly and Lisa slowly turned around. They'd been so excited that they'd forgotten to stop and see if the coast was clear before they walked by the house where Trym and Truls lived. And now the two enormous boys were standing there. They were sporting big sneers, each of them chewing on a matchstick, their jaws moving up and down in their enormous, barrel-shaped heads.

"Good morning, boys," Nilly said. "Sorry, but we

have to hurry. Mrs Strobe doesn't like her geniuses to be late to class."

He tried to say it off-handedly and casually, but Lisa could hear in his voice that Nilly wasn't all that confident. He grasped Lisa by the hand and was about to pull her along after him, but Trym was blocking their way.

Truls was leaning against the picket fence, rolling the matchstick from one corner of his mouth to the other. "We didn't get any powder yesterday," he said menacingly.

"You guys must have joined the queue too late," Nilly said, and gulped. "You can try again this afternoon."

Truls laughed. "Did you hear that, Trym? Join the *queue?*"

Trym hurriedly started laughing.

"Listen up, you freckly anteater," Truls said quietly, grabbing Nilly by the collar. "We're not going to be standing in any queue or paying you anything for that fake powder of yours, you catch my drift? We want that powder right here, right now. Or else . . ." The matchstick flipped up and down in the corner of his mouth as he stared at Nilly grimly.

"Or else what?" Nilly whispered.

Truls looked like he was thinking.

"Or else what?" Lisa repeated dully.

"Come on, Truls," Trym said. "Tell them."

"Shut up!" Truls yelled. "Let me concentrate . . ." He concentrated. Then his face lit up. "Yeah, or else we'll smear honey all over you and tie you to the top of this here oak tree. Then the crows will peck you to pieces."

Truls pointed to an oak tree with a trunk that was as big around as four men the size of Lisa's father. And as big around as two men the size of Truls and Trym's father.

They all looked up.

"Oh," Nilly said.

"Oh," Lisa said.

"Uh-oh indeed," Trym said.

Because the oak tree was so tall it looked like the top branches were brushing against the white cloud that was sweeping past up in the sky.

"In that case," Nilly said, "we'll have to see if we can find some kind of a solution. If you could just let me go for a second . . ."

Truls released his grasp and Nilly started rummaging around in his pockets. When he was done with all six of the pockets he had in his trousers, he started on the six in his jacket.

Truls was getting impatient. "Well?" he said.

"I'm almost certain I have a bag here somewhere," Nilly muttered.

"We don't have time for fakers," Truls said. "Trym, get the honey and the rope."

"Wait!" Nilly yelled desperately.

"Let's get the little girl first," Truls said, grabbing Lisa by the arm.

"Here," Nilly said, holding out a bag of greyish powder. "That'll be five kroner."

"Five kroner" Truls grabbed Nilly's wrist, snatched the bag and spit his half-chewed matchstick into the palm of Nilly's hand. "Here, you can have this. Now you can go home and set yourself on fire."

"Ha, ha," Trym laughed.

Truls eyed the bag suspiciously. "What does this say here?" he said. "D-O-C-T-O-R. P-R-O-C – "

"Doctor Proctor's Fart Powder," Lisa said quickly.

"Shut up, I can read!" Truls yelled.

"Well, excuse me," Lisa said, sounding miffed.

"Hmm," Truls said.

"Hmm," Trym said.

"You first," Truls said to Trym.

"No, you first," Trym said to Truls.

"You guys could share," Nilly said.

"Shut up!" Trym yelled, almost as loudly and nastily as Truls just had.

Then they opened the bag and Truls poured exactly half into Trym's hand and half into his own hand. They looked at each other for a second and then swallowed the powder.

"They'll taste better once we add the jelly flavouring – ," Lisa started.

"Shut up!" Truls and Trym yelled, their mouths full of powder.

"Nothing's happening," Truls said, once he'd managed to swallow.

"Seven," Nilly said.

"What the heck?" Trym said.

"Six," Nilly continued. "Five."

Truls turned to look at Lisa. "What's the puny one babbling about?"

But Lisa was offended and showed with her pursed

lips and crossed arms that *she* of all people was not planning on answering.

"Four," Nilly said.

"Truls . . ." Trym said. "I can feel something happening . . . it's like . . . it's like a tickling in my stomach."

Truls scrunched up his forehead and looked down at his own stomach.

"Three," Nilly said. "Two."

"Hey, now I feel it too," Truls said. A big smile spread over his face as Nilly said, "One. Goodbye."

"Huh?" Truls and Trym said. But no one heard them. Because the only thing anyone could hear was the bang that woke up everyone on Cannon Avenue who wasn't awake already. Lisa rubbed the dust, which had blinded her, out of her eyes, but she still couldn't see anyone besides Nilly.

"Where'd they go?" she asked.

Nilly pointed his index finger towards the sky.

Lisa looked at Nilly in disbelief. "You . . . you didn't give them . . . ?"

Nilly nodded.

"The fartonaut powder? You're crazy, Nilly!" Lisa shielded her eyes, staring up into the sky.

"It was them or us," Nilly said, glancing upwards himself.

"They're gone," Lisa said.

"Vanished into thin air," Nilly said.

"I bet it'll be a long time before we see them again," Lisa said.

"Maybe never," Nilly said. "Or, wait a second."

Now that they were getting their hearing back, they were able to hear the rumble of a long fart. And a pitiful voice.

"Help!" the voice was yelling. "Help, we're falling!" It sounded like the long fart and the voice were both coming from the oak tree.

Lisa and Nilly went over to the tree and there,

way up at the top of the tree, they could see the soles of two pairs of shoes. Truls and Trym were dangling by their arms from a branch at the very top of the tree and the long fart was making all the leaves below them shake.

"Mummy!" Truls yelled.

"Daddy!" Trym yelled.

Nilly started laughing, but Lisa grabbed his arm. "We have to get them down," she said. "They might get hurt."

"Okay," Nilly said. "Let me just finish laughing first."

And then he laughed some more. And when Lisa heard that she couldn't help but laugh, too. And all the neighbours who'd been woken up by the bang now opened their windows to peer out and see what caused it. They heard three things: people yelling "Mummy" and "Daddy", people laughing and a flapping noise that reminded them vaguely of . . . but that couldn't be, could it? . . . yes, it was: a real marathon fart.

"If we're going to rescue you guys, you've got to hurry up before the fart ends and do exactly what I say!" Nilly shouted up to Truls and Trym. "Understand?"

"Just get us down!" Truls yelled.

"Grandma! And Auntie!" screamed Trym.

"Lift your legs so your bum is pointing down towards the ground and let go!" Nilly yelled. "Now, right away!"

Truls and Trym were so scared that they just did what Nilly told them. They let go. And then wafted down between the branches, pulling with them a bunch of leaves and acorns and landing rather hard in a heap in front of Lisa and Nilly.

"Well?" Nilly asked, rolling the matchstick from one corner of his mouth to the other. "Do you guys want some more?"

"N-n-no," Trym said.

"All right," Nilly said. "That'll be five kroner."

"Wh-what?" Trym said. "Did you hear that, Truls?"

But Truls hadn't heard. He was lying on his back on the pavement, staring blankly up at the sky, blinking over and over again.

Trym dug down in his trouser pocket and held out the money, which Lisa accepted.

"Well, gentlemen," Nilly said, stuffing the match-stick into his back pocket. "The clock is ticking and unfortunately Lisa and I have to get going."

Nilly and Lisa started running. They made it onto the playground just as the bell rang.

"Hey, Nilly!" It was a boy whose face was vaguely familiar to Nilly. "Cool powder! You wanna come play football at Kålløkka after school today?"

"Nilly!" someone else yelled. "Børre and I are going to come buy more farters tonight. Do you want to come over to Børre's afterwards and play PlayStation or something?"

A girl came over to Lisa. "Some friends are coming over for pizza tonight. Can you come?"

Nilly and Lisa nodded in all directions and ran towards the door to the school.

"Can you believe it, Lisa?" Nilly whispered. "We're popular. You'll see – you'll have a new best friend in no time."

Lisa nodded slowly.

As they filed into the classroom along with everyone else, she tugged Nilly's sleeve:

"Hey, Nilly, I've been thinking."

"Yeah?" Nilly said.

Lisa smiled and looked down.

Nilly wrinkled his forehead. "What is it?"

Lisa opened her mouth and was about to say something. But then it was like she changed her mind and closed her mouth again. And when she opened it again, it was like she was saying something different from what she'd been planning to say originally.

"Well I was thinking that it was strange that you happened to have a bag of fartonaut powder with you," she said. "And especially strange that it said on the bag that it was regular fart powder."

Nilly shrugged.

"You planned that didn't you?" Lisa said. "You filled one of the regular bags with fartonaut powder when we were sitting in Doctor Proctor's garden yesterday. Because you knew Truls and Trym would stop us someday and when they did, you wanted to have a bag with you so you could trick them."

Nilly just smiled in response.

"Isn't that what happened?" Lisa asked.

But just as Nilly was about to respond, they were interrupted by the loud voice of Mrs Strobe saying, "Good morning, my dear children. Take your seats and be completely quiet, please."

And then they all obeyed. Mostly, anyway.

TRULS AND TRYM didn't go to school that day. They stayed home for four good reasons. The first was that the puny devil might have come up with more dirty tricks. The second was that the other kids at school might have heard about what happened and would forget to be scared of Truls and Trym and laugh at them instead. The third was that when it came right down to it, Truls and Trym were two very lazy guys. But the fourth and most important was that they needed help thinking up a way to get revenge. Because no one was better at revenge than their father, Mr Trane. And now their big, fat father was sitting in a big, fat armchair in their big, fat home, scratching his fleshy belly. "Interesting," he said. "So this professor has a powder that can shoot a person right up into the air? Plus another powder that kids are willing to pay money for?"

"Yeah," Truls said.

"Yeah," Trym said.

"Not such dumb inventions," Mr Trane said with an evil sneer as he jabbed a stick into a cage where a frightened guinea pig was trying to get away. "I think I have a plan, boys. A plan that we can all make some money off of."

"Yippee!" cheered Truls.

"Yippee!" yelled Trym. "What's the plan?"

"A little creative borrowing," Mr Trane said.

"Awesome!" cried Truls.

"How do we start?" asked Trym.

"We start, of course . . ." Mr Trane said, grunting as he reached for the phone, "by calling the police."

A Perfect Day?

NILLY AND LISA danced home from school. It was a perfect day. It had started with Truls and Trym eating the fartonaut powder, which blasted them up into the sky. And continued with everyone wanting to be friends with them. Even Mrs Strobe had been in a good mood and when Nilly had given one of his usual unusual answers, she'd laughed so hard she cried, patted him on the head and said that it was remarkable how many strange things he had room for in there. And this afternoon Lisa, Nilly and Doctor Proctor were going to sell even more fart powder, make even more friends and eat even more jelly, and then just wait for Independence Day. So it wasn't so strange that they were dancing. Because what could go wrong?

Nothing, Nilly thought.

Nothing, Lisa thought.

Which is why they didn't give it a second thought when they noticed a police car parked on Cannon Avenue.

"See you this afternoon," Lisa chirped.

"Definitely," Nilly said, practically jumping over his front gate. He ran up the steps, opened the door and was about to go in when he caught sight of a group of people moving through the tall grass towards Doctor Proctor's front gate. There were two

OFFICER
FU MANCHU

men in police uniforms, one of them with a Fu Manchu moustache, the other with a handlebar moustache. They both looked very determined and between them they were holding Doctor Proctor, who was gesticulating and looked very agitated.

"Stop!" Nilly yelled, leaping down from the porch and running over to the fence. "Stop in the name of the law!" The group stopped and turned toward Nilly.

"We *are* the law," Mr Fu Manchu said, "not you."

"What's going on?" Nilly asked. "What do you want with the doctor?"

"He has broken the law,"

OFFICER
HANDLEBAR

Mr Handlebar said. "And we on the police force don't take that kind of thing lightly."

The doctor groaned. "They claim that I sold a deadly powder to children in the neighbourhood here. As if Doctor Proctor's Fart Powder could hurt so much as a fly!"

The two policemen escorted the professor out his gate towards the parked police car. Nilly ran after them.

"Wait!" Nilly yelled. "Who said Doctor Proctor's Fart Powder is dangerous?"

"The father of two boys who were blasted up into the sky by the powder," Mr Fu Manchu said, opening one of the car's rear doors for Doctor Proctor. "He called and said we had to arrest this crazy professor. And of course he's right. Blasting kids up into the sky like that . . . Watch your head there, Doctor."

"Go home and eat your dinner now, Nilly," Doctor Proctor said, ducking his head and taking a seat in the police car. "I'll get this misunderstanding cleared up down at the police station."

But Nilly didn't back down. "Numbskulls! Doctor Proctor didn't give them the powder that sent them up into the oak tree!"

"Numb-what?" Mr Handlebar said gruffly.

"Well then who *did* give them the powder?" Mr Fu Manchu asked.

"I did," Nilly said, standing in front of them resolutely, his hands on his hips.

First the two policemen looked at Nilly, then at each other and then they both started laughing.

"A tiny little guy like you?" Mr Handlebar laughed. "We're supposed to believe that a pip-squeak like you had something to do with such a serious crime?"

"Well," Nilly said, puffing himself up like a frog, "if you'd been paying attention, then those sharp police brains of yours would have noticed that I said the powder had sent them up into an oak tree. You didn't say anything about a tree, so how else would I have known about that?"

"Hmm," Mr Fu Manchu said, raised his police cap and scratched the bald head he had underneath. "You've got a point. So how did you know that?"

"Because I'm telling you what happened!" Nilly hissed. "I'm the one who sold them the powder. And not the normal, harmless powder that Doctor Proctor sells. No sirree, gentlemen . . ."

Nilly took a deep breath and started a very long sentence: "I sold them Doctor Proctor's Fartonaut Powder, even though the doctor himself decided that it should only be sold to NASA, since the powder is a hyper-explosive special formula that can just be consumed in very small doses by people with at least four years of astronaut training, and even then they need to be wearing padding and be

under the supervision of at least two adults!" As Nilly spoke, he got madder and madder and now he was jumping up and down.

"Hmm," Mr Handlebar said. "And who made this . . . uh, fartonaut powder?"

"I did," sighed Doctor Proctor from inside the car.

"But it's my fault that Truls and Trym got hold of it," Nilly said.

"Hmm," Mr Fu Manchu said. "I don't see any way out of this other than to arrest you both. Do you?"

"I agree," Mr Handlebar said.

And that's how both Doctor Proctor and Nilly got arrested on the day that, until that moment, had seemed like it was going to be perfect.

Three Fishy Fellas with a Plan

WHEN LISA'S DAD came home that day, Lisa was sitting under the apple tree that had no apples.

"I'm so relieved," growled the Commandant, wiping the sweat off his brow. "We thought the whole Independence Day celebration was going to be ruined. You know, Lisa, we've been looking for the special gunpowder for the Big and Almost World-Famous Royal Salute for several days. We were starting to think they'd forgotten to load it on the ship over in Shanghai. But it turns out it was the first thing they loaded onboard, so it's all the way at the bottom. They're going to bring it ashore tomorrow. Phew, imagine what a catastrophe it would have been if the gunpowder hadn't come!"

Only now did he notice that Lisa was hardly paying attention. She was sitting there under the tree with her head in her hands, looking down-hearted.

"Is there something wrong, pal?" he growled.

"Something terrible happened," Lisa said glumly. "They arrested Nilly and Doctor Proctor. Just because Truls and Trym ate a little fartonaut powder."

"I know," the Commandant said.

"You know? How did you find out?"

"Because the police asked if Nilly and Doctor

Proctor could be kept in the most escape-proof cell in all of northern Europe, apart from Finland. And that's where they are."

"You mean . . . you mean . . ." Lisa began, frightened.

"Yup," her dad said. "They're in the Dungeon of the Dead."

"The Dungeon of the Dead! But Nilly and the professor aren't the least bit dangerous!"

"Well, the police don't agree. Mr Trane explained to the police that the professor is a raving lunatic who'll invent an atom bomb if he isn't locked up immediately."

"Mr Trane? And they believed him?"

"Of course they believed Mr Trane," grumbled the Commandant. "After all he's the one who helped us invent the hardest and most secret material in the world. Which is used in the doors of the most escape-proof cell in the world . . ."

"Yeah, yeah, Dad, I've heard of all that," Lisa sighed. "But what do we do now?"

"Now?" The Commandant noisily sniffed the aroma coming through the open kitchen window. "Eat Wiener schnitzel — at least that's what it smells like. Come on."

AS LISA WENT inside the scent of Wiener schnitzel wafted out over the garden, where a light breeze caught it and carried the scent over Cannon Avenue, down to the fjord, to Akershus Fortress, in over the high stone walls and then past the towers and the black, old-fashioned cannons that were aimed out over the fjord. The guards standing outside the Dungeon of the Dead inhaled the scent without noticing it and the part they hadn't inhaled continued in through the bars to a corridor that led to a stone stairway going deep down, down to a *very* thick and *very* locked iron door.

An exceedingly small amount of schnitzel scent seeped through the keyhole into a room that was shaped like the inside of a cannonball. A bridge ran across the centre of the room and led to another iron door, even thicker and even more locked than the first. And with a keyhole so narrow that only a couple of Wiener schnitzel gas molecules made it into the corridor behind. The darkness in that corridor was penetrated only by laser beams that ran back and forth, up and down. The grid of laser beams was so dense that not even a tiny *Rattus norvegicus* could hope to sneak through without triggering the alarm. And the alarm was connected to the guard-room, where the guard on duty was stationed. And also to the main panel at police headquarters. And also to the command centre for the Norwegian anti-terrorism police. And to the command-command centre for the anti-anti-terrorism police. And I'm sure you can understand, triggering an alarm like that would result in a lot of running and yelling and maybe shooting, and definitely the rather rapid arrest of the little rat or spider that was trying to do something so foolish as to break out of the Dungeon of the Dead.

At the far end of the corridor – and by now there was hardly any scent left – was the final door. It was made out of a material that hardly anyone knows exists, but that is so hard, so ingeniously invented and so secret that the author of this book had to promise the Norwegian government that he wouldn't say anything else about the material in this story. The point – as you may already have surmised – is that the Dungeon of the Dead is absolutely impossible to escape from.

And there, behind that last door, sat Doctor Proctor and Nilly. The walls and ceiling were white, windowless and sort of rounded, so it made them feel like they were sitting inside an egg. Each one was sitting on a bed on either side of the egg cell, which was lit by a single lightbulb that hung from the ceiling. There was a small table between the beds, and a toilet and a sink that were both attached to the wall, and a bookshelf with one single book on it: *King Olav – The People's King*. Nilly had already read it four times. The book had a lot of pictures in it and Nilly had gathered from the text that the best thing about Olav was how good-natured he'd been. But there's a limit to how many times anyone sitting in jail wants to read a book about being good-natured. And not just any old jail, but the most escape-proof jail in all of northern Europe, aside from Finland.

As Nilly read, Doctor Proctor scribbled and sketched something on a scrap of paper he had had in his pocket, scratched his head with the pencil stub, mumbled a few passages in Greek and then scribbled some more. He was so absorbed in his work that he didn't notice Nilly sighing loudly several times, trying to draw his attention to how boring it was for a boy like Nilly to be locked up in a place like the Dungeon of the Dead for as long as it had been. Then all of a sudden Nilly stuck his nose up in the air and sniffed. "Do you smell that, Doctor?"

The doctor stopped and sniffed. "Nonsense. There's nothing to smell."

"For those of us with sensitive noses, there is," Nilly said, concentrating. "Hmm. Could it be French bread? No, further east. Goulash? Further south. Wiener schnitzel? Yes, I really think it must be. Fried in margarine."

Right when Nilly said "margarine" he noticed the doctor's shoulders sink and a sad look come over his face. Nilly hopped up onto the doctor's bed and peeked over his shoulder at what he'd been sketching.

"Nice drawing, boring colours," Nilly said. "What is it?"

"An invention," Doctor Proctor said. "A break-out-of-northern-Europe's-most-escape-proof-jail machine. With probability calculations for its chances of working."

"And what do your calculations tell you?"

"Do you see that number?" the doctor asked, pointing to a number that was underlined twice.

"Yes," Nilly said. "That's a zero."

"That means the probability of escape is zero. We're doomed."

"Don't worry," Nilly said. "They'll come let us out soon. Once they've done a little more investigating and found out that the fart powder is basically harmless."

"I don't think so," the doctor said gloomily, rolling up his scrap paper.

"You don't?" Nilly responded. "Sure they will!"

"I wish they would," the doctor said, tossing his papers at the toilet but missing. "I didn't want to mention this before, but when they were questioning me, the police made it pretty clear just what a pickle we're in."

"Why? What did they say, exactly?"

"They said, 'We can't send that little guy named Nilly to jail because he's a kid, but he's looking at at least a year in juvie.'"

"Well, jeez, that wouldn't be so bad," Nilly said. "Maybe that would at least be a place with a band where I could finally do a little trumpet playing. What else did they say?"

Doctor Proctor thought about it, cleared his throat, and continued: "'And you, Doctor, since you're an adult, will be sentenced to up to twelve years behind these walls — or some other walls — and never be allowed to invent anything ever again. Got it?'"

"Yikes," Nilly said. "That's worse."

"A lot worse," Doctor Proctor said. "I can't even bear to think about any part of it — not the twelve years, not the walls and definitely never being allowed to invent anything. I have to escape."

"Hmm," Nilly said. "Where to?"

"To France. I have to find Juliette Margarine. She'll help me, hide me from the police, give me shelter. And Brie. And red wine."

"But how?"

"On my motorcycle, of course. It just needs a little lubrication and then it'll run like, uh, well, like it's been lubricated."

"But how do we get you out of here?"

"I have no clue . . . or, wait a minute!" Doctor Proctor looked like he was lost in thought. "Maybe I made a slight mathematical error. . . ." He leaped up and snatched the crumpled papers off the floor, opened them up, smoothed them out with his hand, let his eyes run up and down the pages, mumbled something and started immediately scribbling and calculating things again. Nilly watched anxiously. Right up until the doctor crumpled the pages up again, threw them over his shoulder and started banging his forehead against the top of the little table.

"It's no use!" he sobbed, covering his head with his arms. "I never make mathematical errors!"

"Hmm," Nilly said, placing his index finger thoughtfully on his chin. "This doesn't look good."

"It looks terrible!" Doctor Proctor yelled. "What are we going to do now?"

"Now?" asked Nilly, who heard the sound of keys rattling and sniffed the air. "It smells like we're going to eat fish cakes."

AFTER DINNER LISA went out into the garden. She needed to think. So she sat down in the grass under the apple tree that had no apples and rested her head in her hands. But the only thoughts that came to her were that the Dungeon of the Dead was completely escape-proof and that Nilly and Doctor Proctor were goners. She burped a Wiener schnitzel burp and mostly felt like crying. So she cried a little and, as usual, crying made her very sleepy, so she yawned a little. And the afternoon sun shone on Lisa and a bird sat on an apple tree branch and sang. But Lisa didn't notice any of it, because she'd fallen asleep. And when something woke her up, it wasn't the birdsong, but voices. The voices were coming from the other side of the fence. There were some people standing in the street talking.

"See that rickety old cellar door there," whispered an adult voice she recognised. "I'm sure it's locked, but you boys won't have any trouble dealing with that."

"Yeah, no problem," said a voice that was even more familiar. "We'll just use a crowbar and pry it open."

"A break-in!" said a third voice and Lisa knew exactly whose it was. "How fun!"

She stood up and peeked cautiously over the fence. And there she saw the backs of three people who were peering cautiously at Doctor Proctor's house.

"Good attitude, boys," whispered Mr Trane's voice.

"And once you're in the doctor's cellar, grab all the fart powder and fartonaut powder you find. Got it?"

"Yes, Dad," Truls said.

"Yes, Dad," Trym said.

"And then, boys, you can sell the fart powder to the kids at school."

Suddenly they turned around, but Lisa was faster and ducked.

"What about the fartonaut powder, Dad?" Truls asked.

"Heh, heh," Mr Trane laughed. "I've already talked to someone in the U.S., in Houston, who's very interested in an invention that can send people right into space without having to build a rocket."

"Who? Who did you talk to, Dad," Trym asked.

"NASA, you idiot," Mr Trane said. "Once we get our hands on the powder, I'm going straight down to the patent office to patent the fartonaut powder. And then, too bad, mister nitwit professor, I'll be the only one who can sell the powder. I'm going to be a millionaire, boys!"

"Aren't you already a millionaire, Dad?"

"Well, sure. But with a few more million, I can buy another Hummer. And an indoor swimming pool. What do you say to that?"

"Oh, yeah, Dad!" Truls and Trym shouted in unison.

"Okay," their father said. "Now we know how it's going to go down. We'll get the crowbar and ski masks and then tomorrow night, we strike! Heh, heh, heh."

Lisa sat motionless, listening as she heard Mr Trane's laughter and all of their footsteps fade into the distance. Then she leaped up and ran inside.

"Dad, Dad!" she shouted.

"What is it, Lisa?" rumbled the Commandant, who was lying on the sofa reading the paper.

She hurriedly told her dad about how she'd been woken up from her sleep and had overhead the Trane family's plans. But as she was talking, a smile spread across the Commandant's face.

"What is it?" she cried when she was done. "Don't you believe me?"

"You never lie, Lisa dear," the Commandant said, chuckling. "But don't you see that you just dreamed it all while you were sleeping, that you weren't actually awake? Mr Trane and his family, breaking into the doctor's house and stealing his invention?" The Commandant laughed so hard he shook. "Can you imagine?"

Lisa slowly realised that if even her own father didn't believe her, who would? Who could help her? And the answer was just as clear: no one. No one except herself.

The sun had just set and tomorrow, Doctor Proctor's Fart Powder would be in the hands of those three fishy fellas. And Lisa was the only one who knew it.

The Dungeon
of the Dead

THAT NIGHT A sound woke Nilly up. He pushed himself up and leaned on his elbow. In the darkness he could hear Doctor Proctor snoring from the other bed. But Nilly knew that wasn't the sound that had woken him up. For a second he wondered if maybe it wasn't the rumbling from his own stomach, because they hadn't had anything to eat since those measly fish balls. But he dismissed that thought. Because Nilly had the distinct sense that he and Doctor Proctor were no longer alone in the cell. . . .

He stared into the darkness.

And all he saw was darkness.

But in the sole strip of light that filtered in through the keyhole, he suddenly saw something. A glimpse of white and fairly sharp teeth. Then they were gone again.

"Hey there!" Nilly yelled, throwing off his wool blanket, jumping off the bed, and running over to the door, where he flipped on the light switch.

Doctor Proctor's snoring had stopped and when Nilly turned around he saw the professor standing on his bed in just his underwear, as white in the face as he was on the rest of his body and pointing at the animal that was now visible.

"It's — it's — it's a . . ." the professor stammered.

"I see what it is," Nilly said.

"I – I – I'm scared to death of – of – of . . ." said the doctor.

"Of that animal there?" Nilly asked.

The doctor nodded, pressing his quaking body against the wall. "Look at the tee-tee-teeth on that beast."

"Beast?" Nilly said, squatting down in front of the animal. "This is a *Rattus norvegicus*, Doctor. A friendly, little *Rattus norvegicus* who's *smiling*. Sure, it's mentioned in a footnote in *Animals You Wish Didn't Exist* by W. M. Poschi, but that's just because it spreads the Black Death and other harmless diseases."

The rat blinked at Nilly with brown rat eyes.

"I can't help it," Doctor Proctor said. "Rats give me the shi-shi-shivers. Where did it come from? How did it get in here?"

SNIFF
SNIFF

RATTUS NORVEGICUS

"Good question," Nilly said, scratching his head and looking around. "Tell me, Doctor, are you thinking what I'm thinking?"

Doctor Proctor stared at Nilly. "I – I – I think so."

"And what are we thinking?"

"We're thinking," said the doctor, totally forgetting to be afraid and hopping down onto the floor and pulling on his professor coat, "that if it's possible to get into a place, it must be possible to get back out of the same place."

"Exactly," Nilly said, holding out a finger that the little rat sniffed out of curiosity. "So I recommend that we pay close attention to our rat friend when he heads home."

The Great Escape

NILLY'S SISTER ANSWERED the door when Lisa rang the doorbell of the yellow house the next morning. Eva gazed at Lisa with her narrow, kind-of-evil eyes, which glowed just as angrily as the two new spots she had on her face and said in a taunting, squeaky voice, "Nilly's not here, Flatu-Lisa."

"I know," Lisa said. "He's in jail."

Eva's eyes got big. "In jail?"

"Yup. In the Dungeon of the Dead."

"Mum!" Eva yelled over her shoulder. "Nilly's in jail!"

They heard someone rummage around, drop several things, fall over and maybe swear a little.

"Haven't you wondered why you haven't seen him for twenty-four hours?" Lisa asked.

Eva shrugged. "It's not easy to spot something so tiny, so I don't think it's that strange if I don't see him for a few days, you know? It's kind of a nice break."

"Well anyway," Lisa said, "prisoners in the Dungeon of the Dead are only allowed to receive visits from people in their immediate family, so I was wondering if you could give him this letter." She held out an envelope.

"We'll have to see," Eva said, snatching the letter. "If we have time."

NILLY AND DOCTOR Proctor were both lying on the floor of the cell, snoring and sleeping, when they were shaken awake by a member of the Royal Guard serving as a prison attendant in a black uniform and hat with a big, silly tassel on it.

"Huh? We must have fallen asleep watching the rat," Nilly said, rubbing his eyes.

"Visitor for prisoner number 000002," the guard said gruffly.

"Is that me?" Nilly asked, still half-asleep. "Or is that him?"

"It's you," the guard said. "He's prisoner number 000001."

Nilly looked around and said, "Where? Where?"

"That guy, right there," the guard said, irritated, pointing at the professor, who was still snoring softly.

"No, not him!" Nilly shouted. "The rat! Did a rat run out the doorway when you came in?"

"Not that I saw," the guard said. "Look, do you want your visitor or not?"

Nilly followed the guard through all the thick, but now open doors, down the corridor where the laser beams had been turned off, over the bridge, up the stairs, through the open door with the metal bars, and into the visiting room. And there was Eva, sitting in a chair chewing gum.

"Hi," Nilly said, surprised, and smiled at his sister. "How nice that you wanted to come visit me."

"As if," Eva said. "I didn't want to. Mum sent me. She didn't feel like she was quite up to a prison visit herself. I brought you a letter. From that weird neighbour girl."

"Lisa?" Nilly said, lighting up and taking the envelope. He could tell right away that it had been opened. "Well, what did she say?" he asked bitterly.

"How should I know?" Eva asked innocently.

Nilly read the letter silently and put it in his pocket.

"What's NASA?" Eva asked.

"Anything else new?" Nilly asked.

Eva snorted and stood up. "I've got to go to school. Have a nice day in jail."

Once Nilly was safely back under lock and lock and lock and key with the professor, he passed him the letter. Doctor Proctor read aloud:

Bad news. The Trane family is going to break into the professor's cellar tonight, steal the fartonaut powder, patent it and sell the invention to NASA. We have to do something. Lisa

"This is hopeless," the professor blurted out. "They're going to rob me! Steal my invention."

"Lisa's right," Nilly said. "We have to do something. We have to get out of here."

"But how?" the professor asked. "The rat is gone. We don't know how it got out."

"Well," Nilly said, "give me the letter. We'll flush it down the toilet so no one finds out that Lisa's working with us. Otherwise they'll put her in jail too."

Nilly crumpled up the letter, tossed it into the toilet and flushed. The toilet made a long, loud gurgling sound, the paper disappeared and then the toilet bowl filled back up with water. Nilly stood there thoughtfully watching the ripples in the bowl where the paper had just been and scratching his scalp through his red hair. And what he was thinking about was how the letter was being carried down through the pipes by the water. Down and down. Until it splashed down into a bigger sewer pipe somewhere way down below them. A sewer pipe that must surely stink and be teeming with . . .

"You know what?" Nilly said. "I think I just figured out where our rat friend went."

"Really?" the professor said.

Nilly pointed down into the toilet.

"It swam up here through the pipes from the sewer. And went back out the same way."

"Pyew!" the professor said, holding his nose.

"Maybe," Nilly said. "But from the sewer pipe, the water keeps going. And going. All the way until it gets to the ocean. Or maybe to a treatment plant. And along the way there are ladders up to the street above, to manhole covers that lead right out onto the streets of Oslo. Do you get where I'm going with this, Professor?"

The professor, who clearly got where Nilly was going, stared at him in disbelief. "You must be crazy!" he exclaimed.

"Not crazy," Nilly laughed. "Just very smart. And

very, very small. We can only hope that I'm small enough."

"You can't!" Doctor Proctor said. "You mustn't!"

"I can, I must and I will," Nilly said.

"The guards look in here all the time — they'll notice that you're gone."

"We'll wait until early evening," Nilly said. "Then we act like we're going to bed early and turn off the light. And then in the dead of night . . ."

THE SUN DRIFTED across the sky and its rays fell on an Oslo that had started preparing for Independence Day, which was only two days away. People were cleaning up their houses and planting flowers in window boxes, ironing flags and the aprons that went with their national costumes, reviewing traditional eggnog recipes and humming the national anthem. And as the sun began to descend towards Ullern

Ridge at the western edge of the city, the men at the wharf carried the last of the crates off the ship from Shanghai.

The rays that penetrated between the planks of the wharf reflected off some seashells. And not just the kinds of shells that are attached to wharf pilings and the rocks that are only visible at low tide. But shells that moved. Shells that were black and attached to the back of something slithering out of the dark opening of a sewer pipe. Shells on the back of something that hadn't eaten anything since the leathery meat on that thirty-five-year-old Mongolian water vole a few days ago . . .

The creature slides through the water. It hears the wharf planks creaking. Sees the soles of a pair of boots. *Food*. It's a man carrying a wooden crate. The creature quickly twists its way up around one of the wharf pilings, up into the blinding sunlight and rises, swaying above the poor guy, and it hears the footsteps on the wharf stop. The creature opens its jaw, the sun shines on its gruesome fangs and it hears a scream. *Yes, yes, this is how food sounds*, it thinks.

The creature gets ready for a bulky mouthful. But the afternoon sun is so low and still so glaring and the creature hasn't seen any light in days. So it strikes blindly, grabs hold of something, seizes it and swiftly vanishes into the water. And then into the sewer pipe. *Food!* The creature can already feel its digestive juices starting to flow from glands throughout its body as it swims its way back into the Oslo sewer system. And then, deep in the sewers, in a strip of light that falls from a little hole for runoff water on a manhole cover in a street way up above, it stops to really enjoy its meal. But . . . what is this? Wood taste? The creature spits the food out. And it isn't food at all. It's a wooden crate. The creature fumes with rage. Blast it! Doggonit! How aggravating!

But then the creature hears something. An echo

from a squeak within the sewer system. A rat squeak? *Rattus norvegicus. Food!* And *whoosh*, the starving creature is swallowed up by the darkness of the sewer, on the hunt again. Leaving the wooden crate floating there, bobbing up and down in the sewer water. And in the strip of light from the manhole cover, one can read the following printed on the lid in red letters: CAUTION! HIGHLY EXPLOSIVE SPECIAL GUNPOWDER FROM SHANGHAI FOR THE BIG AND ALMOST WORLD-FAMOUS ROYAL SALUTE AT AKERSHUS FORTRESS.

THE SUN SANK even further towards Ullern Ridge and started to slip behind it. The last rays cast long, white fingers over the landscape, as if the sun were desperately trying to hang on. And the rays reached all the way to Cannon Avenue. But it lost its hold and then the sun was gone.

It was evening. Truls and Trym stood in one of their three garages on Cannon Avenue, watching

Mr Trane, who had pulled a black crowbar out of the toolbox in his black Hummer. He had already given each of them a ski mask, which would cover their whole heads and faces apart from their eyes and mouth, so they could see and breathe and talk a little. Nice when it's really cold out. Or when you're going to commit a robbery. Because even if someone sees you during the robbery, they're guaranteed not to recognise you afterwards. Unless you're still wearing the ski mask of course.

"Like so," Mr Trane demonstrated, sliding the crowbar in along the edge of a door. "And so and then so."

"Like this," Truls and Trym repeated through their ski masks. "And this and then this."

They repeated and repeated and practised and practised the break-in. But it took some time, because Truls and Trym weren't the smartest boys in the world. And not just not the smartest boys in the world, actually. They were also not the smartest boys in Norway, or the smartest boys in Oslo, or even the smartest boys on Cannon Avenue. Because at that very moment the smartest boy on Cannon Avenue was sitting on a bed in the Dungeon of the Dead, feeling nervous. More nervous than he'd ever been before. Yes, so nervous that he bordered on being scared. And scared was something that Nilly, prisoner number 000002, very rarely was.

"What are you doing?" he asked Doctor Proctor, who'd taken off his professor's coat, turned the pockets inside out and was now carefully brushing the pocket lining over one of his scraps of paper.

"I was thinking," the professor said. "It's going to be awfully dark when you get down there. And you don't have a torch. Then I remembered that there is always residue in my pockets from some of the various powders I've invented. And voilà . . ."

Nilly came over and looked down at the sheet of

paper, where there was a fine layer of light-green powder.

"I've seen that before," Nilly said. "That's Doctor Proctor's Light-Green Powder. You had it in a jar in your cellar. You said it was a phosphorescent powder that makes you glow. And that it was a rather unsuccessful invention."

"Maybe it isn't so unsuccessful after all," the professor said, carefully folding the piece of paper in half so that all the powder slid into the fold. "Open wide!"

With Nilly's mouth open as wide as it would go, the professor poured the powder into the small opening.

"It'll take a little while before it starts working," the professor said. "And meanwhile . . . ." He intensely brushed out the other coat pocket over the sheet of paper.

"Is that what I think it is?" Nilly asked when he

spotted the small, light-blue grains sitting on the professor's mathematical calculations.

"Yup," the professor said. "It's fartonaut powder. Too bad I don't have more here."

"But what do I do with it?"

"The exits to the sewer system are blocked by manhole covers," the professor said. "And they're heavy and hard to move. If you need to get out, you should – "

"Fart one of them up into the sky!" cried the smartest boy on Cannon Avenue.

The professor nodded and poured the fartonaut powder into the envelope that Lisa's letter had come in. "But there's only enough here for one good fart, so don't waste it."

"I won't," Nilly said, folding up the envelope and stuffing it into his trouser pocket.

The professor studied him for a moment. "Your face is green. Are you feeling sick?"

"No," Nilly said, surprised. "Just a little . . . uh, nervous."

"Good, then it's the glowing powder starting to work. Quick, we'd better act now before it stops working."

The professor went over to the door and put his finger on the light switch. He hesitated.

"Come on," Nilly said.

The professor sighed and turned off the light and it got pretty dark. But not completely dark. Because in front of him Nilly could see a glimmering green light, he just couldn't see where it was coming from. Until he looked down at himself.

"Hey!" he yelled. "You can see right through me! I can see my own skeleton!"

"And you're glowing," the professor said. "You're your own torch. Now hurry up!"

Nilly crawled up onto the rim of the toilet and hopped down into the water, making a splash.

"Brr," he said.

"Ready?" the professor asked, looking down at the tiny little, and now phosphorescent, boy who was treading water in the toilet.

"Ready," Nilly said.

"Take a deep breath and hold it," the professor said.

"Roger!" Nilly said, taking a breath and pinching his nose.

And with that, the professor flushed. The toilet gurgled and spluttered and sloshed. And then it turned into a steady rushing noise and the professor peered into the toilet and Nilly wasn't there anymore.

Life in the Sewers

NILLY WAS IN a free fall. He had once tried the waterslide at some water park or other, but this was totally different. His body whooshed like a torpedo towards the centre of the earth until a bend in the pipe flung him to the left. And then to the right. And then straight down again. He felt like a cowboy riding a wild horse of water, and he couldn't help himself — he had to yell, "Yee ha!"

The pipes were exactly big enough with exactly enough water to soften all the falls and turns. He was carried further and further down and, although it was getting both darker and colder, he was having so much fun and things were glowing so green around him that he wasn't thinking about being wet or freezing cold. And he realised why that rat in their cell had swum and scrambled all the way up into their toilet: this was the roller-coaster ride of a lifetime!

It felt like such an adrenaline rush in his stomach every time the pipes turned and Nilly plunged into a new free fall that he hoped the ride would never end. But it did have to, of course. And it did. Rather suddenly too. The walls of the narrow pipe disappeared and he was stretched out in the air as flat as a pancake and saw something black approaching with alarming speed. Then the black thing hit him.

Or to be more precise, Nilly hit the black thing. No one has ever witnessed an uglier belly flop in the Oslo sewer system. Brown, slimy goo sloshed up against the walls. And, boy, did it sting! Nilly felt like he was lying facedown in a frying pan.

He stood up and discovered that the water only came up to his waist. He looked around. Besides the glimmering green light coming from him, it was pitch-black. And once the sloshing had subsided, it was completely quiet too. But, yuck, it stank! It smelled so bad that the author advises you to do the same thing Nilly realised he had to do: Stop thinking about it.

What was I thinking about? Nilly wondered, since he wasn't thinking about the smell anymore. *Right, that I need to find a manhole cover.* And with that Nilly started wading through the sewer system looking for a way out.

Unfortunately it's not as easy as you might think to find a manhole cover in a sewer system after the sun has set. The reason being that the sun is no longer shining through the manhole covers' small holes that are designed to let water in from the street. And although Nilly was glowing the light didn't reach far enough to illuminate the shafts above him. But he didn't give up.

AFTER NILLY HAD waded for a long time and quite some distance, he heard a hissing sound. And he thought the hissing sound must be coming from a manhole cover. Because obviously hissing sounds don't come from the sewer. *Who on earth would be hissing down here?* he thought.

But he wasn't totally sure and noticed that as he approached the location where he thought the sound had come from, his heart beat faster. A lot faster . . .

And as he rounded a corner, he froze and stood still. Completely still. Actually stiller than he'd ever stood before.

Because he thought he'd seen something.

Something that had gleamed at the very edge of the circle of green light. A row of much too white, much too sharp and, most of all, much too big teeth. Because teeth that big and that sharp should not be down in the Oslo sewer system. They should only be found in the Amazon River and thereabouts. Or in a dreadful picture on page 121 of *Animals You Wish Didn't Exist*. More specifically, in the mouth of the world's largest and most feared constrictor. The anaconda.

It had been a long time since Nilly had read the anaconda chapter in the thick, old book from his grandfather, but now he could clearly picture every single dusty word. And Nilly realised that he was in trouble. First of all, because he was standing up to his waist in what, according to his grandfather's book, was the anaconda's favourite element: water. Not very clean water, but water all the same. Secondly, because Nilly was probably the most visible thing in the Oslo sewer world right about now: a transparent, glowing, green boy. And thirdly, because even if he hadn't been a glowing larva, there still wouldn't have been anywhere to hide.

So he kept standing there. And there was that hissing noise again. And there were those teeth gleaming in the light again. And they were attached to the biggest mouth he'd ever seen. On each side of the mouth, an evil anaconda eye was staring at him, and in the middle of the mouth, a split red anaconda tongue was vibrating. And Nilly had to admit that even the dreadful picture on page 121 didn't do

the creature justice. Because this was much, much worse and way creepier. The mouth came towards him relentlessly.

AND NOW AS Nilly is about to be eaten, maybe you hope that something will happen at the last minute, something completely unlikely, the kind of thing that never happens anywhere besides in stories just as the hero is about to meet his demise. But nothing like that happened. All that happened was that Nilly slid right down the gullet of the giant snake, glowing all the way. And only two days before Independence Day.

A FULL MOON hid behind a cloud over Cannon Avenue as if it didn't dare watch. Truls and Trym stood by the fence to Doctor Proctor's garden.

"Breaking in is fun," Truls whispered.

"Breaking in is fun," Trym whispered.

But even though they were whispering they still

made too much noise. The moon emerged from the clouds and cast shadows that ran across the over-grown garden like big men in hats and capes.

"Maybe I should stand watch out here while you go in and get the fart powder?" Truls suggested.

"Shut up," Trym said, staring at the crooked wooden house in front of them, which didn't have any lights on. The house that was so small in the daylight seemed enormous in the dark.

"Are you a tiny bit scared?" Truls asked.

"Nope," Trym said. "You?"

"No way. Just wondering if you were."

"Come on," Trym said, and climbed over the fence. When they were on the inside, they stood still and listened. But all they could hear were a couple of grasshoppers that had lost track of the time and the wind rustling in the pear tree and making the walls of the house creak and groan like an old man telling dusty old ghost stories.

They waded through the grass towards the house. Truls could hear his own heartbeat. And maybe Trym's too. When they got to the cellar door, Trym held the crowbar up.

"Wait!" Truls whispered. "Check if it's locked first."

"You idiot," Trym hissed. "You don't think he'd be so stupid that he would keep a fortune's worth of fart powder in an unlocked cellar, do you?"

"Who knows?"

"You want to bet on it?"

"I'll bet you a bag of fart powder."

"Okay."

Truls pulled down on the door handle and tugged. And do you know what? It turned out that the door was actually . . . was actually . . . *locked*! What did you think? That someone would be so stupid that he would keep a fortune's worth of fart powder in an unlocked cellar?

"Darn it," Truls said.

"Hurray," said Trym, pressing the tip of the crowbar inbetween the door and the frame and pushing on the other end.

It creaked a little. It creaked a little more.

"Wait!" Truls said.

"Not again," Trym groaned.

"Look at the window."

Truls looked at the window. And then eased up on the crowbar.

"Broken," he said. "Must have been some pranksters throwing rocks."

"Or some rotten sneaky thieves who beat us to it."

They climbed in through the window and turned on their torches.

The cones of light from their torches slid over all kinds of strange equipment, test tubes, barrels, drums, tubes, glass containers and an old motorcycle with a sidecar. And stopped on two enormous jars.

"The powder!" Truls whispered.

They moved closer and shone the light on the labels. The writing on them was the kind of swoopy lettering Mrs Strobe had tried to teach them, but that neither Truls nor Trym had really got the hang of.

"*Doctor Proctor's Totally Normal Fart Powder*," Truls read one with difficulty.

"*Fartonaut Powder*," Trym read the other one. "*Keep out of reach of children.*"

"Heh, heh," Truls laughed.

"Ho, ho," Trym laughed. "This'll make Dad happy."

"And then we'll get a swimming pool. Come on, bro."

With that they each grabbed a jar and snuck back out the same way they'd come in. And only the moon saw them as it timidly peeked out from between the hurrying clouds.

And maybe one person in the red house across the street. At any rate, the curtains in one of the windows on the first floor moved a little.

The Even Greater Escape

THE SUN CAME up over Oslo and Akershus Fortress. And there was a great commotion there.

"What do you mean," growled the Commandant, "the gunpowder from Shanghai is *missing*?"

"It disappeared while we were unloading it onto the wharf yesterday afternoon, sir," said the steadfast, but obviously nervous, guardsman in front of him.

"Disappeared? How is that possible?"

"The longshoreman swears it was eaten by a big snake, sir."

The Commandant's growl made the window-panes in his office rattle. "Are you trying to convince me that some snake ate the whole crate of gunpowder?"

"No, sir. The longshoreman is trying to convince me of what I'm trying to convince you, sir."

The Commandant's face was now so red and his stomach so inflated that the guardsman was afraid he might explode at any moment. "Excuses! That butterfingers dropped the crate in the water! Do you know what this means, my dear guardsman?"

And the guardsman knew what it meant. It meant that for the first time in over a hundred years, there wouldn't be any Royal Salute. People from Strømstad, just across the border in Sweden, to Poland, and yes, even all the way to Madagascar, would scoff at their little country way up north, make fun of them and call them things that rhyme with Norway. Gorway and borway and sporway and things that might not sound so bad in English, but that could mean really preposterous things in Madagascarian.

"What do we do now?" asked the guardsman.

And like a big red balloon that suddenly popped, the Commandant sunk down into his chair, thumping his forehead against his desk, and then stopped moving. He tried to say something, but his lips were squashed against the top of the desk so it was impossible to understand him.

"Um, what?" asked the guardsman.

The Commandant raised his head off the desk. "I said, I don't know."

BUT THE SUN kept shining and smiling as if nothing had happened. And it really shouldn't have on a day like this. Because let's review the situation, dear reader. The Commandant's gunpowder is missing. Nilly has been eaten. Doctor Proctor is in jail. And his powder has been stolen by the evil Trane family.

So why does Lisa seem both happy and unconcerned as she plays her clarinet and marches down the streets of the city in the Dølgen School Marching Band at the crack of dawn on the day before Independence Day? Could she have forgotten all of their problems? Is she maybe not who we thought she was? Does she actually not care about her friends at all? Or does she know something we don't?

Perhaps, but we also know something she doesn't know. We know that Nilly was eaten by an anaconda. And the only other one, aside from us and the snake, who knows that is Nilly himself.

I'VE BEEN EATEN by an anaconda, Nilly thought as he sat there in the darkness inside a snake's body that was moving and slithering and dripping from the ceiling and walls. He was still sore from having been kneaded through the snake's jaws and throat, but there was more room in here and he was still more or less in one piece. But, of course, that was just a matter of time. Because he knew from page 121 in *Animals You Wish Didn't Exist* that the stuff dripping on him was a highly corrosive blend of digestive juices. And that in time it would dissolve Nilly's body into its individual components. As it had done to the poor thing that had owned the metal collar Nilly had found when he'd ended up in here the night before.

Nilly'd just had time to read the name engraved on the collar before his phosphorescent powder had stopped working: *Attila*. That was all that was left of the poor thing. The digestive juices had already started eating away at the soles of Nilly's shoes and the scent of burning rubber stung his nostrils. There was little doubt that he was facing a slow and rather gruesome death. There was also little doubt that his hopes that the constrictor would sneeze or hiccup him back out were dwindling rapidly. But there was no doubt that he had to think of something, and he had to do it in a flash.

So Nilly thought of something.

He pulled the envelope of fartonaut powder out of his pocket.

THE CONSTRICTOR ANNA Conda woke up suddenly. It had been dreaming the same dream it always dreamed. That it was swimming with its mother in the delightfully warm waters of the Amazon River among the piranhas, crocodiles, poisonous snakes, and other good friends, and was as happy as a hippo. And that one night it was captured, snatched out of the water and shipped to a freezing-cold country, where it had ended up in a pet store. And that one day a fat little boy had come in with his father, who had yelled at the shop owner and shown her the bite marks on his fat little boy's hand. Then the little boy had discovered the snake. His face had lit up and he had shoved his dad, pointed and yelled: "Anna Conda!" and then that was its name. Even though Anna Conda was a girl's name and he was a boy! Or that's what he thought, anyway.

Anna Conda had been put in a cage in Hovseter and had been fed some pasty white, round, slippery balls that tasted like fish, while the little boy poked it in the side with sticks. And even though this had all happened more than thirty years ago, Anna Conda would still wake up from this awful nightmare and would have been drenched in sweat if constrictors could sweat. And then it sighed in relief, because it wasn't in the apartment in Hovseter, but in the delightfully warm sewer pipes beneath downtown Oslo.

What had happened was that one night the little boy had forgotten to lock the cage and so Anna Conda had managed to escape through the open bedroom window, down along the downspout to the street, where after a great deal of searching and a couple of hysterical women's screams, it had found a loose manhole cover. That first night in the Oslo Municipal Sewer and Drainage System, it had lain curled up in a corner scared to death. But that fear had quickly passed. And by the next day, it had started doing what anacondas do: squeezing things very tightly and then eating them. Because there were lots of *Rattus norvegicus*, bats and regular old mice down there. It wasn't quite the Amazon, perhaps, but it wasn't that bad either. Just the other day it had even come across a genuine Mongolian water vole.

Now that Anna Conda had got so big, it had started easing up on constricting the food first — it just swallowed it, which was so much easier. It was pretty sure it remembered its mother saying that it wasn't good table manners to swallow food without properly squeezing it first, but there wasn't anyone down here to notice. So Anna Conda had just swallowed the tiny, glowing piece of meat with the red hair. But now it had the feeling that that might not have been such a good idea. Because the reason it had woken up was that it suspected something had exploded somewhere inside it and that a massive burp was on its way and wanted out. And Anna Conda suspected that the food was planning to go the same way. So Anna Conda clenched its jaws shut as it felt its long body inflate. And inflate. But it didn't give up; it clenched its jaws harder. Its body began to resemble an enormous sausage-shaped balloon and it was still swelling. But Anna Conda didn't give up; *what's eaten is eaten* he thought. It was so inflated now that its snaky black scales were squished against the sides of the sewer pipe. Its jaws ached. Soon it wouldn't be able to take anymore. And the pressure from within was only getting worse.

Soon it wouldn't . . .

It wouldn't . . .

Wouldn't!

Anna Conda's mouth popped open and out came a burp. We're not just talking about a regular burp, but a thunderclap of a burp that caused all of southwestern Oslo to shake in its foundations. And, just like when you stop pinching together the end of a sausage-shaped balloon, Anna Conda took off like a rocket through the Oslo sewer system. *Vroom!* Just like a cannonball being shot out of a cannon. The speed increased and so when it was shot out of the sewer pipe under the wharf a few nanoseconds later, it kept going quite a long way out over the fjord before it turned and headed straight up into the air. And exactly like a runaway balloon it made sudden, unpredictable turns all over the place, accompanied by a flapping, farting sound.

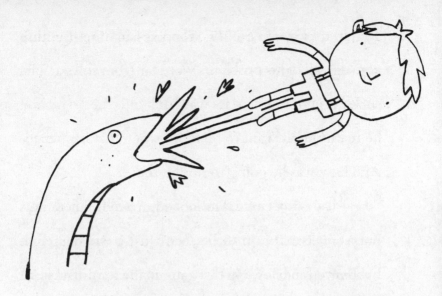

Until it was completely deflated and it landed like a moth-eaten lion pelt in a spruce tree somewhere out on the Nesodden Peninsula.

Nilly, however, was lying on his back, floating in the sewer water like a piece of poop, as he stared up at the ceiling and laughed. His laughter echoed through the network of sewer pipes. He was free! He'd been shot out of the anaconda's jaws like a projectile about one minute after he'd swallowed the fartonaut powder. Who would have thought it would feel so liberating to be in a sewer!

But after a while Nilly stopped laughing. Because actually, all of his problems were far from solved. The snake would soon find its way back into the sewer and he really didn't want to be there when that happened. And how was he going to not be there?

He had to get out. He looked around. There was not a single exit sign to be seen. Just a wooden crate bobbing up and down in the water in the semi-darkness. He clambered up onto it and paddled inwards. Or outwards. He wasn't sure which way. And after he'd paddled around various turns and corners for twenty minutes, he still didn't know where he was or how he was going to find an exit. He stopped paddling. And as he sat there listening to the silence, he thought he heard a faint sound. No, he wasn't imagining it, it *was* a sound. And it was getting louder. A terrible sound. The sound of an explosion, a plane crash and an avalanche. A sound that sends shivers down your spine and sends the devil packing. And Nilly knew that that sound could only be one thing: the Dølgen School Marching Band.

Nilly paddled as fast as he could towards the sound, went around two corners and sure enough he saw a beam of sunlight coming down from something that could only be a shaft leading up to the surface. Nilly paddled over to a metal ladder that was bolted to the side of the shaft and looked up. The ladder led up to where the light was coming from, somewhere way up above him. And, there, up at the top he saw the bottom of a manhole cover. Nilly hopped off the wooden crate and climbed up as fast as he could. When he was halfway up he glanced down, causing his heart to skip a beat and making him immediately promise himself that he wouldn't do that again. Sometimes it's just better not to know how high up you are.

When Nilly had made it all the way to the top and could hear the sound of the Dølgen School Marching Band moving away, he put his shoulder against the manhole cover and pushed as hard as he could. Then he tried one more time. And again. But unfortunately what Doctor Proctor had said was true: the manhole covers in this city would not budge. And there wasn't a single grain of fartonaut powder left for him to blast the iron cover off with.

Nilly shouted as loud as he could: "Help! Help!"

The sound of the most hideous marching band music in the Northern Hemisphere was almost gone now and Nilly's shouts were drowned out by the cars that had started driving on the street above him.

"Help! Help!" Nilly shouted. "There's an anaconda living down here and it's on its way home for lunch!"

Nilly knew probably no one would believe that, but what did it matter since no one could hear him anyway?

Nilly held on until his arms ached and he shouted until he was so hoarse that only a low rasping came from his throat. Resigned, he climbed back down and lay on the crate, exhausted. Then he sat up and started listening for the sound of a snake hissing. And while he was sitting like that, he happened to see a ray of light from the manhole cover shine on some deep holes punched through the lid of the crate he was sitting on. Holes left by large and rather sharp fangs. And some red letters that were printed on the lid:

CAUTION! HIGHLY EXPLOSIVE SPECIAL GUNPOWDER FROM SHANGHAI FOR THE BIG AND ALMOST WORLD-FAMOUS ROYAL SALUTE AT AKERSHUS FORTRESS

Yikes, Nilly thought.

Yeah, yeah, so what? Nilly thought.

Wait a minute, Nilly thought.

Maybe . . . Nilly thought.

He felt around in his back pocket. And there it was. He took it out. It was the half-chewed matchstick he'd got from Truls as payment for the bag of fartonaut powder. Of course it was wet and practically chewed in half, but it still had the red tip made of sulfur.

He held the matchstick in the beam of sunlight,

feeling how the sun warmed the skin on his hand. And two questions occurred to him. Number one: how long would you have to hold a match in the sun on a morning in mid-May before it was dry enough to light? And number two: how long does it take an anaconda to swim across the fjord from somewhere over by the Nesodden Peninsula?

You're going to get the answer right away. It takes *almost* exactly the same amount of time. Which is to say: it takes about one hour and four minutes to dry a match and for an anaconda to swim across the fjord from the Nesodden Peninsula and make its way deep into the Oslo sewer system, it only takes one hour and *three* minutes. So after an hour and three minutes had passed, Nilly had just noticed his hand starting to shake from holding it up in the beam of sunlight for so long, when he heard a familiar hissing noise.

Oh no, Nilly thought, since he felt that getting eaten once in twenty-four hours was more than enough.

He struck the match hard against the metal on the inside of the sewer pipe, but nothing happened.

The hissing noise came closer.

Nilly struck the match against the metal again. The red-tip sparked, but didn't light. And then Nilly was once again staring into the big pink mouth of the largest anaconda anyone had ever seen. The mouth came around the corner and Nilly thought, *This time that's it, a red-haired boy only gets so many chances.*

He pulled the match along the wall one last time.

It sparked. It sizzled. It ignited.

Nilly acted very quickly now. He set the match down in one of the holes left by the fangs in the crate with the burning end up. Then he dove into the water and swam away underwater as fast as he could. And for the first time he was glad that this was really nasty, filthy sewer water, where it would be impossible to see or smell much of anything besides nasty, filthy sewer water. And the match burned. From the top of the crate down into the highly explosive special gunpowder from Shanghai.

And for the second time that day, the foundations of downtown Oslo shook. On Sverdrup Street a manhole cover shot up into the air. Drivers slammed on their brakes and pedestrians froze on the pavement, staring at the hole in the street. The manhole cover was followed by a spray of wooden splinters and sewage. And then nothing. And then finally a tiny, red-haired and soaking wet boy climbed up out of the hole. He bowed politely to the frightened onlookers before rolling up his shirtsleeves.

Then he leaned over the manhole, spit some sewer water into it and yelled, "Take that, you earthworm!"

Before turning around to face the pedestrians, the shopkeepers who had come out of their shops to see what was going on and the drivers who had rolled down their windows.

"I'm Nilly!" the little boy yelled with his hands on his hips. "Anyone have anything to say about that?"

But the people on Sverdrup Street just stared, their mouths hanging open, at this strange being that had emerged from the inside of the earth.

"Nope, that's what I thought," said the boy, spitting one more time and walking away.

The Patent Office

WHEN THE BELL rang for first lesson Lisa was still wearing her band uniform. Everyone in the band was standing in the playground, talking about the weird thing that had happened that morning when they were marching through downtown Oslo. About the two band members who'd been knocked unconscious, about the ambulances that had come and about Conductor Madsen, who'd been so upset that they thought he might pass out as well.

Lisa pushed her way towards the classroom, through a crowd of children who were all pestering her and pulling on her and asking when they would be able to buy some fart powder because, after all, Independence Day was tomorrow!

Lisa was sitting at her desk when Mrs Strobe entered the classroom, pushed her glasses way out to the tip of her nose and peered at the one empty desk.

"Lisa, do you know if Mr Nilly is out sick today?"

Lisa just shook her head.

Mrs Strobe eyed her with suspicion. "Is something wrong, Lisa?"

Lisa really wanted to say no, but she knew that Mrs Strobe had a special kind of X-ray vision that could see through kids' skulls and into their brains, to where their thoughts were. So Lisa just came right out with it:

"Nilly's in jail."

A gasp ran through the classroom and Mrs Strobe raised one of her eyebrows so high that it totally vanished up into her hair.

"I'm sorry, could you please repeat that, Lisa?"

"Yes, Mrs Strobe. Nilly's in jail. Actually, to be specific, he's in the Dungeon of the Dead."

And then Mrs Strobe lowered both eyebrows and pulled them together so that it looked like she had a moustache on her forehead. "You used to be someone I could rely on to tell the truth, Lisa," she said. "But you've obviously been spending too much time with Mr Nilly."

"But I am telling the truth!" Lisa cried.

"Nonsense," Mrs Strobe scoffed. "Nilly is not in jail. Let's pick up reading where we left off. Page seventeen, everyone."

"He's in jail!" Lisa said.

"No!" said Mrs Strobe.

"Yes!" Lisa said.

"No," said a voice. "Not anymore."

Everyone in class turned around and looked at the door. And there was Nilly. He was soaking wet and the ends of his hair were a little singed, but otherwise he was exactly the same.

"Been swimming in the drinking fountain again, Mr Nilly?" Mrs Strobe asked sarcastically.

"Just a little altercation with a relatively large anaconda in the sewer, Mrs Strobe. And we handled it just fine with a couple of explosions."

The students all gasped, but were interrupted by Mrs Strobe slapping her desk with the palm of her hand.

"That's enough nonsense for today. Take your seat, Mr Nilly."

Nilly did as he was told, but as soon as he was seated he leaned over to Lisa. "I got your message," he whispered. "Sorry I wasn't able to get out sooner.

I was unfortunately delayed by a visit to the digestive system of a large snake. What's the situation?"

"Truls and Trym broke into Doctor Proctor's house last night," Lisa whispered. "And stole both jars of powder, as far as I could see."

"See? You just watched them do it?"

"Yup," Lisa said. "To make sure everything went according to plan."

"Plan? What plan?"

"Oh, just a teensy-weensy little emergency plan," Lisa said. "There's not really much to say."

AT THAT VERY moment five serious men were sitting behind a long table at the Oslo Patent Office. They were looking at Mr Trane, who was standing on the floor in front of them, going on and on about the amazing fartonaut powder that was in the jar he had set on the table in front of them.

"It's faster than a race car or a rocket," Mr Trane

said. "It's a better and cheaper fuel than a gazillion gallons of gasoline," he continued. "It can move men to the Moon, Mars and maybe Mercury."

As he talked and talked, the chairman – who was the most serious of the serious men – stared intently at Mr Trane. Because wasn't there something familiar, both about Mr Trane's name and his fat, pear-shaped body? Yes, he definitely reminded him of a boy in the neighbourhood he'd grown up in more than thirty years ago. A place called Hovseter. And this boy was always getting new pets since his old ones went crazy, kicked the bucket, or escaped. He vaguely remembered a Mongolian water vole. And a nice little snake from the Amazon. Could this be that same boy, all grown up?

When Mr Trane was done, the chairman cleared his throat and said, "All this is well and good, Mr Trane, but we here at the Oslo Patent Office cannot grant you a patent on what you call the . . . uh, fartonaut powder you say you invented if you don't know what it's made of. So, as the chairman of the Industrial Property Office, I am asking you for the third time. What is your invention made of?"

Mr Trane smiled as graciously as he could. "As I've already explained twice, I simply don't remember at all. It sort of happened by accident. I just tossed in a little of this, a little of that and stirred it up over a low heat. And then it turned into this powder that you see before you."

"Hmm," the chairman said seriously.

"Hmm," the four other serious men chimed in.

"We need proof," the chairman said.

"Yes, proof," the other four said.

"What kind of proof?" Mr Trane asked, looking at the clock. The men from NASA had said they'd be arriving on the two o'clock flight from Houston and he'd been hoping to have the patent signed and ready before he met with them at three.

"A test," the chairman said.

"Exactly!" the others said. "A patentable patent test."

Mr Trane looked at them uncertainly.

"You must demonstrate for us," the chairman said. "A very small dose, of course. Just so that we can see that what you're claiming appears reasonably likely."

"Of course," Mr Trane said, obviously nervous. "Of course, my dear gentlemen of the patent office."

"By all means, please wear that," said one of the serious men, pointing to a helmet hanging on a hook on the wall. "Although it didn't help the last man who used it very much."

"Who was that?" Mr Trane asked meekly.

"A man who thought he'd invented a new special gunpowder for the cannons at Akershus Fortress," the chairman said gravely. "It turned out it was far too explosive."

The other four shook their heads somberly and crossed themselves.

Then Mr Trane put on the helmet, walked up to the table and stuck the teaspoon down into the jar of powder, making sure he took only a tiny little bit. Then he swallowed, squeezed his eyes shut tight, and waited. And waited. And waited.

But nothing happened.

Nothing that he noticed, anyway.

But then he heard the five men starting to murmur back and forth to each other.

"Remarkable," one of them said.

"Highly unusual," the second one said.

"But haven't we seen this before?" the third one said.

Mr Trane cautiously opened one eye and saw the fourth man leafing through a big book.

"Here it is," the man said, pointing to the book. "This invention has already been patented."

The chairman cleared his throat and became even more serious. "Mr Trane, you are a fraud, who's trying to steal Doctor Proctor's invention."

Mr Trane stared and then sputtered, "Has that crazy professor already patented the fartonaut powder?"

"Fartonaut powder? Certainly not. We're talking about an essentially rather unsuccessful invention called Doctor Proctor's Light-Green Powder. Look for yourself, my dear sir!"

Mr Trane looked down at himself. And he emitted a shriek of disbelief. Because he was glowing a phosphorescent green and was partially transparent, like some kind of see-through larva.

AT THAT SAME instant, in Mrs Strobe's classroom, Nilly leaned over to Lisa's desk and whispered skeptically, "You did what?"

"I broke the cellar window, snuck in, and glued

a new label over the old one on the jar containing Doctor Proctor's Light-Green Powder."

"And on the new label you wrote . . . ?"

"Fartonaut Powder," Lisa giggled. "Keep Out of Reach of Children!"

They ducked as they saw Mrs Strobe's eyes sweeping across the classroom, searching for the source of the whispering.

"And then?" Nilly whispered.

"I put a jar full of regular fart powder next to the other jar. So Truls and Trym wouldn't suspect anything," she whispered. "Then I put all the fartonaut powder and the regular fart powder that were left in my backpack."

"And what did you do with all that powder?"

"Hid it in my wardrobe."

"And then you watched Truls and Trym . . . ?"

"Yup! I watched them from my bedroom window. They broke in and took both of the jars."

"I wonder where they are now. I didn't see them on the playground before the bell rang."

"Oh, I know where they are all right!" Lisa said, forgetting to whisper. "There was actually this very weird accident when the band was marching in downtown Oslo this morning. Something fell out of the sky – "

"Lisa!" Mrs Strobe slapped her desk. "Mr Nilly! What are you two talking about?"

Nilly cleared his throat. "We were just discussing why women like Lisa and yourself are so much smarter than men, Mrs Strobe," Nilly said. "I think women ought to take over the world."

Mrs Strobe looked at him, bewildered.

"But it was just a thought," Nilly said. "And since I'm a man, it was probably a very dumb thought. So I say let's forget the whole thing and thanks for your interest, Mrs Strobe. Please, just pick up where you left off."

The corners of Mrs Strobe's eyes twitched. Her prominent nose and the corners of her mouth twitched. But before she managed to say anything, there was a loud knock on the door.

"Come in!" she yelled quickly, actually sounding like she was relieved to have the interruption.

The door opened and there was a man standing there with a pair of dark sunglasses perched on a short, thick nose with black pores.

"Good day, Mrs Strobe," he said. "Pardon me for interrupting."

"Come in, Mr Madsen. What can we do for you?"

The director stepped into the classroom and cleared his throat. "We have a little crisis. Or to be more precise: a big crisis. As some of you know, there was a freak accident as our marching band was practising downtown this morning. Something very heavy and very hard and very unexpected fell out of the sky and hit two of our musicians on the head. They're in the hospital with mild concussions. The two students are Truls and Trym Trane."

A murmur spread through the classroom. And a couple of almost inaudible hurrahs could be heard. Mr Madsen cleared his throat again.

"And now the crisis is that the two of them will not be able to play with us in the Independence Day parade tomorrow. In other words, I'm looking for someone who can stand in for them at extremely short notice. Someone who plays the . . . uh, trumpet."

Lisa looked at Nilly, who was sitting there with his mouth hanging open, staring at Mr Madsen.

Mr Madsen shuffled his feet and looked like he was feeling sort of uncomfortable, but then he continued: "And if I'm not mistaken, there's someone in this class who plays the . . . uh, trumpet. A boy with . . . uh, perfect pitch. A boy named . . . uh, Nilly."

Everyone turned to look at the red-haired, tiny little guy who was now studying his nails with a distant, aloof expression.

"Nilly?" Mrs Strobe asked.

"Yes, Mrs Strobe?"

"Aren't you beside yourself with happiness, son? You're going to get to play with Mr Madsen in the Dølgen School Marching Band in the big parade on the seventeenth of May!"

Nilly squeezed one eye shut and stared thoughtfully off into space. "The seventeenth of May, May seventeenth, that date sounds familiar . . . oh, yeah, now I remember! Isn't that Norwegian Independence Day? Because first of all I already have a lot of plans for Independence Day. I was planning to drink some traditional eggnog. Then there are a few sack races I'm signed up for. And then of course I have to defend my title as the reigning champion of the Great Egg-Rolling Race in Eggedal. And that's even in the toughest group, the hard-boiled egg group."

The kids started laughing, but an extraordinarily powerful palm-against-teacher's-desk slap shut them all up again immediately. Apart from Nilly, of course.

"In short," he said. "It may be difficult for me to squeeze any trumpet playing in on that particular day."

Mr Madsen grimaced and groaned in despair.

"Unless . . . ," Nilly said.

"Yes?" Mr Madsen lit up. "Yes, tell me!"

"Unless I'm asked very nicely, of course . . ."

"Yes, yes, I'm asking nicely!" Mr Madsen cried out.

"Or even better, unless I'm begged."

"I'm begging, I'm begging!" Mr Madsen wailed.

"On your knees?" Nilly asked.

And Mr Madsen dropped to his knees and begged while Mrs Strobe's glasses slid twenty inches down her nose at this unusual sight.

"All right!" Nilly said, leaping up onto his desk. "I'll play. Just make sure you have a uniform that's small enough."

And then all the kids cheered. So did Mr Madsen. And although it was hard to tell, even Mrs Strobe did, a little bit, on the inside. And while they were cheering, Lisa whispered a few words into Nilly's ear. And then he stuck two fingers into his mouth and whistled so loudly that the keyhole in the door made a squeaking sound and suddenly it got totally quiet again.

"Now, a message for all children!" Nilly yelled. "This afternoon we'll be selling fart powder in Lisa's garden. Right, Lisa?"

"Yeah," Lisa said, jumping up on her desk. "And we're lowering the price to two kroner, since . . . well, since it's cheaper."

"Isn't she smart?" Nilly smiled.

And with that the cheering started again and since

the bell rang right then, Lisa and Nilly were carried out of the room in triumph.

Mrs Strobe and Mr Madsen were left standing there in the classroom watching them go, shaking their heads and laughing.

"Those two are quite a pair, aren't they?" Mr Madsen said.

"They sure are," Mrs Strobe said. "But there was just one thing I was wondering about."

"Yeah?"

"What hit Truls and Trym?"

"That's the most mysterious part of the whole thing," Mr Madsen said. "Believe it or not, it was a manhole cover."

The Confession

EVENING HAD FALLEN and in just one night it would be the seventeenth of May, Norwegian Independence Day, when all children and grown-ups put on their traditional costumes and march in parades until they get blisters and their feet swell up so much they can't get their brand-new shoes off. They yell

"Hurrah" until their voices are so hoarse they wouldn't even be able to whine when they stuff themselves way too full of hot dogs and ice cream and their stomachs feel like they are crammed full of barbed wire. In other words, it was the evening before the day that all children and grown-ups were really looking forward to.

And on this evening Truls woke up and discovered that he was lying in a hospital bed. He looked around and discovered Trym lying awake in the bed next to him.

"What happened?" Truls asked. "Why do you have a bandage around your head?"

"A manhole cover," Trym said. "And you have a bandage around your head too."

"We were supposed to sell fart powder to the kids and make a fortune today!" Truls said. "Independence Day is tomorrow!"

"And we were supposed to play the trumpet," Trym said, dazed.

Right then the door to the room opened and a nurse came in.

"Hi, boys," she said. "There are two people here to see you."

"Daddy!" Truls yelled, on the verge of tears, he was so relieved.

"And Mummy!" Trym whimpered.

"Not quite," the nurse said, stepping to the side.

Truls and Trym stiffened in their beds. Before

them stood two men that we have met before. They were wearing their police uniforms and tucked under their arms each of them was holding a jar that we've also seen before.

"Good evening, boys," Mr Fu Manchu said. "I trust your head injuries won't be permanent."

"And," Mr Handlebar added, "that you'll be able to confess right away that you were the ones who broke into Doctor Proctor's cellar."

"And stole these jars," Mr Handlebar continued.

"It wasn't me," Truls blurted.

"Or me," whimpered Trym.

"We followed a tip and found them in your garage," Mr Handlebar said.

"And we also found two pairs of shoes there with glass shards in the soles. Like the glass shards that came from the broken glass in the cellar. You're done for."

"But if you'll give us a confession now, you may

be able to avoid winding up in the Dungeon of the Dead."

"It was me," Truls blurted.

"No, it was me," Trym whimpered.

"And Dad," Truls said.

"Yes, Dad," Trym said. "He . . . he . . . tricked us."

"We were duped." Truls sniffed.

"We're so easily tricked," Trym sobbed. "Poor us!"

"Hmm," Mr Fu Manchu said. "Mr Trane, you say. Just as we thought. We should put out an A.P.B."

"Yeah," Mr Handlebar said. "And fast. Neither he nor that dreadful Hummer of his were home when we checked."

Mr Fu Manchu got out his police radio and called the police station. "Put out an A.P.B. for all cars to stop any black Hummers they see. We're looking for a man named Mr Trane. He's incredibly dangerous. I repeat: incredibly dangerous."

And with that he started the biggest car chase in

Oslo's history. We won't go into details, but more than one hundred police cars chased Mr Trane's black Hummer as it raced through the streets of Oslo, spewing out more carbon dioxide than two locomotives. Every time the police blocked off a street and thought they had him, Mr Trane just gave the Hummer more gas and broke through the barricades, speeding past the police cars, the police horses and the policemen all over downtown Oslo.

And that's what they were still doing when the sun rose and Independence Day was finally here.

Independence Day

FOR THE LAST time in this story the sun rose in a cloudless sky. It had already shone for a while on Japan, Russia and Sweden, and now it was starting to shine on the very small capital city, of a very small country called Norway. The sun got right to work shining on the yellow, and fairly small, palace that was home to the king of Norway, who didn't rule over enough for it to amount to anything, but who was looking forward to waving at the children's parade as it marched by and to listening to the Big and Almost World-Famous Royal Salute in his honour. And of course the sun shone on Akershus Fortress, on the old cannons that were aimed out over the Oslo Fjord and onto the most remote of doors. The door that ultimately led to the city's most feared jail cell, the Dungeon of the Dead.

And just at this moment the door to the Dungeon of the Dead opened and out onto the grassy embankment stepped Doctor Proctor, who had to squint in the sunlight. He was followed by two guards.

"Hip hip hurrah!" yelled Nilly and Lisa, who were standing there waiting for him. They jumped up and down and waved their Norwegian flags.

"Freedom, sunshine, Independence Day and my assistants," Doctor Proctor said, laughing and hugging them. "Could the day get any better?"

"For some," mumbled the Commandant, who was standing a few steps behind Lisa and Nilly and rocking back and forth on his heels.

"But nobody's told me why I was set free," the doctor said after he set Nilly and Lisa down.

"Truls and Trym admitted everything," Lisa said. "That they bullied Nilly into giving them the fartonaut powder that day."

"And that you never sold fartonaut powder to children," Nilly said.

"And the police are going to have Mr Trane in custody soon," Lisa said. "For stealing the powder and passing it off as his own. They just have to finish racing around the city first."

"Good heavens!" the professor said. "Then all of the problems are solved!"

"Not quite all," Lisa said, nodding towards the Commandant. "Dad?"

"Of course, of course," rumbled the Commandant, stepping forwards. He seemed embarrassed; maybe that was why he spoke a little louder and more commandingly than necessary. "Yes, well, we are so sorry for this idiotic imprisonment, Doctor Proctor. It won't happen again. Unless you do something very illegal, of course. Like stuffing bananas in exhaust pipes, for example. Or hoisting infants up to the tops of flagpoles. Or . . ."

"Get to the point, Dad," Lisa said sternly.

"Of course, of course, the point," rumbled the Commandant, his neck reddening a little. "As you can see, we have some old cannons over there. And as you can't see, we don't have any special gunpowder from Shanghai, which we need for the Big and Almost World-Famous Royal Salute that was supposed to be fired off from these cannons later today. It has never happened before in the modern era that the Royal Salute hasn't been fired off, and we're afraid the whole world will laugh at us. Well, all of northern Europe, anyway . . . except maybe Finland . . . and . . . and . . ."

"Dad!"

"Of course, of course. The question is – "

"The question is," Doctor Proctor interrupted, "whether I can help you with the Royal Salute. And the answer, my dear Commandant and neighbour, is: *yes!*"

And with that, cheering broke out for the second time in a very short period. But Lisa and Nilly weren't able to cheer for very long, because of course they were about to go perform in the Dølgen School Marching Band in the Independence Day parade.

THE DØLGEN SCHOOL Marching Band marched and played like never before. They hit a lot of the right notes and had never been closer to playing in time. And Nikolai Amadeus Madsen led the way in his sunglasses and grinned his biggest smile as he dreamed of the marching band competition at Eidsvoll that summer.

Lisa played the clarinet and every once in a while she glanced over at Nilly, who practically had to do splits to keep up with everyone else. But he played amazingly well as his fingers danced over the keys, his eyes hurriedly scanning the music.

The band had reached Sverdrup Street and Nilly was concentrating so hard that he didn't hear the wailing sirens of the police cars approaching. And he didn't see the big, roaring Hummer turn the corner on squealing tyres and slam on its brakes when it saw that its path was finally blocked by something it couldn't just run over or push out of the way: a whole Independence Day parade marching towards it. And the noise the school marching band was making sent shivers down the driver's spine, because it was the sound of an explosion, a plane crash and an avalanche all at the same time.

Following the Hummer roughly a hundred police cars turned the corner with their blue lights and sirens.

A man jumped out of the Hummer.

Lisa stopped playing. "But isn't that . . ." she said. "It is! It's Mr Trane."

Nilly stopped playing too and looked up.

Mr Trane was standing in the middle of the street, looking around frantically. There was nowhere to hide. It seemed like this was the end of the chase.

"Ha!" Mr Trane yelled. "You'll never get me, you idiots, you worthless turds!" And with that he yanked up the manhole cover next to him and jumped down into the hole.

"Hey!" Lisa said.

The policemen ran over, peered down into the hole, scratched their heads and discussed the situation. Nilly and Lisa could make out a few random snippets of the conversation:

"I'm wearing my Independence Day uniform today, and I don't want to go down in the sewer and get it all dirty."

"Well I have asthma; the smell of excrement just isn't good for me."

"And I'm signed up for a sack race."

So they shoved the manhole cover back into place, checked to make sure it was on tight, cancelled the whole police chase and waved the Independence Day parade on.

ANNA CONDA WAS lying in the pipe, listening to its stomach rumble with hunger. It could hear the noise of a marching band and smell the scent of boiled hot dogs from up on the street. And now, suddenly, it heard a huge splash in the Oslo sewer system. The creature was so hungry that it was only just barely able to swim towards the sound. But when it got there, it saw something it recognised. Two-legged food glowing a faint green. The last time it had eaten something like this, Anna Conda had been blasted all the way out to the Nesodden Peninsula. But that wasn't the only thing it recognised. There was something about this two-legged food, something familiar from when the creature was a little anaconda snakeling in a cage in Hovseter. Because wasn't there a certain similarity between this fat, fleshy, sausagelike man and that fat boy who used to poke sticks in his side back then? Yes, that was it! And now the anaconda could see that the man had noticed it and that the recognition was mutual. And that the man had opened his mouth to scream. That his mouth was as far open as it would go. Which was very far. But of course nowhere near as far open as Anna Conda's mouth now was.

"WOW, THAT WAS good!" Nilly yelled as he chewed. He was holding a steaming hot dog in a bun.

"Really good!" Lisa said, taking a bite of her hot dog.

They were sitting on the grassy embankment at Akershus Fortress, watching the seven brave guardsmen who were pacing nervously in front of the table where Doctor Proctor was standing with a big jar of Doctor Proctor's Totally Normal Fart Powder. The seven of them had signed up as volunteers for this honourable assignment.

"Assistant Nilly!" Doctor Proctor yelled, glancing up at the clock on the tower at City Hall, which was approaching the time of the Big and Almost World-Famous Royal Salute. "Can you help me dole out the portions?"

"Of course," Nilly said and scarfed down the rest of his hot dog, ran over to the table, grabbed the wooden ladle that was lying there and stuck it down into the jar.

"I'm Nilly," he told the guardsmen. "What do you have to say about that?"

One of the guardsmen started swaying back and forth and singing "Silly Nilly." Two more quickly joined in.

"Shut up," Nilly said, looking at the clock. "Or rather, open up. And bend over. Quick, we only have seconds to go."

"Is it dangerous?" one of the guardsmen asked nervously, opening his mouth.

"Yes," Nilly said and stuck a whole ladleful of powder into the guardman's mouth. "But it tastes like pears. Nine . . . eight . . ."

"Thanks, assistant," the professor said, adjusting his motorcycle goggles. "My dear guardsmen, please assume your positions."

The guardsmen, who were not used to following commands that included words like "dear" and "please," looked at each other in confusion.

"I feel a little tickle in my stomach," one of them said.

"Listen up!!" the small, red-haired boy bellowed. "Point your rear ends in the same directions as the cannons, now! And bend over!"

This was a language the guardsmen understood and they followed the orders immediately.

And right then the clock on the Town Hall tower started to toll twelve times.

IT WAS SUCH a funny sight that Lisa had to laugh out loud. Seven guards bending over with their rear ends aimed out over the wall of the fortress at the Oslo Fjord as the Town Hall clock chimed.

But after the third chime neither Lisa nor any of the other inhabitants of Oslo and the surrounding area heard the clock anymore. Because both it and Lisa's laughter were drowned out by a bang so loud that frost formed on people's eardrums and their eyes were pressed quite a way back into their heads. The next bang sent a rush of air up Rosenkrantz Street to Karl Johan Street, where it made all the flags stand out straight. The third bang shattered three window-panes on the Nesodden Peninsula and made the grand, old apple trees in the Ulleval Garden town burst into bloom out of sheer fright. The fourth bang caused a girl Lisa knew in Sarpsborg to look up at the cloud-less sky and wonder if a thunderstorm was approach-ing. The fifth wasn't that loud at all; it just sounded like a fart and made the people in Oslo look at each other in surprise. But the sixth nearly caused a cruise ship in the middle of the fjord on its way to Denmark to capsize and a flight of swallows on their way to Norway changed their minds and decided to fly back to Africa. The sound wave reached all the way to Trafalgar Square in London, where it bent the spray from the fountain so that all the tourists standing

around it got wet and children laughed with glee.

When the final and seventh bang rang out, the king in his castle nodded in satisfaction at the farting and thought he had never heard a finer salute. And before the last echo had faded into silence, the king's adjutant was already on the phone to the Commandant of Akershus Fortress to tell him that the king would like to award him and his cannoneers the Royal Medal of Merit, a promotion to honourary cannoneer and a long and happy life.

"Can he really give us a long life?" the Commandant asked skeptically.

"He's the king," the adjutant said, hanging up the phone, offended.

The Commandant walked out onto the embankment again, where seven guardsmen with rips in the seats of their trousers, two policemen with their eyes wet from tears of laughter and Lisa, Nilly and the professor were still dancing around in joy.

The Last Chapter

IT HAD BEEN a loooooong Independence Day and there was still a little of it left.

The afternoon sun shone lazily on the pear tree in Doctor Proctor's garden, and Lisa and Nilly sat underneath it, each in their own chair, clutching their stomachs. Along with the professor, they'd polished off a three-metre-long jelly, and now they were so full that the professor had gone inside to rest a little.

"You did great today," Lisa said.

"You didn't do so badly yourself," Nilly admitted. "It was all thanks to you."

"You think?" Lisa smiled, closing her eyes to the rays of sunlight that filtered through the leaves.

"Yeah," Nilly said. "You're the smartest girl I know. And even more important, you're the best . . ."

It got quiet and Lisa opened her eyes and was surprised to see that Nilly's face had become really red. And she thought he might have got something stuck in his throat because he had to clear it three times before he was able to continue in a slightly hoarse voice.

"You're the best friend anyone could have."

"Thanks," Lisa said, her whole body feeling warm. "So are you."

And then neither of them knew what to say, so maybe it was just as well that there was a *bang*. Because there was. There was a final bang on this loooooong Independence Day and they both turned towards Doctor Proctor's cellar. Because this didn't sound like Doctor Proctor's Totally Normal Fart Powder.

"Oh, no," Lisa said, dismayed.

"Not the fartonaut powder . . ." Nilly said.

"No," said Doctor Proctor, appearing in the cellar doorway. His face was black with soot and oil. "Just a faulty muffler on a motorcycle that hasn't been started in twelve years. But that just needed a little lubrication to run, well, like it had been lubricated."

And with that the professor drove his motorcycle and sidecar out of the cellar and through the high grass, stopping in front of them. There was a brown, worn leather suitcase in the sidecar.

Nilly and Lisa stood up.

"Where are you going?" Nilly asked.

"Where do you think, my fartonaut assistant?" the professor asked, beaming under his hockey helmet and motorcycle goggles.

"You're going to Paris," Lisa said. "You're going to try to find Juliette Margarine."

"Wish me luck," Doctor Proctor said. "And lock the cellar and keep an eye on my house until I get back."

"Good luck," Nilly said.

They walked ahead of the motorcycle and opened the gate.

The professor revved the engine and it gave a satisfying growl.

"And if you go through Sarpsborg . . ." Lisa said.

"Yes?"

"Then you can say hello to my *second* best friend."

And the last rays of sunlight shone on the pear tree, on Nilly's red hair, on Lisa's smile and maybe on a tiny tear, as Doctor Proctor's motorcycle and sidecar drove away down Cannon Avenue.